Fast Ion-Atom and Ion-Molecule Collisions

Interdisciplinary Research on Particle Collisions and Quantitative Spectroscopy

ISSN: 2315-4233

Series Editor: Dževad Belkić

Vol. 1: Fast Ion-Atom and Ion-Molecule Collisions
edited by Dž. Belkić

Forthcoming

Vol. 2: Fast Collisions of Light Ions with Matter: Charge Exchange and Ionization
by Dž. Belkić

Interdisciplinary Research on Particle Collisions
and Quantitative Spectroscopy

Volume 1

Fast Ion-Atom and Ion-Molecule Collisions

Editor

Dževad Belkić

Karolinska Institute, Stockholm, Sweden

NEW JERSEY • LONDON • SINGAPORE • BEIJING • SHANGHAI • HONG KONG • TAIPEI • CHENNAI

Published by

World Scientific Publishing Co. Pte. Ltd.
5 Toh Tuck Link, Singapore 596224
USA office: 27 Warren Street, Suite 401-402, Hackensack, NJ 07601
UK office: 57 Shelton Street, Covent Garden, London WC2H 9HE

British Library Cataloguing-in-Publication Data
A catalogue record for this book is available from the British Library.

Interdisciplinary Research on Particle Collisions and Quantitative Spectroscopy — Vol. 1
FAST ION-ATOM AND ION-MOLECULE COLLISIONS

ISBN 978-981-4407-12-0

Printed in Singapore.

Editorial

Vision of interdisciplinarity within this book series

It is instructive to inquire about the real meaning of interdisciplinary research within a given context, such as a book series. In general, besides obvious obstacles such as evident differences in the attitudes to problem solving and terminology, there is also the more subtle issue of mentality and barriers when experts from different research fields envisage to join forces. It would not be a rare event if a theoretical physicist would wish, from the start, to venture to tell, e.g. a potentially collaborating biologist what is actually important to study in biology. A biologist might attempt to view physicists or mathematicians alike as those who merely provide the needed service to biology. Both approaches are wrong, as they evidently stem from the lack of genuine appreciation of the special aspects and potential of the other branches of research. Some research physicists might even be tempted to think of going to another hardly necessary extreme of studying, e.g. all of biomedicine or vice versa, as if this were a pre-requisite for a successful interdisciplinary cooperation.

A more appropriate approach to a good cooperation of experts from different fields would seem to be striking a balance in acquiring the necessary minimum information about each other's expertise within the concrete problems under study, then to ask the right questions and finally to know what is at stake. It is within this general and realistic realm of viewing interdisciplinarity that this series is planned to encourage cooperative approaches to complex problems to be reviewed through different volumes of varying levels of complexity.

Specifically, this series is aimed at giving expert coverage to cutting-edge advances in studies of collisions and spectroscopy with a focus on prospects for interdisciplinary applications in the sense to which we alluded above. These two main strategies for investigating the structure of matter on various fundamental levels are deeply and intrinsically intertwined through a panoply of similar or common concepts as well as via mathematical and computational methods, of both a deterministic and stochastic nature.

The notion of spectroscopy naturally emerges from resonant scattering phenomena when unstable, decaying states are involved. Regarding collisions, the series will cover both light and heavy particle laser-free and laser-assisted collisions from low to high energies. This is rooted in the key interdisciplinary significance of such collisions in wide applications, including plasma physics, astrophysics, thermonuclear fusion research, radiotherapy, etc. The series will encourage contributions from basic as well as applied sciences, the joint home of both collisions and spectroscopy.

Over a long period of time, collisions and spectroscopy from basic research in physics and chemistry have made gigantic strides across interdisciplinary fields, including the life sciences. For example, nuclear magnetic resonance spectroscopy from fundamental research in physics and chemistry is viewed by experts to possess the potential of revolutionizing particularly early cancer diagnostics, molecular image-guided surgery and radiotherapy as well as follow-up.

Likewise, energetic collisions involving multiply-charged heavy particles are of paramount importance, as they are nowadays going through a veritable renaissance in several leading strategies, such as hadron therapy, fusion research, etc. High-energy light ion beams from protons to carbon nuclei, as a powerful part of hadron therapy of deep-seated tumors, are increasingly in demand worldwide, and this highly motivates construction of hospital-based accelerators. Positron-emitting secondary ions generated from collisions of primary projectiles with the treated tissue provide the possibility for special tomographic verifications of dose depositions at the targeted sites. Such information is of utmost importance, since any inaccuracy in dose delivery to the tumor relative to the prescribed dose planning system can be used to update the input data to the algorithmic codes for energy losses in order to provide the corrected doses in the subsequent fractionated treatments of the patient. This adaptive dose delivery and the proper inclusion of the so-called relative biological effectiveness are among the major constituents of biologically-optimized and, indeed, personalized radiotherapy which is adapted to the specific needs of each patient.

Collision physics involving ions is topical again in the quest for new energy sources, as greatly boosted by the International Thermonuclear Reactor (ITER) which is currently being built in France and shall be put into operation in 2020. This type of energy stems from high-temperature fusion of light nuclei and the main significance of this achievement is not just in a commercially profitable energy gain, but, most importantly, in having an incomparably safer energy production than the one from nuclear fission, due to minimum radioactive waste and no runaway nuclear reactions. This was made possible by tremendous intra- and inter-disciplinary cooperative efforts involving ion collision physics, plasma physics, engineering and technology. For example, the most critical for steady energy production is plasma stability. Plasma, being comprised of hot charged ions, can readily be neutralized and, as such, lost for generation of the current. Neutralization can occur through collisions of ions with the reactor walls via electron capture from the material contained in the tokamak. Knowledge from atomic collision physics about the largest cross sections for electron capture for certain elements enabled elimination of such materials from the walls of the tokamak and thereby contributed substantially to plasma stabilization.

This series explores both direct and inverse problems in scattering and spectroscopy. The primary emphasis is on methodologies of proven validity simultaneously based on fundamental theoretical concepts and principles as well as on indispensable and thorough comparisons with experiments. There is a growing concern among e.g. medical physicists that critical input from theories of basic sciences is needed for going beyond phenomenological and empirical, data-driven modeling with no predictive power. Such models are still overwhelmingly in use in, e.g. radiobiology of cell survival as well as in cross section data bases for Monte Carlo simulations of energy losses of ions during their passage through tissue. Critical and expert state-of-the-art reviews of the existing progress and stringent assessments of the available methods in comparative analyses of theories versus experiments will be strongly encouraged.

Although physics was the main branch for studying particle collisions as well as to the major three spectroscopies (sub-nuclear, nuclear and atomic), mathematics was and still is pivotal to furthering these scientific strategies. This coupling resulted in the creation of a special field in mathematics *per se*, called "Scattering Theory". The themes of the series will nurture the ever needed inspiration from mathematics, both computational and fundamental, ranging from integral transforms through differential equations to inverse problems. The most interesting/important areas in theoretical and experimental studies across inter-disciplinary fields are within inverse problems of finding the causes from the observed effects. Only the simplest experiments in, e.g. physics deal with direct measurements of the sought observables. By contrast, in more fundamental physics experiments such observables are directly inaccessible, so that resorting to the concept of inverse problems is unavoidable by making inferences about the causes for the recorded effects. Likewise, and practically by definition, virtually the whole of medicine is about solving inverse problems. This is of paramount significance in practice and the series will embrace inverse scattering and spectroscopy problems that are plagued by notorious mathematical ill-conditioning and ill-posedness due to the lack of the models' continued dependence upon the analyzed data. Such inherent difficulties are further exacerbated by the presence of unavoidable noise.

Ubiquitous noise in experimental data and computationally generated spuriousness are invariably considered as a nuisance and detrimental to analysis and synthesis. This series will pay special attention to the modern advances in inverse collisional and spectroscopic problems where, surprisingly, noise can advantageously be exploited to reliably identify the sought physical and biochemical information and, thus, separate noise from true signal (stochastic resonances, Froissart doublets, etc.).

The series' interest in basic structure and overall interactions is aimed to offer the possibility for perceiving kindred relationships among apparently diverse systems in vastly different fields. This naturally includes system theory and optimization. System theory perceives the system's interactive dynamics as a whole, rather than merely its isolated constituent parts. Optimization comes into play to synergize the favorable and minimize the adverse influences of the surrounding environment of the system. This approach has already brought fruits to biology and medicine. For instance, optimization is a key to conformal radiotherapy which selectively targets the diseased tissue, while simultaneously sparing the healthy parts of the organ. In the like spirit of cross-fertilization, the self-contained field of signal processing is itself currently benefitting from the recently imported advances in quantum-mechanical spectral analysis. Here, regarding, e.g. medical diagnostics, mathematical optimization is also the most important, since experimentally encoded data are frequently uninterpretable in the direct domain in which measurements are performed.

There is not a single book or book series or journal which unifies spectroscopy and collisions into a common framework, as envisaged here. This series is designed to bridge this gap as a publication of the outlined unique profile. Such a scope is indispensable, given the inevitable scatter of information across interdisciplinary fields. The need to make bridges across various sources of spectroscopy and collisions is urgent, given the demonstrated potential of these universal methodologies for studying the structure of matter on vastly different levels/complexities. One of the rationale for establishing this type of interdisciplinary book series is to avoid duplication of methods and results. The lack of cross-disciplinary fertilization often leads to "discovering" some methodologies or techniques in one branch only to subsequently realize that it was merely a reinvention of the already known results from another branch of research and development.

The logo of the series is a symbol consisting of three intertwined rings that are known as Borromean rings in Mathematics and Effimov states in Physics. The latter symbolize a bound system of, e.g. three atoms where none of the three two-body sub-systems is bound. As such, the compound tripartite system breaks apart as soon as any of the sub-systems is removed. In the logo, the three rings represent symbolically Mathematics, Physics and Medicine, as the main backbone of the interdisciplinarity of this series. What is meant by this symbolism is that both Physics and Medicine are substantially enriched by their linkage to Mathematics, without which progress of either science would be most severely hampered, to say the least. Modern hospitals could not function without X-rays, computerized tomography, ultrasound, positron emission tomography, magnetic resonance imaging and magnetic resonance spectroscopy, all of which were brought to Medicine by Physics and Mathematics. The logo of the series places emphasis on the synergism achievable when these three disciplines are tightly bound together. This triple link is beneficial to Mathematics, as well. For example, the most difficult problems and questions raised particularly within Physics are already known as being capable of spurring fundamental progress in Mathematics. Critical advances on deeper levels in biomedical sciences also rely heavily upon mathematics due to the complexity of the problems and systems in the life sciences. It is likely that in the future, biomedical problems might also initiate research in yet unexplored pure mathematical avenues.

Overall, this series will focus on the multifaceted interdisciplinary character of collisions and spectroscopy. This is motivated by the existing wide consensus that the most profound progress in modern sciences of strategic importance for the society at large is critically guided by inter- and multi-disciplinary approaches to complex research problems.

By its versatile nature, the intended audience for this series is vast, including researchers and graduate students in physics, chemistry, biology, physicians of various levels of training, and the associated mathematical/computational sub-branches alongside the separate field of signal processing. The series will be published twice per year through independent books by one or more invited authors or edited volumes with chapters by different expert contributors.

Editor-in-Chief: Dževad Belkić
Professor of Mathematical Radiation Physics
Nobel Medical Institute, Karolinska Institute
Stockholm, Sweden

Preface to Volume 1

Volume 1 of this series is a collection of reviews on fast ion-atom and ion-molecule collisions. Here "fast" refers to the initial velocity or energy of the given projectile. Energetic ion beams are those with projectiles of speed considerably larger that the Bohr orbiting velocity of the target electron which is undergoing a transition from an initial to a final state. A typical reference energy of an ion beam is 25 keV/amu for a projectile nucleus of velocity $v = 1$ au, which matches the speed of the electron in the ground state of atomic hydrogen target. Relative to this reference energy, impact energies such as 25–400 keV/amu could be viewed as being in the intermediate region, whereas those above, e.g. 400 keV/amu might be considered as high incident energies. Volume 1 covers intermediate and high impact energies 25–7000 keV/amu of light ions (protons, alpha particles, lithium and carbon nuclei) colliding with atomic (helium) and molecular (water) targets. The investigated collisions are those that cause one- and two-electron transitions, such as single and double electron capture, single and double ionization, transfer ionization as well as simultaneous electron transfer and excitation of the residual target ion. Angular distributions of scattering projectiles and ejected electrons as well as total cross sections are among the main observables in both experiments and theories from Volume 1. Direct collaboration of experimentalists and theoreticians is always welcome and the results of such joint efforts occupy the first half of Volume 1 through chapters 1–4. The remaining chapters 5–8 are by theoreticians who likewise place their major attention on validation of their models against experimental data.

The experimental parts of chapters 1 and 2 are based upon one of the most powerful measuring devices from atomic and molecular collision physics called "cold target recoil ion momentum spectroscopy". This apparatus is a veritable "reaction microscope" which is capable of peering most deeply into the interactive dynamics of atomic and molecular collisions by way of kinematically complete experiments. Here, all the momenta of every actively participating particle can, in principle, be available, some of them by direct measurements and the others by reliance upon the momentum conservation law. This was made possible by exploiting the recoil target ion kinematics, which is usually left unexplored by, e.g. translational spectroscopy, due to conventionally unmeasurable small momenta. The rescue was in introducing the concept of target cooling, which enabled precise measurements of small recoil momenta, and this paved the road to kinematically complete experiments. With these achievements, theoreticians were offered the possibility for highly stringent testing of their methods on, e.g. fully differential cross sections and other similar observables.

As to theories from Volume 1, the challenges are multifaceted, especially at intermediate impact energies where the otherwise powerful perturbation methods are of limited applicability. On the other hand, non-perturbative methods are expected to be more adequate for this particular energy region, as also reviewed in Volume 1. The other level of challenges for theoreticians encompasses collisions with participation of two and more active electrons. Such collisions are characterized by electron-electron correlation effects that are known to play an important role at low and high energies, but have been reported in the past to be of minor significance at intermediate energies, as also analyzed in Volume 1. Yet another challenging aspect of the theories presented in this book is the area of molecular targets, among which water was chosen for illustration because of its pivotal role in medical physics. In medicine, and more precisely in radiotherapy, water is considered as a tissue-like matter because of its abundance in the human body. For this reason, most benchmark computations on ion energy losses in tissue, as needed for dose planning in radiotherapy, are carried out with water as a prototype tissue-like target. Here, the challenges are heightened by a large number of open channels, on top of the difficulties in handling the accompanying particle transport phenomena, such as energy as well as range fluctuations called straggling. One of the endpoints in this endeavor is to evaluate precisely the dose to be delivered by energetic ions at a given position located deeply in the treated body where the tumor resides. This is a very complex problem indeed, which demands many-layered expertise from nuclear and atomic collision physics. Part of Volume 1 is an initial contribution, which needs to be further pursued.

In light of the above-outlined theme, our aim in Volume 1 is to present some of the leading classical, semi-classical and quantum-mechanical methods from ion-atom and ion-molecule collisions that could be relatively readily extended to a wider class of biomolecular targets of direct relevance to cancer therapy by energetic light ions. This is motivated by the intention of encouraging a systematic participation of a larger number of atomic and molecular collision physicists to the important problem of comprehensive and accurate determination of energy losses of ions when they traverse biomolecular targets, including tissue. This motivation is supported by the advantageous circumstance that the existing atomic collision physics methods can be adopted to biomolecular targets without undue difficulty, by using molecular wave functions in terms of linear combinations of Slater-type orbitals. This has indeed already been shown with the boundary-corrected first Born approximation and a simplified continuum distorted wave method. The next step would be to incorporate these methods into several powerful Monte Carlo simulations of energy losses of ions transported through tissue-like materials alongside their secondary particles (electrons and lighter nuclei). Very recently the first results along these lines have been reported.

The need for ions in radiotherapy stems from the most favorable localization of the largest energy deposition, precisely at the tumor site with small energy losses away from the target. Such a dose conformity to the target is due to heavy masses of ions that scatter predominantly in the forward cone and lose maximal energy only near the end of their path in the vicinity of the so-called Bragg peak, as illustrated in Fig. 1. The heavy masses of nuclei preclude multiple scattering of the primary ion beam. This occurrence is responsible for only about 30% of ion efficiency in killing tumor cells. However, ionization of targets by fast ions yields electrons that might be of sufficient energy to produce further radiation damage. These so-called delta-electrons can accomplish the missing 70% of tumor cell irradication mainly through multiple scattering enabled by the small electron mass. Therefore, energy depositions by both heavy (nuclei) and light (electrons) particles need to be simultaneously transported in Monte Carlo simulations. This is yet to be implemented in the existing codes.

Ions, electrons and photon beams are traditionally viewed as ionizing radiation modalities. This term "ionizing radiation" for cell kill is actually misleading, as it necessarily implies that only ionization is of relevance to radiotherapy as far as electromagnetic interactions are concerned. This is hardly the case, however, since radiation damage can be made by

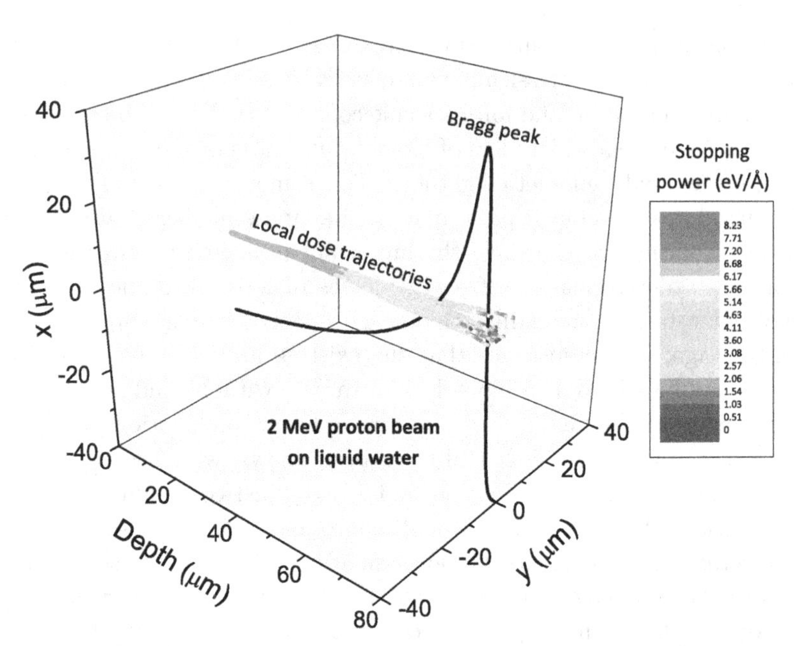

Fig. 1. Simulated trajectories and the associated local dose deposited by a 2 MeV proton beam colliding with liquid water target. To avoid clutter, only a few trajectories are shown and the displayed depth-dose curve with the Bragg peak is qualitative. The figure is from chapter 8 with permission of the authors (Garcia-Molina *et al.*) who are hereby kindly thanked.

many other electronic processes such as excitation, electron capture, etc. Ionization and excitation are included in the Bethe-Bloch formula for stopping power, but not charge exchange. However, processes that govern charge states of projectiles, such as electron capture, are important when initially fast ions suddenly slow down near the Bragg peak. For example, 2 MeV protons from the highlighted figure steadily keep the initial charge 1 for most of their track, but in the vicinity of the Bragg peak they easily become neutralized by electron capture from liquid water targets. The stopping power of the formed hydrogen atoms differs from that of protons. This implies that the Bethe-Bloch formula needs to be improved due to the missing charge exchange channel, which is important near the Bragg peak. The situation is actually more complicated because electron capture is promptly counteracted by electron loss, which is ionization of the formed hydrogen atoms in this case (or dressed ions in a general case). Only within the last few centimeters of the proton path, do literally electron capture and electron loss continuously and interchangeably take place thousands of times. Such a dynamic charge state of projectiles needs to be properly taken into account through, e.g. the appropriate rate equations. Instead, the common practice in Monte Carlo simulations is to use an empirical velocity-dependent screened nuclear charge of projectiles via the so-called Barkass effect. A number of other empirical formulae are employed in these stochastic algorithms for cross sections that are the most important input data for evaluation of ion energy losses. One of the goals of Volume 1 is to offer alternatives to these phenomenolgical approaches by presenting the perturbative and non-perturbative methods, both deterministic and stochastic (as well as their mixture) from atomic and molecular collision physics, with no adjustable parameters.

As per these remarks, there is plenty of room for improvement of the description of ion energy losses in matter, relative to the current situation encountered in, e.g. therapy by high-energy light ions. Part of the reason for relying heavily upon empirical cross section formulae in Monte Carlo codes for ions in hadron therapy is the lack of cross-talk between theory of atomic/molecular collisions and particle transport physics, when it comes to applied physics. One of the goals of Volume 1 is to bridge this gap and to initiate this type of cross-talk on a deeper and more fruitful level for the benefit of everyone involved, as per the spirit of this series.

When it comes particularly to hadron therapy, the interdisciplinarity of this series is inspired by the seminal work of physicist Robert Wilson who in the 1940s was the first to suggest in the medical journal *Radiology*, that protons, alpha particles or carbon nuclei and other light ions with their Bragg-peak-tailored dose distribution should provide optimal target conformity and tumor control. The significance of that proposal was in having been conveyed in such a convincing manner that physicians were quick to accept it. Moreover, physicians wasted no time in beginning patient treatments by light ions already in the 1950s. Such a unique example is admittedly hard to match, but this series will nevertheless strive whenever feasible to present the concepts and methodologies from one field in a way which should be relatively readily grasped and used by other research branches, including the life sciences.

- We will now go to the brief outlines of each of the chapters from Volume 1 with their most salient features.

Chapter 1 reviews recent studies on differential cross sections for double electron capture, as well as for simultaneous electron transfer and excitation of the residual target ion in proton-helium collisions at intermediate impact energies. Experimental data are compared to computations based upon perturbative and non-perturbative theories. The reported comparisons between experiment and theory show that the most interesting/important electron-electron correlation effects are difficult to identify in double capture and in transfer accompanied by target excitation in this energy region, because they are weak and masked by the projectile-residual target ion interaction.

Chapter 2 is on state-selective electron capture in collisions of alpha particles with helium targets at intermediate energies. State-selective capture mechanisms are identified from longitudinal momentum spectra. The recorded experimental data are compared with the classical trajectory Monte Carlo theory and good agreement is obtained at all the investigated energies.

Chapter 3 addresses the problem of double ionization as well as of single and double electron capture at intermediate energies. Two non-perturbative theories are employed, the Monte Carlo event generator and the coupled-channel basis generator methods. The independent particle formulation of the latter method is shown to work well even for double capture at the studied intermediate energies.

Chapter 4 presents several first-order four-body quantum-mechanical theories for fully differential cross sections in proton-helium collisions at intermediate energies involving single as well as double electron capture and also electron capture with target excitation. It was found that the electron-electron correlation effects in either the initial or final bound state have no appreciable influence on the computed angular distributions of scattered protons.

Chapter 5 deals with approximate three-body scattering states for the system of three charged particles that satisfy the correct asymptotic Coulomb boundary conditions for single electron capture as well as for direct and resonance ionization. It is shown that the interference of the distorted plane and spherical waves in scattering states can advantageously be used to analyze post-collisional interaction effects for spectra of auto-ionized electrons. Moreover, within the continuum intermediate state approximation, an illustration is given demonstrating that the classical Thomas double scattering could alternatively be described by interference between the distorted plane and spherical waves from the scattering Coulomb function.

Chapter 6 reviews total cross sections for several four-body collisional processes involving lithium nuclei and helium targets at intermediate as well as high impact energies. The possible reasons for the discrepancies among a number of theoretical predictions based upon classical and quantum mechanics are thoroughly discussed by reference to the available experimental data. Special attention has been paid to single and double electron capture, simultaneous transfer and ionization as well as to single and double ionization.

Chapter 7 is concerned with computations of cross sections for one- and two-electron processes in collisions of protons, alpha particles and carbon nuclei with water molecules. Studied are single electron capture, single ionization, double capture, transfer ionization and water fragmentation. Two methods are used, improved impact parameter classical trajectory Monte Carlo simulations and a semi-classical treatment with expansions in terms of molecular wave functions. A scaling law with respect to the projectile charge is derived at high energies and cross sections for molecular fragmentation after electron removal by single capture as well as single ionization are estimated.

Chapter 8 reports on the depth-dose profile and the spatial distribution of proton beams colliding with liquid water molecules. A combination of molecular dynamics and Monte Carlo simulations is used to investigate electronic stopping power, target electronic excitations, energy-loss straggling, elastic scattering as well the projectile charge-state due to electronic capture and electron loss processes. It is shown that electron capture becomes significant near the Bragg peak. Moreover, computations demonstrate that the position and the width of the Bragg peak is predominantly determined by the stopping power and energy straggling, respectively.

• Overall, as per chapters 1–8, this book, like any other book for that matter, is not a panacea for a straightforward knowledge transfer from one discipline to another. This is the case because, quite expectedly, there is no single theory, which would work universally by covering all the impact energies and all the collisions governed by electromagnetic interactions of light ions with atomic and molecular targets. Such theory is yet to come and this is the subject of active research in atomic and molecular collision physics. Nevertheless, for the purpose of, e.g. Medical Physics, Monte Carlo simulations of ion transport can still have a promising prospect from a judicious combination of the presented methods with their respective validity domains to cover most of the collisional phenomena of relevance to hadron therapy regarding electronic stopping power. This would be possible under the provision that a systematic extension to molecular targets, such as water molecules, will be soon accomplished e.g. along the lines discussed above or in some other alternative directions. Once this is achieved, pre-computations of cross sections from the best performing methods could be done to cover all the cases of interest in practice. This, in turn, would provide an invaluable modern recommended data base stored as a tabular interface from which fast and direct sampling could be made for Monte Carlo simulations. Such a procedure would make obsolete the ubiquitous reliance upon empirical formulae with fitting parameters adjusted to a limited set of the available experimental data. Some of the methods presented in Volume 1 are already near this sought operational stage, whereas others are to reach this status in the nearest future. It is hoped that Volume 1 of this series will spur a wider interest in the community of atomic and molecular collision physics for a more proactive participation to solving acute problems in Medical Physics with a particular focus on therapy by energetic light ions by using the available methodologies from Physics' first principles instead of phenomenological approaches.

Editor-in-Chief: Dževad Belkić
Professor of Mathematical Radiation Physics
Nobel Medical Institute, Karolinska Institute
Stockholm, Sweden
July 14, 2012

Acknowledgments

We are grateful for the support from the Swedish Cancer Society Research Fund (Cancerfonden), the Radiumhemmet Research Fund, the Karolinska Institute Research Fund and the COST Action MP1002 "Nano-scale insight in ion beam cancer therapy (Nano-IBCT)".

Contents

Chapter 1

Electron Capture Processes in Ion-Atom Collisions at Intermediate Projectile Energies

M. Schulz*
Department of Physics and LAMOR,
Missouri University of Science & Technology,
Rolla, Missouri 65409, USA
schulz@mst.edu

A. L. Harris
Department of Physics, Henderson State University,
Arkadelphia, Arkansas 71999, USA

T. Kirchner
Department of Physics and Astronomy,
York University, Toronto, Ontario, Canada M3J 1P3

D. H. Madison
Department of Physics and LAMOR,
Missouri University of Science & Technology,
Rolla, Missouri 65409, USA

Recent studies on differential cross sections for double capture as well as transfer and target excitation of intermediate energy p + He collisions are reviewed. Experimental data are compared to both perturbative and non-perturbative calculations. One of the main interests in two-electron processes in general is the role of electron-electron correlations. The comparison between experiment and theory reveals that such effects are difficult to identify in double capture and transfer and target excitation in this energy regime because they cannot be easily disentangled from features related to the projectile-residual target ion interaction.

*The author to whom corrspondence should be addressed.

1. Introduction

Of the fundamental forces acting in nature the Coulomb force is the one which is by far the best understood. As a result the properties of the most basic atom, the hydrogen atom, consisting of only two particles, are essentially completely known since the Schrödinger equation can be solved exactly for two charged particles interacting through the Coulomb force. In contrast, the nuclear counterpart to the hydrogen atom, the deuteron, also consisting of only two particles, is not nearly as well understood because the underlying strong force is not fully explored yet. However, even if the underlying force is precisely known, the Schrödinger equation is not analytically solvable for more than two mutually interacting particles. Therefore, properties of more complex atoms containing 2 or more electrons can only be calculated using iterative methods, such as the Multi-Configuration Hartree-Fock procedure (e.g. Ref. 1). Such numerical methods often provide accurate binding energies and wave functions of stationary atoms.

More dynamic systems, such as collisions of charged particles with atoms, represent a significantly bigger challenge to theory (e.g. Refs. 2–5). To accurately describe such systems requires much more modeling than for stationary single atoms. The approximations and assumptions entering in the theoretical models have to be tested by detailed experimental data, ideally by kinematically complete measurements. In such experiments the momentum vectors of all collision fragments are determined, from which fully differential cross sections (FDCS) can be extracted. For processes in which the active electrons remain bound to one of the two collision partners, i.e. for excitation and capture processes, the measurement of the scattered projectile momentum already constitutes a kinematically complete experiment because the momentum of the recoiling target atom is given by momentum conservation. In fact, due to energy conservation, it is sufficient to measure just two of the projectile momentum components. Therefore cross sections differential in the projectile solid angle, which is comprised of two momentum components in spherical coordinates, already represent FDCS. A rich literature on such measurements exists (e.g. Refs. 3, 6–14).

Ionization processes leave at least three unbound particles in the final state so that the momentum vectors of at least two particles have to be measured directly in order to extract FDCS. For electron impact, such experiments have been performed for several decades (for reviews see e.g. Refs. 15, 16) since the pioneering work of Ehrhard *et al.*[17]

However, for ion impact these measurements are much more difficult and have become feasible only with the development of Cold Target Recoil Ion Momentum Spectroscopy (COLTRIMS) (e.g. Refs. 18, 19). Since then an extensive literature on FDCS measurements has been generated (e.g. Refs. 4, 5, 20–24). One important advantage of studying FDCS for ionization is that the final state possesses a higher degree of freedom than in excitation or capture processes so that an experiment completely determining the final state space offers a more sensitive test of the theory.

For collision systems containing more than one electron, one aspect of the collision dynamics that has attracted particularly strong interest is the role of electron-electron correlations. Such effects tend to be rather unimportant in processes in which only one electron undergoes a transition, but they can be crucially important in two-electron processes like e.g. double ionization (e.g. Ref. 25 and references therein) or double excitation.[26] The degree of freedom of the final state is even further enhanced compared to the corresponding one-electron process so that, from this point of view, studies especially of double ionization should be ideally suited to test the theory. However, this high degree of freedom, combined with the usually very small cross sections, make measurements of FDCS for double ionization very difficult. Such studies for electron impact have been reported by two groups only[27,28] and for ion impact the literature is even more sparse: there, only one nearly fully differential measurement was reported, which was differential in the energy difference between the ejected electrons, but not in the individual energies.[29] Although valuable information on the role of electron-electron correlations has been obtained from these measurements, and more recently from studies applying innovative analysis techniques to less differential data (e.g. Ref. 30), the role of such effects in double ionization is not fully understood yet. This is largely due to the fact that the high degree of freedom of the final state, which is rather advantageous for single ionization, actually becomes too large in double ionization, which not only makes fully differential measurements very difficult, but also theoretical analysis very complex.

Additional information about the role of electron-electron correlations in atomic collisions can be obtained from studies of other two-electron processes. Double excitation is kinematically much simpler than double ionization because the final state involves three unbound particles (the doubly excited state decays almost exclusively through autoionization leading to a free electron). However, the identification of the process is more difficult and it requires measuring either the electron energy (e.g. Ref. 31) or the projectile energy-loss[32] in high resolution. Furthermore, the interpretation of the data is rather complicated. In addition to double excitation followed by autoionization, direct ionization also contributes to the electron flux. Since these two channels are indistinguishable, the electron energy spectra exhibit pronounced interference effects.[33]

An appealing alternative to study the role of electron-electron correlations is offered by two-electron processes involving capture of at least one electron. Transfer ionization (TI; i.e. one target electron is captured by the projectile and a second electron is ejected to the continuum) has been studied extensively (e.g. Refs. 34–38). There, the final state has the same degree of freedom as in double excitation followed by autoionization, but it is not subject to the complications introduced by the interference effects in the latter process described above. The theoretical analysis is even less complicated for transfer and target excitation (TTE; i.e. one target electron is captured by the projectile and a second target electron is excited), especially when p + He collisions are studied. There, the final electronic state can be expressed in terms of a simple product of two hydrogenic wave functions, which makes the incorporation of the electron-electron interaction in the transition amplitude much easier. Furthermore, the final state involves only two unbound particles so that a fully differential study is also much easier from an experimental point of view than for the two-electron processes mentioned above. Total cross sections for TTE were already measured two decades ago (e.g. Refs. 39–41), but FDCS have only become available relatively recently.[42,43]

Studies of double capture (DC) not only share the advantages of TTE, but an additional benefit for p + He collisions is that the DC process is intrinsically state-selective since the ground state is the only bound state of the H^- ion. Furthermore, this final state is highly correlated, in sharp contrast to TTE. Therefore DC and TTE studies are complementary as they probe different aspects of electron-electron correlation effects. Measured total DC cross sections have been available for a long time (e.g. Refs. 44–50). Several data sets on differential cross sections for He^{2+} + He collisions were also reported (e.g. Refs. 13, 46, 51, 52), but only one differential measurement has been performed for p + He collisions so far.[11]

Theoretically, capture processes involving two active electrons have been treated by perturbative (e.g. Refs. 53–56), such as 4-body distorted wave (4DW) approaches, and non-perturbative methods (e.g. Refs. 12, 57–60). One of the more recent non-perturbative models is based on the

Basis Generator Method (BGM).[61] The advantage of the BGM is that it uses dynamically adapted basis sets so that couplings to states not representing the process of interest are accounted for during the time propagation of the system. On the other hand, the distorted wave approaches incorporate electron-electron correlations, while the BGM represents an independent electron model. In the case of DC in p + He collisions, this means that the highly correlated final state of the H^- ion cannot be realistically described and the BGM approach is thus not feasible to treat this process. In this chapter we review recent studies on TTE and DC. Because of the advantages mentioned above, we will focus on experimental and theoretical work performed for p + He collisions.

2. Experimental Methods

A sketch of the ion-atom collision laboratory at Missouri University of Science & Technology is shown in Fig. 1. The processes of interest in this chapter are

$$\text{(a) } p + He \rightarrow H^- + He^{2+} \quad \text{(DC)} \quad \text{and}$$
$$\text{(b) } p + He \rightarrow H^0 + He^{+*}(nl) \quad \text{(TTE)}.$$

In the case of DC, the process is already unambiguously determined by a charge analysis of the projectile because this process leaves the residual target ion without any electrons. Furthermore, as mentioned above, there is only one bound state in the H^- ion so that the charge analysis even determines the final electronic state. The only other information that is required to extract FDCS from an experiment is the projectile scattering angle. This can be accomplished by detecting the H^- ions (selected by the switching magnet, see Fig. 1) with a position-sensitive detector, which has been done for intermediate projectile energies (25 to 150 keV).[11]

For large projectile energies it is very difficult to achieve a sufficient angular resolution because typical scattering angles are of the order of tens of micro-radians or even smaller. An alternative is to deduce the scattering angle from the transverse recoil-ion momentum p_{rt} measured using COLTRIMS (e.g. Refs. 18, 19). Since DC kinematically represents two-body scattering, the transverse projectile momentum component must have the same magnitude as the transverse recoil momentum component p_{rt}. The scattering angle is thus given by

$$\theta = \sin^{-1}(p_{rt}/(Mv_p)), \tag{1}$$

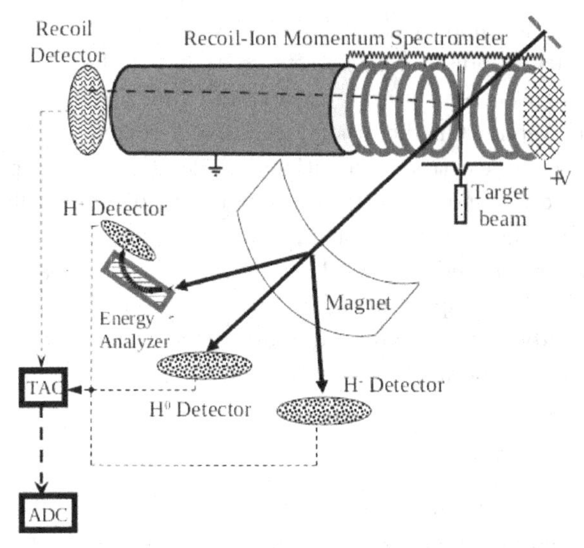

Fig. 1. Sketch of ion-accelerator laboratory at the Missouri University of Science & Technology. The actual accelerator is not shown. The key components are a recoil-ion momentum spectrometer, a switching magnet and several projectile detectors located at various ports of the magnet, so that the projectiles can be detected in various charge states. The proton component of the charge analyzed beam can also be energy-analyzed; however, this feature is not used in the studies reviewed here. The recoil-ion detector is set in coincidence with one of the projectile detectors, depending on the process of interest.

where M and v_p are the projectile mass and speed, respectively. The angular resolution achievable with this method actually improves with increasing projectile speed (e.g. a few micro-radians at a collision energy of 5 MeV). One disadvantage is that the detection of the recoil ion alone is not sufficient to unambiguously identify DC. Rather, the recoil ions have to be measured in coincidence with the charge-analyzed projectiles. This has been done for He^{2+} + He collisions,[13,52] but no fully differential DC data measured with COLTRIMS have been reported for p + He collisions yet.

To measure FDCS for TTE is significantly more involved than for DC. A charge analysis of the projectile only determines whether one target electron was transferred to the projectile, but it gives no information as to whether the residual target ion and the projectile are in the ground state or in an excited state. There are two possibilities to identify the final state of the target ion: first, the photon emitted by the decay of the excited state can be detected. However, one important disadvantage is that the photon detection efficiency is very small, making a coincidence measurement with the charge-exchange projectiles very challenging. Without such a coincidence set-up, TTE cannot be distinguished from ionization plus excitation and no differential cross sections can be obtained.[39–41] Furthermore, to also determine the final state of the projectile would require a coincidence set-up between the photons emitted by both collision partners, which is even more difficult than a photon-projectile coincidence.

The second possibility to experimentally determine the final state of both collision partners is to measure the Q-value of the collision, i.e. the difference in total binding energies between the final and initial state of both collision partners. For a process involving the capture of one target electron in which no electron is ejected to the continuum, the Q-value is to a very good approximation related to the longitudinal recoil-ion momentum component p_{rz} by (e.g. Refs. 18, 19)

$$p_{\mathrm{rz}} = \frac{Q}{v_{\mathrm{p}}} - \frac{v_{\mathrm{p}}}{2}. \tag{2}$$

At the same time, the projectile scattering angle is determined by the transverse recoil-ion momentum component (see equation (1)). Therefore, experimental differential cross sections for TTE can be extracted from a coincidence measurement between the momentum-analyzed recoil ions and the charge exchanged projectiles.

3. Theory

The central task of (atomic) scattering theory is to formulate, approximate, and calculate the transition (T) -matrix elements T_{fi} for the reaction of interest. This can be done in a number of different ways. Here, we consider two approaches for the particular case of p + He collisions: (i) 4-body distorted-wave theory; (ii) the nonperturbative basis generator method on the level of the independent electron model (IEM) within the impact-parameter picture. Atomic units will be used in this section.

3.1. *4-Body Distorted Wave (4DW) Theory*

For single capture, DC, and TTE, the fully differential cross section is differential only in the projectile solid angle, and can be written as

$$\frac{d\sigma}{d\Omega} = (2\pi)^4 \mu_{pa}\mu_f \frac{k_f}{k_i}|T_{fi}|^2, \tag{3}$$

where μ_{pa} is the reduced mass of the projectile and target atom, μ_f is the reduced mass of the outgoing particle and the residual ion, and $\mathbf{k}_i$ ($\mathbf{k}_f$) is the wavevector of the incident (scattered) projectile. The 4DW models[55,56] use a distorted wave approach and two potential formulation, where the final state interaction potential is written as $V_f = U_f + W_f$. In the two potential formulation, U_f is the part of the potential that is used to calculate the distorted waves and W_f is the perturbation. The exact T-matrix can be written in the post form as[62]

$$T_{fi} = \langle \eta_f^{(-)} | V_i | \Phi_i \rangle + \langle \eta_f^{(-)} | W_f | (\Psi_i^{(+)} - \Phi_i) \rangle, \tag{4}$$

where $\eta_f^{(-)}$ is the distorted wave for the potential U_f, V_i is the initial state interaction potential, Φ_i is the unperturbed initial state wave function, W_f is the final state perturbation, and $\Psi_i^{(+)}$ is the exact initial state wave function. The T-matrix for any four-body process is a 9-dimensional integral, and within the 4DW models, this integral is calculated completely numerically. In equation (4), the first term represents the contribution from first order perturbation theory, and the second term represents contributions from all higher order terms. Note that if the exact initial state wave function $\Psi_i^{(+)}$ is approximated as the unperturbed wave function Φ_i (a plane wave times the helium atom wave function), then equation (4) becomes the first order Born approximation

$$T_{fi}^B = \langle \eta_f^{(-)} | V_i | \Phi_i \rangle. \tag{5}$$

The initial state projectile-atom interaction V_i is given by

$$V_i = \frac{Z_\mathrm{p} Z_\mathrm{nuc}}{r_1} + \frac{Z_\mathrm{p} Z_\mathrm{e}}{r_{12}} + \frac{Z_\mathrm{p} Z_\mathrm{e}}{r_{13}}, \tag{6}$$

where Z_p, Z_nuc, and Z_e are the charges of the projectile, nucleus, and electron respectively, and r_1, r_{12}, and r_{13} are the magnitudes of the relative coordinates for the projectile-target nucleus and projectile-target electrons.

If the initial state projectile wave function is approximated as something more sophisticated than a plane wave, the second term of equation (5) does not vanish, and parts of all higher order terms remain in the perturbation series.

In either case, the exact initial state wave function $\Psi_i^{(+)}$ is approximated as a product of a projectile wave function times an atomic wave function

$$\Psi_i^{(+)} = \chi_i^{(+)} \xi_{\text{He}}, \tag{7}$$

where $\chi_i^{(+)}$ is the incident projectile wave function and ξ_{He} is the ground-state helium atom wave function. The final state distorted wave function $\eta_f^{(-)}$ is also approximated as a product of wave functions for the final state particles. For single capture and TTE,

$$\eta_f^{(-)} = \chi_f^{(-)} \psi_{\text{He}^+} \phi_{\text{H}}, \tag{8}$$

where $\chi_f^{(-)}$ is the scattered hydrogen wave function, ψ_{He^+} is the final state He^+ wave function, and ϕ_{H} is the captured electron wave function. Both ϕ_{H} and ψ_{He^+} are hydrogenic wave functions, and thus known exactly. For DC,

$$\eta_f^{(-)} = \chi_f^{(-)} \psi_{\text{H}^-}, \tag{9}$$

where ψ_{H^-} is the wave function for the outgoing H^- ion, and $\chi_f^{(-)}$ is the scattered projectile wave function.

3.2. *The Basis Generator Method within the Impact-Parameter Picture*

If one considers the special case $W_f = V_f$ (i.e., $U_f = 0$) in equation (4) one arrives at the standard expression

$$T_{fi} = \langle \Phi_f^{(-)} | V_f | \Psi_i^{(+)} \rangle \tag{10}$$

for the T-matrix, in which $\Phi_f^{(-)}$ is the unperturbed final state wave function. Equation (10) can be used as the starting point for the introduction of the impact-parameter picture. One assumes (again) that electronic and heavy particle motions can be separated and one describes the latter in terms of plane waves. Equation (10) then takes the form

$$T_{fi} = \frac{1}{(2\pi)^3} \int \langle \phi_f | \hat{V}_f | \Psi_i \rangle e^{i\mathbf{q}\cdot\mathbf{r}_1} d^3 r_1 \tag{11}$$

with the momentum transfer $\mathbf{q} = \mathbf{k}_i - \mathbf{k}_f$. If one further assumes that the wave function Ψ_i solves the *time-dependent* Schrödinger equation (TDSE) for the Hamiltonian

$$\hat{H} = \hat{H}_{\text{He}} + V_i, \tag{12}$$

where $\hat{H}_{\text{He}}$ is the Hamiltonian of the helium atom and $\mathbf{r}_1$ in V_i (equation (6)) is treated as a classical straight-line trajectory $\mathbf{r}_1(t) = \mathbf{b} + \mathbf{v}_p t$ with the impact parameter **b** and the perpendicularly directed constant projectile velocity $\mathbf{v}_p$ one arrives[a] at the expression

$$T_{fi} = \frac{iv_p}{(2\pi)^3} \int e^{i\mathbf{q}\cdot\mathbf{b}} A_{fi}(\mathbf{b}) d^2b. \tag{13}$$

A_{fi} is the impact-parameter-dependent transition amplitude to the asymptotic electronic state ϕ_f

$$A_{fi}(\mathbf{b}) = \lim_{t\to\infty} \langle \phi_f | \Psi_i \rangle \tag{14}$$

and is obtained from the solution of the TDSE.[2]

As a consequence of treating $\mathbf{r}_1$ classically the projectile-target nucleus interaction in equation (6) is a merely time-dependent term. Hence, equation (14) can be written as

$$A_{fi}(\mathbf{b}) = a_{fi}(\mathbf{b}) \exp\left[-2i \int_0^\infty \frac{Z_\mathrm{p} Z_\mathrm{nuc}}{r_1(t)} dt\right], \tag{15}$$

where $a_{fi}(\mathbf{b})$ is a transition amplitude of the TDSE from which the projectile-nucleus interaction has been removed. Still, the nonperturbative solution of this TDSE is a formidable problem because of the presence of the electron-electron interaction in $\hat{H}_\mathrm{He}$. While it is not impossible to address it explicitly (see e.g. Refs. 63, 64), we have not attempted this, but have adhered to the IEM. This means that the electron-electron interaction is replaced by an effective potential such that the TDSE separates into a set of single-particle equations

$$i\partial_i \psi_j(\mathbf{r}, t) = \hat{h}\psi_j(\mathbf{r}, t) \quad j = 1, \ldots, N. \tag{16}$$

For the results presented below we have used the Hamiltonian

$$\hat{h} = -\frac{1}{2}\nabla^2 + V_\mathrm{eff}(r) - \frac{Z_\mathrm{p}}{|\mathbf{r} - \mathbf{r}_1|} \tag{17}$$

that includes the kinetic energy and (effective) target and projectile potentials. For V_eff we employ a helium ground-state potential obtained from the so-called exchange-only limit of the optimized potential method (OPM) of density functional theory.[65] The ground state of the spin-singlet helium atom is realized by populating the K shell with two equivalent electrons. As a consequence, the exchange-only OPM potential is equivalent

---

[a]If one exploits the large mass ratio of the heavy particles and the electrons and the fact that the scattering angles are small.

to the Hartree-Fock ground-state potential and only one (initial $1s$) orbital has to be propagated.

This is done in terms of a basis representation obtained from the two-center version of the basis generator method (TC-BGM). The general ideas of the (TC-) BGM have been explained at length in previous publications[61,66] and will not be repeated here. Let us just describe the TC-BGM basis set used for the results presented below. It includes finite sets of bound target and projectile states, which are multiplied by appropriate electron translation factors to ensure Galilean invariance. This two-center atomic orbital basis is augmented by time-dependent BGM pseudostates which account for ionization channels and for quasimolecular effects at low collision velocity. The TC-BGM ansatz translates equation (16) into a set of ordinary differential equations for the expansion coefficients that can be solved by standard methods.

Considering two-electron processes such as TTE within the IEM involves the calculation of appropriate products of those coefficients. This yields the amplitudes a_{fi}, which are subsequently multiplied by the projectile-target nucleus phase factor according to equation (15) and inserted in the T-matrix (13). The azimuthal integration over the impact parameter can be carried out and the two-dimensional Fourier integral reduces to a one-dimensional Bessel integral. For heavy-particle collisions the magnitude of the momentum transfer can be related to the polar scattering angle θ according to

$$q \approx 2k_i \sin\frac{\theta}{2} \approx k_i\theta \tag{18}$$

and the square of the T-matrix element is directly proportional to the projectile angular differential cross section in the center-of-mass system. This procedure for the calculation of differential cross sections is known as the eikonal approximation.[67]

4. Discussion

4.1. *Transfer and Target Excitation*

In Fig. 2, a two-dimensional recoil-ion momentum spectrum for 25 keV p + He collisions is shown.[42] The transverse and longitudinal recoil-ion momentum components p_{rt} and p_{rz} are plotted on the vertical and horizontal axes, respectively. The spectrum was measured in coincidence with the neutralized projectiles. Discrete lines in the p_{rz} dependence are

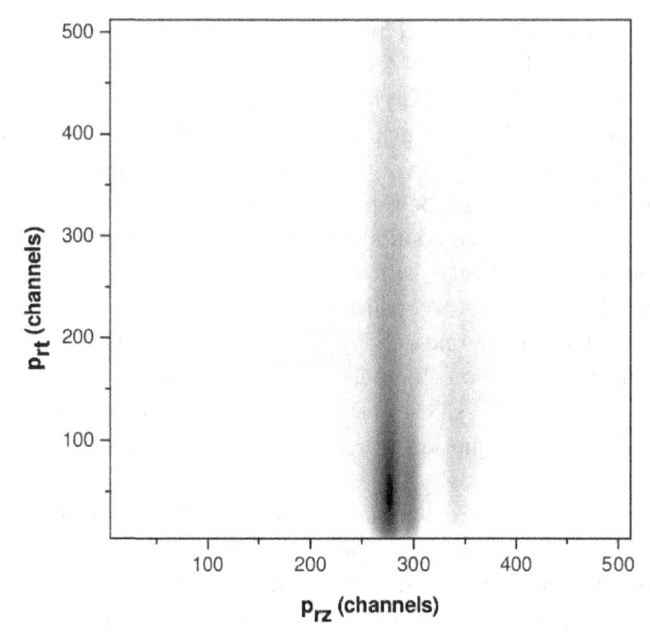

Fig. 2. Two-dimension recoil-ion momentum spectrum. The horizontal axis represents the longitudinal component, which is essentially determined by the Q-value (i.e. the inelasticity) of the process (see equation (2)). The transverse recoil-ion momentum component, plotted on the vertical axis, determines the projectile scattering angle (see equation (1)).

clearly visible reflecting the Q-value for the various capture processes and for different final states according to equation (2). The two main lines between channels 250 and 300 are due to single capture to the $n = 1$ and 2 states of the projectile leaving the residual target ion in the ground state. Capture to higher n-states cannot be resolved. TTE is represented by only one resolved line near channel 350. The experimental resolution was sufficient to separate TTE for capture to $n = 1$ from $n = 2$ and higher. Therefore, the presence of only one line demonstrates that TTE is dominated by capture to the ground state of the projectile. In contrast, the various excited n-states of the target ion could not be resolved and only a tail, extending beyond channel 350, representing $n > 2$ is barely visible.

The projection of the TTE line onto the p_{rt} axis represents the scattering angle dependence (see equation (1)) of the TTE cross sections differential in the scattering angle $\mathrm{d}\sigma/\mathrm{d}\theta$. The corresponding cross sections differential in the projectile solid angle $\mathrm{d}\sigma/\mathrm{d}\Omega = (2\pi \sin\theta)^{-1}\mathrm{d}\sigma/\mathrm{d}\theta$ are shown in Fig. 3 as a function of θ for projectile energies of 25, 50, and 75 keV.

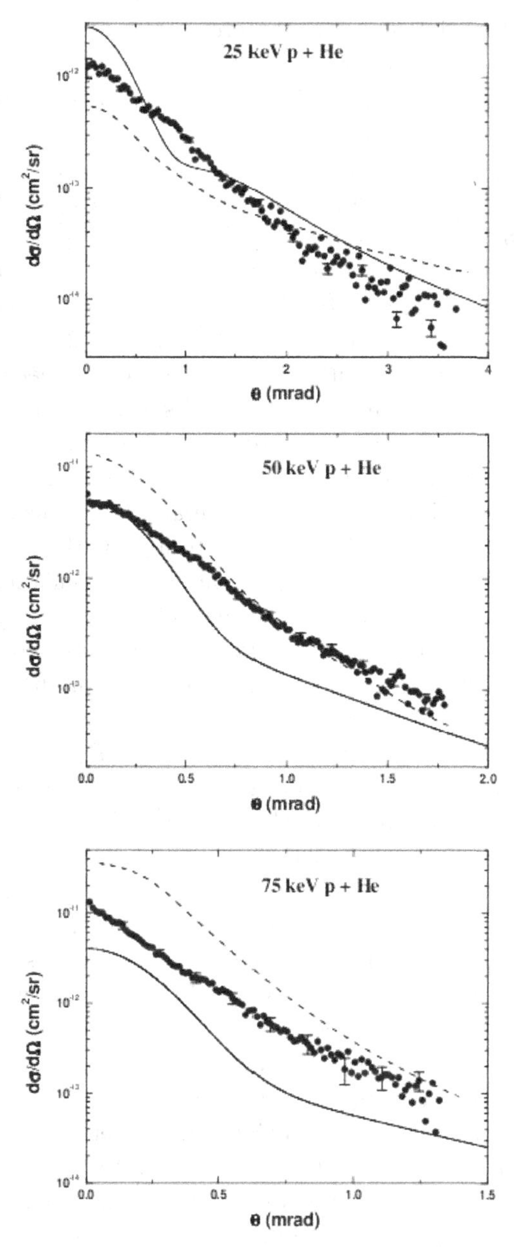

Fig. 3. Differential transfer and target excitation cross sections as a function of projectile scattering angle for collision energies of (from top to bottom) 25, 50, and 75 keV. The dashed curves show 4DW calculations and the solid curves BGM calculations (see text for detail).

The solid and dashed lines show the cross sections computed with the BGM[12] and 4DW[55] models, respectively. Neither calculation is in good agreement with the experimental data, nor do they agree with each other. This comparison between experiment and theory does thus not help much in understanding the collision dynamics underlying the TTE process. However, using the 4DW model, the relative importance of the various interactions within pairs of particles of the collision system was systematically studied by removing one interaction from the calculation at a time. It was found that within the 4DW model, electron-electron correlations are completely insignificant at a projectile energy of 25 keV, but become somewhat more important with increasing energy, especially at small scattering angles. At the same time, the results of the calculations changed sensitively when the interaction between the projectile and the residual target ion was removed. Much better agreement with the experimental data in the magnitude was obtained with that interaction included. Furthermore, at small projectile energies it is important to include the distortion of the incoming projectile wave by the target atom in the calculation.

For other two-electron processes, like e.g. TI or double target excitation (DE), the role of electron-electron correlations was studied by analyzing the ratios of the differential cross sections to those for the corresponding one-electron process as a function of scattering angle (e.g. Refs. 32,34,68). In the ratios of TI to single capture cross sections R_{TI}, pronounced peak structures were found at a scattering angle of about 0.5 mrad.[34] These were interpreted as due to the so-called Thomas mechanism of the second kind. In this process, the projectile first undergoes a binary collision with one of the target electrons such that the electron is initially moving at an angle of 45° relative to the projectile. Neglecting the initial momentum of the electron in the target ground state and applying the classical kinematic conservation laws, the electron comes out of this collision with a speed of $\sqrt{2}\, v_{\mathrm{p}}$ and the projectile gets deflected by 0.55 mrad. In the second step the electron then scatters off the second target electron. If it gets deflected by 45° in this second step as well, it will move in the direction of the projectile and, again as a result of the classical kinematic conservation laws, its speed must be equal to v_{p} so that capture will occur with high probability. For large v_{p} the energy transfer to the second electron is much larger than its initial binding energy so that it is lifted to the continuum. Classically, the requirement that the first electron should move at the same speed and in the same direction, in order to be captured with large probability, thus entails a critical scattering angle of the projectile, which leads to the peak structure in R_{TI}. In this Thomas mechanism of the second kind, electron-electron correlation plays a central role.

A similar peak structure was also observed in the double to single excitation cross section ratios R_{DE} at slightly larger scattering angles.[32,68] This was interpreted as due to interference between the first-order and second- (or higher) order double excitation amplitudes. The second-order mechanism proceeds through two independent interactions with both target electrons and does not require electron-electron correlation. However, in the first-order process, the projectile only interacts with one of the two electrons directly. The second electron is excited through electron-electron correlation which thus again plays a crucial role in the peak structure.

A similar analysis was performed for TTE. In Fig. 4 the ratios between TTE and single capture cross sections R_{TTE} are shown as a function of scattering angle. At all projectile energies a steep increase of R_{TTE} is found at small scattering angles followed by a plateau at large angles but no peak structure is observed. This is not really surprising. In TI the Thomas peak of the second kind is only significant at much larger projectile energies than discussed here. More importantly, it cannot contribute to TTE at all because here the second electron remains bound to the residual target ion. Therefore, any second scattering of the first electron from the target, following the interaction with the projectile, can only transfer momentum to the entire target ion, but not to only the second electron. In this case the process is known as the Thomas mechanism of the first kind and it can occur for single capture, leaving the target ion in the ground state, as well and it leads to a critical scattering angle of 0.47 mrad. In fact, for pure single capture this mechanism should even be more pronounced than for TTE because at these projectile energies the first electron is not fast enough to transfer sufficient energy to the second electron to lift it to an excited state. Therefore, no peak structure (or perhaps even a minimum) due to a Thomas-type of mechanism near 0.5 mrad is expected in the scattering angle dependence of R_{TTE}.

Interference between first- and second-order amplitudes is not likely to lead to a pronounced peak structure in R_{TTE}, either. This would require that both amplitudes are of similar magnitude. While the first-order amplitude is probably not unimportant, at these relatively small projectile energies it is nevertheless expected to be significantly smaller than the second-order amplitude. Furthermore, any structure could be further

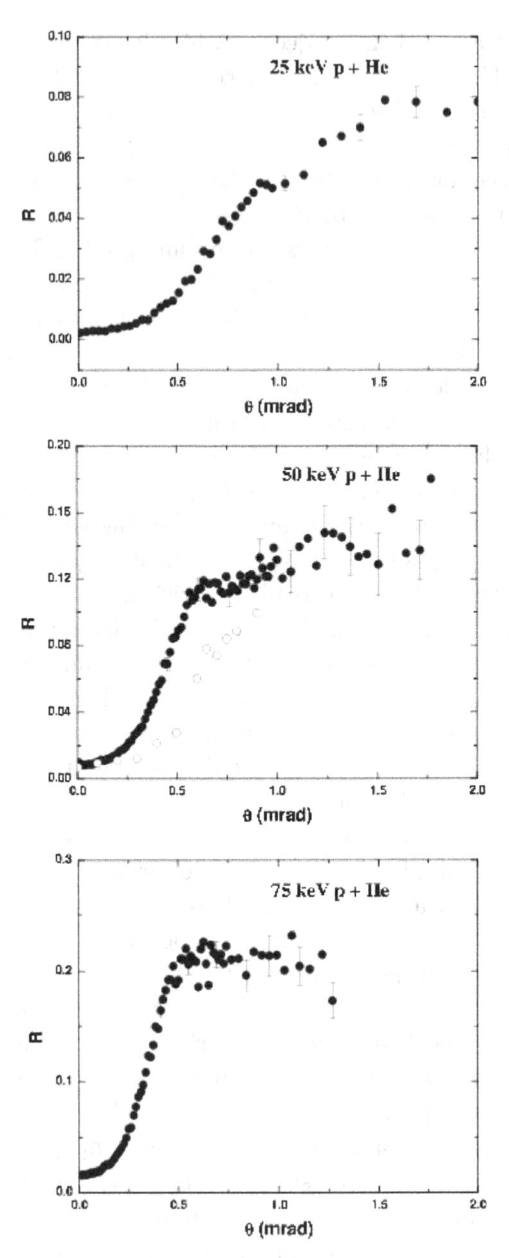

Fig. 4. Differential ratios between transfer and target excitation and single capture cross sections as a function of projectile scattering angle for the same energies as in Fig. 2. For 50 keV differential double to single excitation cross sections are also shown as open symbols.

"smeared out" by deflection of the projectile from the target nucleus, which becomes increasingly important with decreasing projectile energy. Even the ratio R_{TTE} may thus not be sensitive enough to reveal an observable peak structure.

A more sensitive analysis of the collision dynamics in TTE is offered by a comparison between the ratios R_{TTE} and R_{DE}. The latter have not been measured yet for projectile energies of 25 and 75 keV, but experimental values are available for 50 keV, which are shown as open symbols in Fig. 4. Within the IEM, and assuming that the elastic scattering between the projectile and the target core can be described classically, the differential TTE and single capture cross sections are given by:

$$\left(\frac{\mathrm{d}\sigma}{\mathrm{d}\Omega}\right)_{\mathrm{TTE}} = 2P_{\mathrm{SE}}(\theta)P_{\mathrm{SC}}(\theta)\left(\frac{\mathrm{d}\sigma}{\mathrm{d}\Omega}\right)_{\mathrm{el}} \tag{19}$$

$$\left(\frac{\mathrm{d}\sigma}{\mathrm{d}\Omega}\right)_{\mathrm{SC}} = 2P_{\mathrm{SC}}(\theta)P_{\mathrm{NT}}(\theta)\left(\frac{\mathrm{d}\sigma}{\mathrm{d}\Omega}\right)_{\mathrm{el}} \tag{20}$$

so that the ratio is given by $R_{\mathrm{TTE}} = P_{\mathrm{SE}}/P_{\mathrm{NT}}$. Here, P_{SE}, P_{SC}, and P_{NT} are the single electron probabilities for excitation, capture, and no electronic transition at all, and $(\mathrm{d}\sigma/\mathrm{d}\Omega)_{\mathrm{el}}$ is the differential elastic scattering cross section. Likewise, the differential DE and single excitation cross sections in the IEM are given by

$$\left(\frac{\mathrm{d}\sigma}{\mathrm{d}\Omega}\right)_{\mathrm{DE}} = P_{\mathrm{SE}}^{2}(\theta)\left(\frac{\mathrm{d}\sigma}{\mathrm{d}\Omega}\right)_{\mathrm{el}} \tag{21}$$

$$\left(\frac{\mathrm{d}\sigma}{\mathrm{d}\Omega}\right)_{\mathrm{SE}} = 2P_{\mathrm{SE}}(\theta)P_{\mathrm{NT}}(\theta)\left(\frac{\mathrm{d}\sigma}{\mathrm{d}\Omega}\right)_{\mathrm{el}} \tag{22}$$

and R_{DE} is given by $P_{\mathrm{SE}}/(2P_{\mathrm{NT}})$. Therefore, the double ratio $R = R_{\mathrm{TTE}}/R_{\mathrm{DE}}$ within the IEM is constant at 2.

Ratio R is plotted as a function of θ in Fig. 5. The dotted line indicates the constant value of 2 expected from the IEM. Although the experimental data averaged over θ is consistent with 2, the measured values are clearly not constant. Rather, a pronounced maximum is seen between 0.4 and 0.5 mrad. This at first glance may raise some hope that the double ratios R may indeed be more sensitive to electron-electron correlation effects. However, before such a conclusion can be drawn, other potential causes for the peak structure and for a deviation from 2 have to be considered. One aspect that is not properly accounted for in the simple analysis above is that the screening of the nuclear charges by the electrons depends on the specific

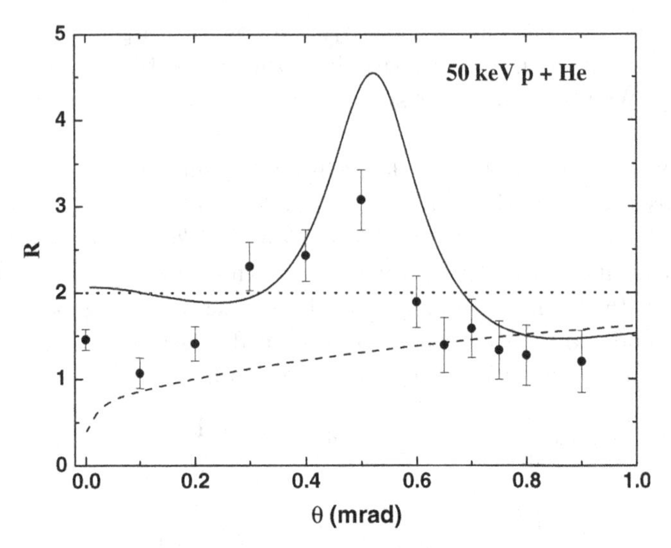

Fig. 5. Differential double ratio between transfer and target excitation to single capture and double to single excitation cross-section ratios (shown in Fig. 3) as a function of scattering angle. All theoretical curves are based on the BGM model. The dotted curves further use an independent electron model in which the effective nuclear charges of both collision partners are constant an independent of the process (equations (19) to (22)). For the dashed curves a variation of the independent electron model was used in that different effective nuclear charges were used for different processes. Both models treat the projectile as a classical particle. Finally, the solid curves show a BGM calculation in which the projectile is treated quantum-mechanically within the eikonal approximation (for details see text).

process, which could affect both $(\mathrm{d}\sigma/\mathrm{d}\Omega)_{\mathrm{el}}$ and the one-electron transition probabilities in equations (19) to (22). For example, if the excitation step in TTE follows the capture step, it occurs in the He^+ ion while for DE both excitation steps occur in neutral He. The dashed curve in Fig. 5 shows a BGM calculation which is based on a modified IEM, which accounts for the varying nuclear screening for different processes. Indeed, the screening has a significant impact on the θ-dependence of R, especially at small θ, but it does not lead to a peak structure.

Another approximation in the BGM calculation represented by the dashed curve is that the projectile is treated classically within an impact parameter formulation. The solid curve also shows results from the BGM approach, but this time elastic scattering of the projectile from the residual target ion is described quantum-mechanically using the eikonal approximation (BGM-EA) as described in Sec. 3.2. This calculation leads to significantly improved qualitative agreement with the experimental data in so far as a pronounced peak structure near 0.55 mrad is obtained. It should be kept in mind that the BGM model does not account for electron-electron correlations. Therefore, the mere existence of a peak structure is no evidence yet for a prominent role of such correlation effects. On the other hand, considerable discrepancies with the measured data remain both in the location of the peak and in the magnitude of R. It can thus also not be ruled out that R is significantly affected by electron-electron correlations. Calculations of R which treat the projectile quantum-mechanically and at the same time account for electron-electron correlations have not been reported yet. Until such theoretical results become available, the role of electron-electron correlations in TTE at such relatively small projectile energies cannot be conclusively evaluated.

4.2. *Double Capture*

Differential DC cross sections for p + He collisions at energies ranging between 15 and 150 keV are plotted as a function of θ in Fig. 6. The solid curves show 4DW calculations divided by 100 for selected energies.[56]

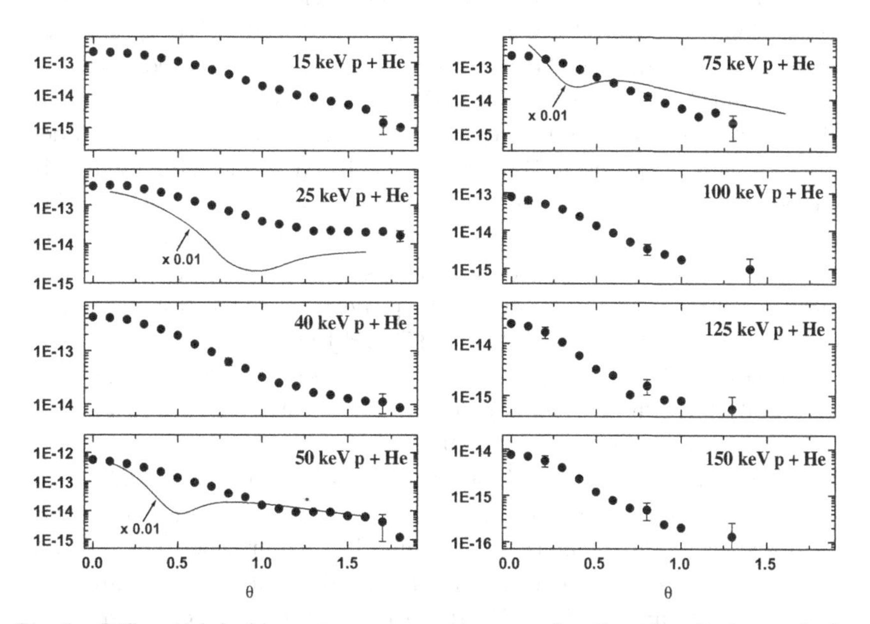

Fig. 6. Differential double capture cross sections as a function of scattering angle for projectile energies as indicated in the legends. The curves represent 4DW calculations divided by 100 (see text).

Here, both the initial He^0 and the final H^- states are described in terms of a Hylleraas wave function.[69] Severe discrepancies are quite apparent: pronounced minima occurring in the calculation at scattering angles between 0.3 and 0.9 mrad (depending on projectile energy) are not present in the data and the magnitude of the cross sections is overestimated by about two orders of magnitude. Since without correlation the H^- ion would not even form a bound state, the mere existence of non-zero cross sections already establishes the overpowering importance of electron-electron correlations in the final state in DC. One possible explanation for the large overestimation of the magnitude by theory is that such correlation effects might even be too strong in the calculation. On the other hand, another possible cause for these discrepancies could be higher-order effects, in which an electron captured by the projectile is recaptured by the target ion, which are not accounted for by the 4DW model. The influence of correlation in the initial state on the cross sections was also tested by replacing the Hylleraas wave function by a Hartree-Fock wave function. The difference between both calculations was found to be smaller than 20%.

The minimum obtained in the scattering-angle dependence of the DC cross sections with the 4DW model could be traced to the projectile-target nucleus interaction. When this interaction was removed from the perturbation this minimum disappeared. A similar structure was also found in the differential single capture cross sections, both with the 4DW model and the BGM-EA approach. However, in the BGM model the minimum is not present when the projectile is treated classically.[12] It is actually this minimum which leads to the peak structure in the calculated double ratio $R = R_{\mathrm{TTE}}/R_{\mathrm{DE}}$ discussed above. This suggests that a quantum-mechanical description of the scattering between the projectile and the residual target ion plays a similarly sensitive role in TTE and DC.

Similar to TTE, the DC to single capture cross section ratios may reveal more detail about the reaction dynamics. These ratios R_{DC} are shown in Fig. 7 for projectile energies of 25, 50, 75, and 100 keV as a function of θ. Except for 25 keV, a clear peak structure is found at around 0.7 mrad. Like in TTE, here too, the Thomas mechanism of the second kind cannot contribute because of kinematic restrictions. The requirement that the first electron must eventually move at the same speed and in the same direction as the projectile can only be satisfied in this process if the second electron is ejected from the target at an angle of 90° relative to the initial projectile beam axis,[70] i.e. if the second electron is not captured.

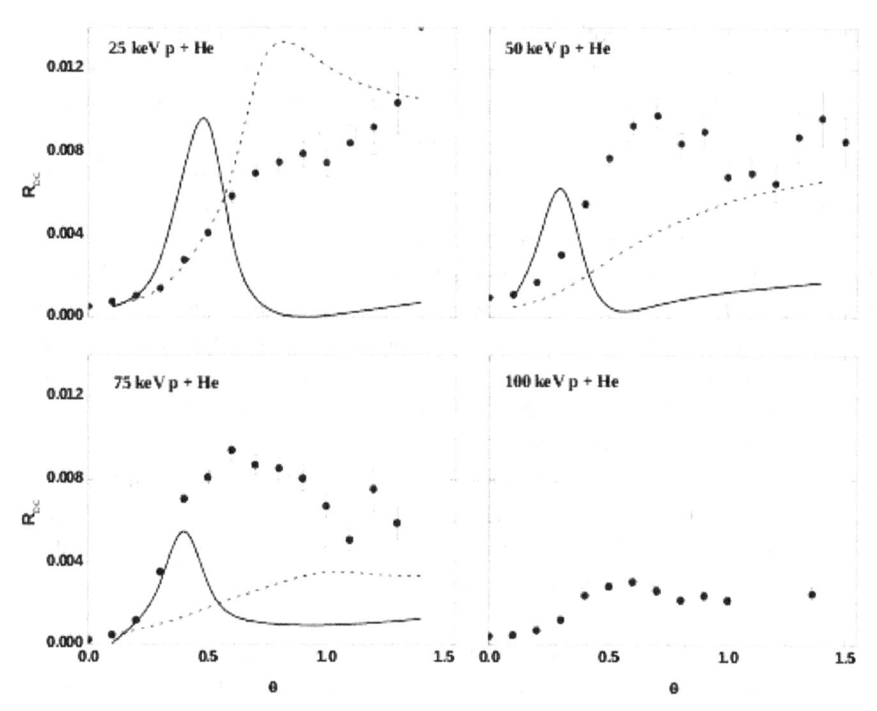

Fig. 7. Differential double to single capture cross section ratios as a function of projectile scattering angle for projectile energies as indicated in the legends. The solid and dashed curves are 4DW calculations with and without the projectile-residual target ion interaction, respectively.

DC channels involving electron-electron correlations other than the Thomas mechanism of the second kind could nevertheless be important. As mentioned above, interference of such processes with higher-order mechanisms are believed to lead to a peak structure in double to single excitation ratios.[32,68] However, there is one important difference to the double to single capture ratios: while in double excitation the peak structure was not observed at projectile energies below 150 keV, in DC it was most pronounced between approximately 50 and 75 keV. This can be explained by considering that the relative importance of higher-order transitions, involving two independent interactions of the projectile with both electrons, varies with projectile energy quite differently for both processes. The total single capture cross sections maximize near 25 keV (i.e. the smallest projectile energy considered here) and then fall off steeply.[71] The cross sections for DC proceeding through two independent projectile-electron interactions should fall off even steeper. At the same time the total cross sections for all contributions to DC maximize around 35 keV, implying that the contributions involving electron-electron correlations fall off much slower and that they constitute a significant fraction of total DC already at this energy. As a result, the amplitudes for the mechanisms with and without correlation become similar at relatively small projectile energies, which is a requirement for pronounced interference effects. In contrast, the total cross sections for single excitation maximize near 100 keV and vary rather slowly with projectile energy.[72] At the same time, the relative importance of the process involving electron-electron correlations increases with increasing projectile energy, so that here both amplitudes become similar at much larger projectile energies than in the case of DC.

Apart from electron-electron correlations, one has to consider the possibility that the peak structures in R_{DC} are, like in the double ratio R for TTE, related to the projectile-residual target ion interaction. In the 4DW model, indeed this interaction seems to play an overwhelming role. The solid and dashed curves in Fig. 7 show calculations based on this model with the projectile-residual target ion interaction included or not included, respectively. If this interaction is included, R_{DC} is dominated by a very strong peak structure around 0.3 to 0.4 mrad. Without that interaction, the peak structure is much weaker, and even absent for 50 keV, and it has moved to significantly larger scattering angle. This comparison shows that, within the 4DW model, the peak structure is mostly due to the projectile-residual target ion interaction, but that other factors, possibly electron-electron correlations, also contribute. Neither calculation is in good agreement with the experimental data. In particular, the role of the projectile-target ion interaction appears to be drastically overestimated by theory. On the other hand, without that interaction included, the peak structures in the measured data are either not present at all in the calculation or appear at significantly different scattering angles. The calculation does thus not provide any evidence that electron-electron correlations play an important role in the peak structures. Although strongly overestimated by the theory, it is nevertheless quite possible that, like in TTE, these structures can to a large extent be associated with the projectile-residual target ion interaction.

5. Conclusions

Two-electron processes involving capture of at least one electron have been widely studied to investigate the dynamics of reactions involving two active electrons. Double capture and transfer and target excitation are particularly well suited because both the initial and final states only entail bound states. From an experimental point of view this is advantageous because a single differential measurement in projectile solid angle already constitutes a fully differential experiment, providing a maximum of information about the process. For the theory, these processes are somewhat easier to treat than e.g. double ionization because the complications introduced by the rather complex two-electron continuum are avoided.

In this chapter we have reviewed recent experimental and theoretical work on double capture and transfer and target excitation in p + He collisions at intermediate projectile energies. The interest in two-electron processes is often focused on the role of electron-electron correlations. Undoubtedly, such effects are important in transfer and target excitation and especially double capture as well. However, in the projectile energy regime covered here, manifestations of electron-electron correlations are very difficult to identify because they are to a large extent masked by effects due to the projectile-residual target ion interaction. Theoretical work shows that, at intermediate and small projectile energies, the cross sections are very sensitive to how this interaction is described. For example, a fully quantum-mechanical treatment results in pronounced interference effects in the scattering angle dependence of the cross sections, which are absent when the projectile is treated classically.

To study the role of electron-electron correlations without complications introduced by the projectile-residual target ion interaction, it would be advantageous to study double capture and transfer and target excitation at large projectile energies. Not only does the relative importance of higher-order processes diminish with increasing projectile energy, but also the width of the projectile wave packet becomes very small compared to the target dimension. It was recently demonstrated that this can have drastic effects on differential cross sections[73] like e.g. the disappearance of interference structures present at small energies. However, because of the rapidly dropping cross sections, experiments become increasingly difficult with increasing projectile energy and, to the best of our knowledge, experimental differential data for p + He are currently not available above 300 keV (for He^{2+} + He data have been reported for 375 keV/amu). In order to extract information on electron-electron correlations from the existing differential data at intermediate energies, it is crucially important to advance our understanding of the role played by the projectile-residual target ion interaction.

Acknowledgements

This work was supported by the National Science Foundation under grant nos. 0969299 and 0757749, by the Natural Sciences and Engineering Research Council of Canada, and by TeraGrid resources provided by the Texas Advanced Computing Center, Grant No. TG-MCA07S029.

References

1. C. Froese-Fischer, *Adv. At. Mol. Opt. Phys.*, **55**, 235–291 (2008).
2. M. R. C. McDowell and J. P. Coleman, *Introduction to the Theory of Ion-Atom Collisions*, North-Holland Publishing Company, Amsterdam (1970).
3. Dž. Belkić, *J. Comp. Meth. Sci. Eng.*, **1**, 1–74 (2001).
4. M. Schulz and D. H. Madison, *Int. J. Mod. Phys. A*, **21**, 3649–3672 (2006).
5. M. Schulz, *Comments on Atomic, Molecular, and Optical Physics, Phys. Scr.*, **80**, 068101(1–9)2009.
6. J. T. Park, J. E. Aldag, J. M. George, and J. L. Peacher, *Phys. Rev. Lett.*, **34**, 1253–1256 (1975).
7. P. Martin, D. M. Blankenship, J. J. Kvale, E. Redd, J. L. Peacher, and J. T. Park, *Phys. Rev. A*, **23**, 3357–3360 (1981).
8. E. Horsdal-Pedersen, C. L. Cocke, and M. Stöckli, *Phys. Rev. Lett.*, **50**, 1910–1913 (1983).
9. H. Vogt, R. Schuch, E. Justiniano, M. Schulz, and W. Schwab, *Phys. Rev. Lett.*, **57**, 2256–2259 (1986).
10. L. K. Johnson, R. S. Gao, R. G. Dixson, K. A. Smith, N. F. Lane, R. F. Stebbings, and M. Kimura, *Phys. Rev. A*, **40**, 3626–3631 (1989).
11. M. Schulz, T. Vajnai, and J. A. Brand, *Phys. Rev. A*, **75**, 022717(1–6) (2007).
12. M. Zapukhlyak, T. Kirchner, A. Hasan, B. Tooke, and M. Schulz, *Phys. Rev. A*, **77**, 012720(1–9) (2008).
13. M. S. Schöffler, J. Titze, L. Ph. H. Schmidt, T. Jahnke, N. Neumann, O. Jagutzki, H. Schmidt-Böcking, R. Dörner, and I. Mančev, *Phys. Rev. A*, **79**, 064701(1–4) (2009).
14. Dž. Belkić, I. Mančev, and J. Hanssen, *Rev. Mod. Phys.*, **80**, 249–314 (2008).
15. H. Ehrhardt, K. Jung, G. Knoth, and P. Schlemmer, *Z. Phys. D*, **1**, 3–32 (1986).
16. A. Lahmam-Bennani, *J. Phys. B*, **24**, 2401–2442 (1991).
17. H. Ehrhardt, M. Schulz, T. Tekaat, and K. Willmann, *Phys. Rev. Lett.*, **22**, 89–92 (1969).
18. R. Dörner, V. Mergel, O. Jagutzski, L. Spielberger, J. Ullrich, R. Moshammer, and H. Schmidt-Böcking, *Phys. Rep.*, **330**, 95–192 (2000).
19. J. Ullrich, R. Moshammer, A. Dorn, R. Dörner, L. Ph. H. Schmidt, and H. Schmidt-Böcking, *Rep. Prog. Phys.*, **66**, 1463–1545 (2003).
20. M. Schulz, R. Moshammer, A. N. Perumal, and J. Ullrich, *J. Phys. B*, **35**, L161–L166 (2002).
21. M. Schulz, R. Moshammer, D. Fischer, H. Kollmus, D. H. Madison, S. Jones, and J. Ullrich, *Nature*, **422**, 48–50 (2003).

22. A. B. Voitkiv, B. Najjari, R. Moshammer, M. Schulz, and J. Ullrich, *J. Phys. B*, **37**, L365–L370 (2004).
23. N. V. Maydanyuk, A. Hasan, M. Foster, B. Tooke, E. Nanni, D. H. Madison, and M. Schulz, *Phys. Rev. Lett.*, **94**, 243201(1–4) (2005).
24. M. Schulz, M. Dürr, B. Najjari, R. Moshammer, and J. Ullrich, *Phys. Rev. A*, **76**, 032712(1–8) (2007).
25. J. H. McGuire, *Electron Correlation Dynamics in Atomic Collisions*, Cambridge University Press, Cambridge, England (1997).
26. M. Schulz, *Int. J. Mod. Phys. B*, **9**, 3269–3301 (1995).
27. I. Taouil, A. Lahmam-Bennani, A. Duguet, and L. Avaldi, *Phys. Rev. Lett.*, **81**, 4600–4603 (1998).
28. A. Dorn, A. Kheifets, C. D. Schröter, B. Najjari, C. Höhr, R. Moshammer, and J. Ullrich, *Phys. Rev. Lett.*, **86**, 3755–3758 (2001).
29. D. Fischer, R. Moshammer, A. Dorn, J. R. Crespo López-Urrutia, B. Feuerstein, C. Höhr, C. D. Schröter, S. Hagmann, H. Kollmus, R. Mann, B. Bapat, and J. Ullrich, *Phys. Rev. Lett.*, **90**, 243201(1–4) (2003).
30. M. Schulz, M. F. Ciappina, T. Kirchner, D. Fischer, R. Moshammer, and J. Ullrich, *Phys. Rev. A*, **79**, 042708(1–7) (2009).
31. J. Giese, M. Schulz, J. K. Swenson, H. Schöne, S. L. Varghese, R. Vane, P. F. Dittner, M. Benhenni, S. M. Shafroth, and S. Datz, *Phys. Rev. A*, **42**, 1231–1244 (1990).
32. W. T. Htwe, T. Vajnai, M. Barnhart, A. D. Gaus, and M. Schulz, *Phys. Rev. Lett.*, **73**, 1348–1351 (1994).
33. U. Fano, *Phys. Rev.*, **124**, 1866–1878 (1961).
34. E. Horsdal, B. Jensen, and K. O. Nielsen, *Phys. Rev. Lett.*, **57**, 1414–1417 (1986).
35. S. W. Bross, S. M. Bonham, A. D. Gaus, J. L. Peacher, T. Vajnai, H. Schmidt-Böcking, and M. Schulz, *Phys. Rev. A*, **50**, 337–342 (1994).
36. V. Mergel, R. Dörner, M. Achler, Kh. Khayyat, S. Lencinas, J. Euler, O. Jagutzki, S. Nüttgens, M. Unverzagt, L. Spielberger, W. Wu, R. Ali, J. Ullrich, H. Cederquist, A. Salin, C. J. Wood, R. E. Olson, Dž. Belkić, C. L. Cocke, and H. Schmidt-Böcking, *Phys. Rev. Lett.*, **79**, 387–390 (1997).
37. A. L. Godunov, C. T. Whelan, H. R. J. Walters, V. S. Schipakov, M. Schöffler, V. Mergel, R. Dörner, O. Jagutzki, L. Ph. H. Schmidt, J. Titze, and H. Schmidt-Böcking, *Phys. Rev. A*, **71**, 052712(1–4) (2005).
38. Dž. Belkić and I. Mančev, *Phys. Rev. A*, **83**, 012703(1–7) (2011).
39. K.-H. Schartner, B. Lommel, and D. Detleffsen, *J. Phys. B*, **24**, L13–L17 (1991).
40. S. Fülling, R. Bruch, E. A. Rauscher, P. A. Neill, E. Träbert, P. H. Heckmann, and J. H. McGuire, *Phys. Rev. Lett.*, **68**, 3152–3155 (1992).
41. W. C. Stolte and R. Bruch, *Phys. Rev. A*, **54**, 2116–2120 (1996).
42. A. Hasan, B. Tooke, M. Zapukhlyak, T. Kirchner, and M. Schulz, *Phys. Rev. A*, **74**, 032703(1–5) (2006).
43. M. S. Schöffler, J. N. Titze, L. Ph. H. Schmidt, T. Jahnke, O. Jagutzki, H. Schmidt-Böcking, and R. Dörner, *Phys. Rev. A*, **80**, 042702(1–6) (2009).

44. L. I. Pivovar, X. Tubavaev, and M. T. Novikov, *Sov. Phys. JETP*, **15**, 1035–1037 (1962).
45. N. V. de Castro Faria, F. L. Freire, and A. G. de Pinho, *Phys. Rev. A*, **37**, 280–283 (1988).
46. R. Schuch, E. Justiniano, H. Vogt, G. Deco, and N. Grün, *J. Phys. B*, **24**, L133–L138 (1991).
47. U. Schryber, *Helv. Phys. Acta*, **40**, 1023–1052 (1967).
48. L. H. Toburen and M. Y. Nakai, *Phys. Rev.*, **177**, (191–196) (1969).
49. Ya. Fogel, R. V. Mitin, V. F. Kozlov, and N. D. Romashko, *Sov. Phys. JETP*, **8**, 390–395 (1959).
50. J. F. Williams, *Phys. Rev.*, **150**, 7–10 (1966).
51. W. C. Keever and E. Everhart, *Phys. Rev.*, **150**, 43–47 (1966).
52. R. Dörner, V. Mergel, L. Spielberger, O. Jagutzski, J. Ullrich, and H. Schmidt-Böcking, *Phys. Rev. A*, **57**, 312–317 (1998).
53. Dž. Belkić, *Phys. Rev. A*, **47**, 3824–3844 (1993).
54. Dž. Belkić and I. Mančev, *Phys. Scr.*, **47**, 18–23 (1993).
55. A. L. Harris, J. L. Peacher, D. H. Madison, and J. Colgan, *Phys. Rev. A*, **80**, 062707(1–6) (2009).
56. A. L. Harris, J. L. Peacher, and D. H. Madison, *Phys. Rev. A*, **82**, 022714 (1–7) (2010).
57. W. Stich, H. J. Lüdde, and R. M. Dreizler, *J. Phys. B*, **18**, 1195–1207 (1985).
58. M. Kimura, *J. Phys. B*, **21**, L19–L24 (1988).
59. W. Fritsch, *J. Phys. B*, **27**, 3461–3474 (1994).
60. K. Roy, S. C. Mukherjee, and D. P. Sural, *Phys. Rev.A*, **13**, 987–991 (1976).
61. O. J. Kroneisen, H. J. Lüdde, T. Kirchner, and R. M. Dreizler, *J. Phys. A*, **32**, 2141–2156 (1999).
62. S. Jones and D. H. Madison, *Phys. Rev. A*, **65**, 052727(1–9) (2002).
63. W. Fritsch and C. D. Lin, *Phys. Rep.*, **202**, 1–97 (1991).
64. M. S. Pindzola, F. Robicheaux, and J. Colgan, *Phys. Rev. A*, **82**, 042719 (1–8) (2010).
65. E. Engel and S. H. Vosko, *Phys. Rev. A*, **47**, 2800–2811 (1993).
66. M. Zapukhlyak, T. Kirchner, H. J. Lüdde, S. Knoop, R. Morgenstern, and R. Hoekstra, *J. Phys. B*, **38**, 2353–2369 (2005).
67. R. McCarroll and A. Salin, *J. Phys. B*, **1**, 163–171 (1968).
68. M. Schulz, W. T. Htwe, A. D. Gaus, J. L. Peacher, and T. Vajnai, *Phys. Rev. A*, **51**, 2140–2150 (1995).
69. J. F. Hart and G. Herzberg, *Phys. Rev.*, **106**, 79–82 (1957).
70. J. Pálinkás, R. Schuch, H. Cederquist, and O. Gustafsson, *Phys. Rev. Lett.*, **63**, 2464–2467 (1989).
71. C. F. Barnett, *Atomic Data for Fusion*, Oak Ridge National Laboratory, Tennessee, **A-32**, (1990).
72. C. F. Barnett, *Atomic Data for Fusion*, Oak Ridge National Laboratory, Tennessee, **C-60**, (1990).
73. K. N. Egodapitiya, S. Sharma, A. Hasan, A. C. Laforge, D. H. Madison, R. Moshammer, and M. Schulz, *Phys. Rev. Lett*, **106**, 153202(1–4) (2011).

Chapter 2

COLTRIMS Experiments on State-Selective Electron Capture in Alpha-He Collisions at Intermediate Energies

M. Alessi[1], S. Otranto[2] and P. Focke[1,*,†]

[1]*Centro Atómico Bariloche, Av. E. Bustillo 9.500, San Carlos de Bariloche (8400), Argentina*

[2]*IFISUR and Departamento de Física, Universidad Nacional del Sur, Bahía Blanca (8000), Argentina*

[†]*focke@cab.cnea.gov.ar*

The cold target recoil-ion momentum spectroscopy (COLTRIMS) technique has settled in recent years as a very valuable tool to perform kinematically complete experiments. As a result, several laboratories worldwide have implemented COLTRIMS devices during the last decade. In this work, we provide a survey over the fundamental concepts of this technique and focus on how it allows one to perform state-selective electron capture studies. Results for the reaction $^3He^{2+}$ + He explored at the Atomic Collisions Laboratory in Bariloche are shown and discussed. The energy range considered is that corresponding to our Kevatron accelerator (40–300 keV). We find this energy range to fit just between those previously explored at the University of Lanzhou and University of Frankfurt laboratories. The present experimental results are found to be in agreement with classical trajectory Monte Carlo simulations.

1. Introduction

The experimental investigation of atomic collision processes of excitation, charge transfer and ionization provides valuable information for the understanding of the dynamics involved in different collision systems.

*The author to whom corrspondence should be addressed.

In many cases, the experimental advance has also pushed the development or improvement of existing theoretical methods that refine our description of the underlying physics. Besides these fundamental aspects, experimental studies are also relevant for practical applications in controlled fusion research and other areas in plasma physics.[1] The most desirable study of a collision process would consist of the detection of all the resulting reaction products, i.e. all the emitted particles, projectile and recoil ion, either charged or neutral, and the products: electrons and photons. These many particle reactions are characterized by fully differential cross sections, i.e. distribution in terms of the kinetic energy and emission angle of the different particles, and internal excitation of the reaction partners. Experiments that provide this almost-complete information about a collision process are known as kinematically complete, in the sense that the energy and emission angle of each particle involved is measured, with the exception of the spin. The spin observable that is difficult to detect and include in today's experiments provides only averaged information over the other detected observables.

Experiments aimed at obtaining such detailed information on collision processes, which go beyond the determination of total cross sections, were performed in the 60s and 70s. These used spectroscopic techniques consisting of recording with dispersive spectrometers the projectile energy and scattering angle, the recoil-ion energy and emission angle, or performing coincidences between projectile and recoils.[2–11] Electron[12] and photon[13] spectroscopies were also used and are characterized by their high resolution, but provide only information concerning the deexcitation process. Many of these experimental techniques are grouped under the general name of collision spectroscopy.[14] By making use of these techniques, it was possible to determine the energy and angle of the emerging projectile and recoil-ion. It was also used to measure differential cross sections as well as the inelastic energy defect Q of the reaction (that is the change in potential energy during the collision that is transformed into translational energy of the colliding particles). The measurement of Q is an important observable because it allows the identification of the excitation processes. The branch of collisional spectroscopy related to the study of Q for collision channels involving small impact parameters is usually denoted scattered particle spectroscopy.[15] In the special case that the scattering angle θ of the projectile is small, ($\theta \ll 1$), that is for nearly forward scattered particles, we are concerned with translational spectroscopy.[14,15] In the latter, the change of the kinetic energy of the projectile, ΔE, is well approximated by $\Delta E = Q$. This technique using electrostatic spectrometers allows one to perform high resolution measurements with projectiles at intermediate to low impact energies (i.e. projectiles reaching the keV energy region). The usable projectile energy limit represents the main limitation of this technique. Important studies can be performed with the use of coincidence measurements between the emitted projectile and recoil ion.[3,4,8,9] In this last category we may also mention as a prominent example the (e-2e) and (e-3e) experiments.[16,17] The use of dispersive spectrometers, with low coincidence efficiency and scanning procedures for energy and angle, represents the main drawback of these coincidence measurements, due to the large measuring times involved. Apart from the mentioned coincidence measurements at low projectile energies, measurements on the recoiling target ion have practically not been considered at those times.[10,11] The reason is that, in general, at intermediate to large collision energies the transferred momentum is very low and practically shadowed by the initial thermal movement of the target atoms.[15,18] These recoil-ion energies are typically in the meV to μeV energy regime. Pioneering successful attempts in this direction were performed by Levin *et al.*[19] who obtained information on the mean energies, and Ullrich *et al.*[20,21] who succeeded to measure the transverse momentum (in the perpendicular direction to the incident projectile beam) of the recoil-ions with a target at room temperature. Those experiments evolved with the use of cooled static-targets in order to reduce the thermal motion, resulting in the RIMS spectrometers developed at the University of Frankfurt, by Dörner *et al.*[22] and Ullrich *et al.*[23,24] A rapid progress developed right after those studies with the implementation of localized gas-jet targets, ion projection techniques and position-sensitive detectors.[24–27] With these improvements all the momentum components of the recoil-ion could be measured. Warm effusive jets[28] and cold supersonic gas-jets were used as targets. The latter resulted in an important increase in momentum resolution and the technique may be identified at this stage as COLTRIMS. By precooling the gas target, prior to supersonic expansion, the resolution in energy of the detected recoil ions reached the μeV range.

The cooling of the gas target together with the use of position-sensitive detectors represented an important advancement in experimental techniques, that made possible with COLTRIMS to measure the momentum of the recoil-ions resulting from collision processes. Together with the ion-projection technique, the developed spectrometers had the capability to detect ions emitted in the full 4π angular range. These features made the study of momentum distributions with COLTRIMS accessible and represented an interesting approach, in contrast to the traditional spectroscopic techniques for which those studies were so elusive. With an attainable energy (or momentum) resolution that ranges into the μeV, COLTRIMS is fully competitive with the previously mentioned translational spectroscopy (the last frequently named energy gain or energy loss spectroscopy). Besides, there is an additional advantage that now Q measurements can be performed for any projectile impact energy.

The COLTRIMS technique is not limited to the detection of recoil-ions only. With an additional positional detector it is possible to measure the momenta of electrons emitted in ionization reactions.[29,30] In this case, only a partial detection of the electrons is possible. In the simpler versions the spectrometers can only detect low energy electrons (usually less than 10 eV) at all emission angles. The range of accessible emission angles is reduced as the electron energy increases. With the development of a spectrometer which superposes a solenoidal magnetic field, the spectrometer is able to detect electrons emitted at 4π angles with energies reported as high as 30 eV,[31] and an even higher limit is possible.[27] These spectrometers are known as projection microscopes. With these devices it is possible to perform complete kinematical experiments for several collisional processes.

The state of the art in cooling the target with the use of the COLTRIMS technique has been reached with laser cooled targets in a magneto-optical trap.[32–34] This high resolution technique named MOTRIMS was applied to perform charge exchange experiments with reported momentum resolution of an order of magnitude improvement over conventional COLTRIMS.

2. The COLTRIMS Concept

In the following section we are going to give only a brief description of the COLTRIMS method. Excellent, full detailed descriptions and plenty of examples concerning applications can be found in the references[25–27,31] from the main laboratories that were involved in the development of the technique.

The high resolution recoil-ion momentum spectrometers are based essentially on the following features: (a) a well localized cold atomic target; (b) an ion-projection system; and (c) the use of position-sensitive detectors.

A cold target is essential for recoil-ion momentum resolution since the momenta resulting from the collision are of the order or even smaller than the typical thermal motion of the target atoms at room temperature. The cooling of the gas is achieved by means of an adiabatic expansion of the gas through a small nozzle (typical diameter 30 μm). The features of the adiabatic expansion of a gas to produce a free-jet molecular beam are described in Ref. 35. The coolest inner part of the jet passes through a skimmer into the scattering chamber to the collision region. The skimmer is used to provide a collimation of the gas-jet. This target gas beam crosses the projectile beam (realized by a fast ion, electron, or photon beam) perpendicularly in the collision region. The expansion converts the random internal kinetic energy of the gas atoms into an ordered directed kinetic energy in the gas-jet defined by the free enthalpy H of the gas. As an example, for He atoms at room temperature of 300 K this directed movement corresponds to a momentum of 5.9 a.u. (for a precooled gas at 30 K it is lower at 1.8 a.u.). This directed momentum implies that the collision is not performed with a target at rest but with one with a significant initial velocity perpendicular to the projectile beam. This effect is easily visible as a shift in the recoil-ion momentum distributions.

The quality of the jet is expressed by the speed ratio, S, which is the mean jet velocity divided by the thermal spread in velocities. For a mono-atomic gas we have $S = \sqrt{5T_i/2T}$,[31] where T_i is the initial gas temperature before expansion, and T is the leftover temperature of the jet after expansion. The speed ratio is significantly improved by cooling the gas (typically to 30 K) before expansion, reaching reported speed ratios of $S = 200$;[31] while for gas at room temperature, S may be between 3 to 8. The speed ratio concept is useful for considerations of the remaining random movement of the atoms in the jet direction or longitudinal (respect to the jet) momentum spread; it is the parameter relevant for recoil-ion spectroscopy resolution considerations. For the transverse directions the remaining random momentum spread is more conveniently described by the collimation of the jet, defined by the geometry of the nozzle-skimmer arrangement and known as geometric cooling, that is very effective in most cases.

The ions produced in the collision region, resulting from the overlap volume of the incident projectile beam with the target gas-jet, are extracted and guided onto a two-dimensional position-sensitive detector. From the impact position on the detector and the ion time-of-flight the momentum in the collision region can be calculated. The extraction is realized with a static electric field that guides the recoil ions, emitted in any direction, onto the detector surface. The simplest configuration is a homogeneous field.[24,28,36] The extraction field is oriented usually perpendicularly to the gas-jet and projectile beam. A configuration with the extraction field along the direction of the projectile beam has also been used.[31] Adding to the accelerating field a field free drift-region assures that the ions starting at slightly different positions along the electric field arrive at the same time at the detector. For a homogeneous electric field the ratio of acceleration length to drift length is 1:2 and results in a first order focusing.[37] This time focusing geometry reduces the degrading effects of the finite collision region size, along the field direction, on the momentum resolution. In the direction perpendicular to the extraction field a configuration consisting in addition of an electrostatic lens to the accelerating section, and a properly adjusted drift region length, has been reported to produce space focusing.[26,38] This improves the momentum resolution along the two directions perpendicular to the electric field, and reducing the degrading effect on the resolution due to the finite extension of the interaction volume. For the time-of-flight measurement it is necessary to have a trigger signal which defines the time of interaction of the projectile with the target atom. This can be done using a pulsed beam of projectiles, or single projectiles of a continuous beam with a time-sensitive detector. The time-of-flight not only allows the determination of one of the momentum components but also to distinguish the different ion species produced in the collision, based on the mass to charge ratio. The position-sensitive detector should also be time-sensitive for good time-of-flight measurements. If it has a multi-hit capability, the spectrometer can be used for studies of fragmentation processes from molecular targets.

Electrons may also be detected using the projection technique as done for recoil-ions. The same electric field that guides the ions in one direction can be used to accelerate electrons in the opposite one. Placing a large area position-sensitive detector close to the collision region allows the detection of low energy electrons ($\lesssim 10\,\mathrm{eV}$) with nearly 4π emission solid angle.[29,39] In this case a drift region is not used and the electron momentum component along the electric field is not measured. To increase the electron energy acceptance at all emission angles, and maintain at the same time a good resolution in the recoil-ion branch, a magnetic solenoidal field is superimposed parallel to the electric field.[31] This allows one to measure all the three momentum components of the electrons and results in the realization of an instrument capable of a complete kinematical investigation for several collision processes. This more evolved concept of the COLTRIMS technique is identified as a reaction microscope.

The COLTRIMS technique, as briefly outlined here, has the capability of measuring the initial momenta of the charged particles (recoil-ions and emitted electrons) present in a collision. Only the projectile momentum after the collision is not measured. In some cases the scattering angle is measured, but in principle it is not needed. The projectile momentum after the collision can be deduced from energy and momentum conservation once the momenta of the other emitted particles are known. This feature of the technique makes it a powerful tool in the study of the complete final kinematics of the collision. Another interesting feature is that the data are taken in an "event mode" whereby a computer registers in a file a list of all detector signals for every collision event. In this way, the complete data of the experiment can be examined later, as many times as needed, without repeating the measurement.

There are also some limitations for the COLTRIMS technique. The detection of particles is limited to charged ones. It works well only for detection of low energy particles. This is particularly a limiting feature with regard to the electron detection. The resolution on the electron momentum can be considered poor if compared to the traditional methods that make use of dispersive spectrometers when electron energies $>100\,\mathrm{eV}$ are considered. On the other hand, the possibility to measure electron distribution for energies below $5\,\mathrm{eV}$ with good resolution may provide a promising tool for the study of collisionally emitted low energy electrons, which are difficult to detect applying the conventional electron spectroscopy.

3. Experiment

A COLTRIMS spectrometer was assembled in our laboratory in a simple version that can measure the recoil-ion momenta. The design follows the lines of development performed by the COLTRIMS group at the University of Frankfurt. A schematic view of the spectrometer is displayed in Fig. 1 together with the coordinate system used. A description of the experimental setup, spectrometer and method of measurement is given in Ref. 40. The spectrometer consists of two sections, an accelerating section where a homogeneous electrostatic field is established, followed by a field free drift-region. The accelerating section is made of a series of square shaped electrodes with large circular openings in their center. These are closely spaced along 8 cm, and electrically connected by a resistive chain. At each end electrode there are meshes of high transparency. This configuration

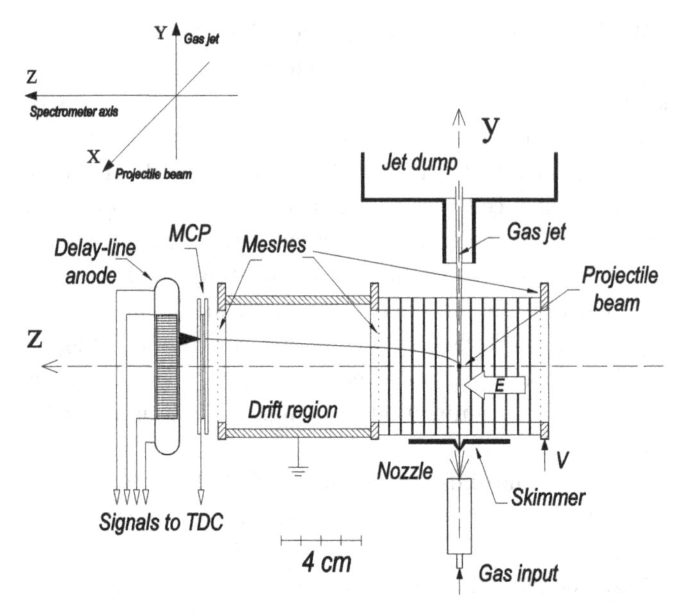

Fig. 1. Schematic of the COLTRIMS spectrometer with two dimensional position-sensitive detector and supersonic gas-jet arrangement. Also shown is the coordinate system used. MCP are microchannelplates, TDC is the time to digital converter electronics, and $\boldsymbol{E}$ the electric field.

provides a properly homogeneous electric field to extract the recoil-ions. The drift region is a metal cylinder 8 cm long with a mesh at its end, followed by a position-sensitive detector. This last mesh helps to define a field-free region inside the drift tube, shielding it from electric fields originating from the detector. The projectile beam and the target gas-jet from a supersonic nozzle intersect perpendicularly inside the spectrometer along the center plane. The recoil-ions produced in the small collision region are accelerated along 4 cm toward the drift region. This configuration satisfies a first order time focusing condition.[37]

The detector consists of a stack of two microchannel plates with a 4 cm diameter active area and a delay-line anode with two dimensional position-sensitive capability.[41–43] The position resolution of our detector with the associated processing electronics has been found to be of nearly 1 mm. This limit is due mainly to electronic noise in the pulse processing electronics, which in the end defines the instrumental limit in the recoil-ion momentum resolution. The target is supplied by a supersonic gas-jet. It is produced by an adiabatic expansion of a gas contained in a reservoir through a nozzle, consisting of a $30\,\mu$m diameter opening, followed by a skimmer of 0.3 mm diameter. The collimation attained with this geometry results in an interaction length in the target consistent with the detector position resolution. Improving the collimation may reduce appreciably the count rates without much significant gain in momentum resolution. The collimation provides also a geometrical cooling as has been mentioned before. For He the present settings may result in mean momentum spread of ± 0.04 a.u. The highest driving pressure used was 2 bar, achieving a target density estimated in 7.8×10^{11} atoms/cm^3 for He gas. The jet enters the spectrometer perpendicular to the electric field. At the exit of the spectrometer the jet is caught in a dumper chamber. This arrangement helps to keep the pressure inside the collision chamber low. Usually the background gas density in the collision chamber with the jet on was found to be 6×10^9 atoms/cm^3. Our gas-jet is produced without a precooling of the reservoir gas, using only the adiabatic cooling effect. For He, the expansion starting at room temperature (300 K) produces a jet with a terminal momentum, along the jet direction, of 5.9 a.u. The added random motion along this direction can be estimated from the speed ratio S as having a mean momentum spread of 1 a.u. that corresponds to a residual temperature of 10 K.[44]

The projectiles leaving the interaction region inside the spectrometer pass through an electrostatic deflector that separates the different charge-states. After the deflector the projectiles are detected with two channeltron detectors working as a secondary emission mode devices. One of these detectors collects the direct non-deflected component, and the other a selected charged component.[40] The signal from the charge-selected detector is used as the trigger signal for the time-of-flight measurement of the recoil-ions, and at the same time allows by a coincidence condition the identification of the collision process.

The incident projectile beam delivered from a 300 kV Cockroft-Walton accelerator, before entering the collision chamber with the COLTRIMS spectrometer is collimated, by two sets of adjustable slits, to a beam cross section of $0.7 \times 0.7\,\text{mm}^2$ at the target location. This additionally provides, by geometry considerations, along the direction perpendicular to the jet and perpendicular to the projectile beam a geometrical cooling for the momentum, estimated at ± 0.02 a.u. for He.

The incident projectile beam, target gas-jet and electric extraction field intersect mutually inside the spectrometer at right angles. We define a right handed coordinate system for our measurements with the x-axis along the projectile beam, the y-axis along the gas-jet, and the z-axis parallel to the electric field coincident with the spectrometer axis and pointing towards the detector, as depicted in Fig. 1. The origin of the coordinate system is fixed at the interaction region. Measuring the position of arrival of the recoil-ion on the detector and the time-of-flight, we can find the initial momentum of the particle from a trajectory analysis. For a homogeneous electric field we use for the momentum components p_x, p_y and p_z the expressions[40]:

$$p_x = \frac{M}{T_o} x, \tag{1}$$

$$p_y = \frac{M}{T_o} y, \tag{2}$$

$$p_z = (T_o - t) qE. \tag{3}$$

Here M is the recoil-ion mass, q its charge, and E the intensity of the electric field. x, y are the hit coordinates on the detector and t is the time-of-flight. T_o is the time-of-flight of a recoil-ion that was initially at rest. In the momentum space the origin of coordinates is shifted along the y direction by p_j, the terminal momentum of the gas-jet particles after expansion; and in the x directions it does not change (placed at the crossing of the gas-jet trajectory). Along the z-axis the origin is located at $T_o qE$. In order to apply equations (1)–(3), T_o is obtained from an analysis of the time-of-flight spectra, and the x and y positions are found performing a calibration of the position-sensitive detector using a mask.[40]

In Fig. 2 we show an image of recoil-ions obtained from the collision of a He^{2+} beam incident on a He target. The figure shows the recoils that are in coincidence with emerging He^{+} projectiles. This corresponds to a single capture process. The figure is represented in x, y coordinates and with help of equations (1)–(3) can be converted into p_x, p_y coordinates. One observes the recoil-ions produced resulting from the localized target as a group of two patches displaced in the y-axis by an amount of the jet momentum p_j. In the present case, the two groups of recoil-ions observed is an example of a collision reaction where different reaction channels and Q are involved. Also visible is a broad trace resulting from the incident projectile beam interacting with the background gas. In Fig. 3 we display a x projection, equivalent to a p_x distribution, integrated over all the p_y and p_z momentum components. Also shown is a distribution resulting from an independent background measurement when turning the gas-jet off.

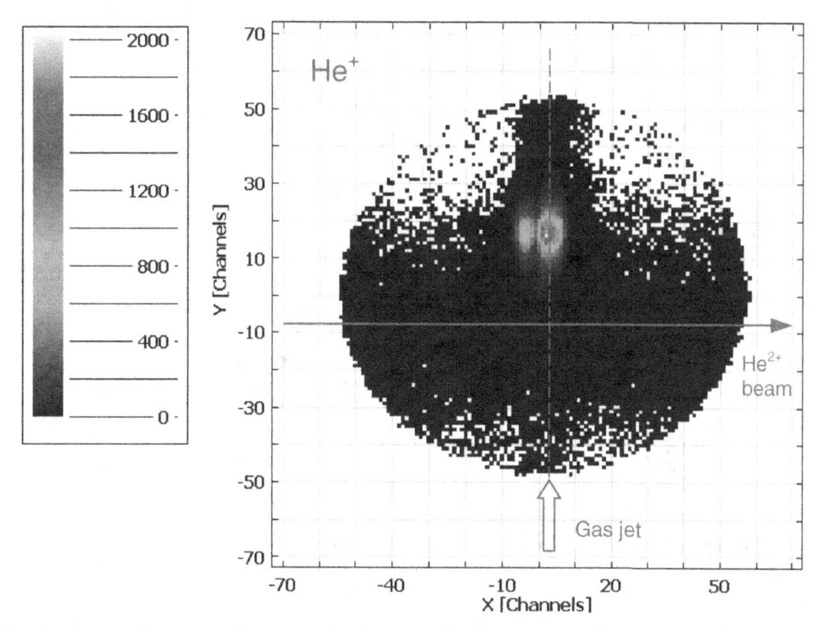

Fig. 2. Image from position-sensitive detector, in the x, y coordinates, of He^+ recoil-ions for 100 keV $^3He^{2+}$ projectiles incident on a He target. Image resulting from coincidences with emerging $^3He^+$ projectiles. Accelerating electric field was 2.4 V/cm. The figure shows the trajectory of the gas-jet and the projectiles. The He^+ produce two patches associated to different Q reactions. Ions from the background gas form a wide trace along the projectile trajectory.

4. Electron Capture in $^3He^{2+}$ + He

The incorporation of the COLTRIMS technique in several laboratories renewed the interest in the study of atomic and molecular collision processes concerning charge exchange, transfer ionization, single and double ionization among others.[25–27,45,46] The subject of charge exchange processes involving atomic targets has been thoroughly investigated theoretically and experimentally in the past decades.[13,15,47,48] For light projectiles, most of the experiments were focused on obtaining total cross sections[49–52] and only in a few cases were studies made to determine the different state-selective capture cross sections.[36,53–57] For the latter the COLTRIMS method with its high resolution capability provided a convenient tool to perform such studies. Many of the studies using this technique were also involved with heavy highly ionized projectiles at low collision velocities. State-selective single capture for He^{2+} incident on a He target was investigated by Mergel *et al.*[36] for collision energies in the range from 60 to 250 keV/amu,

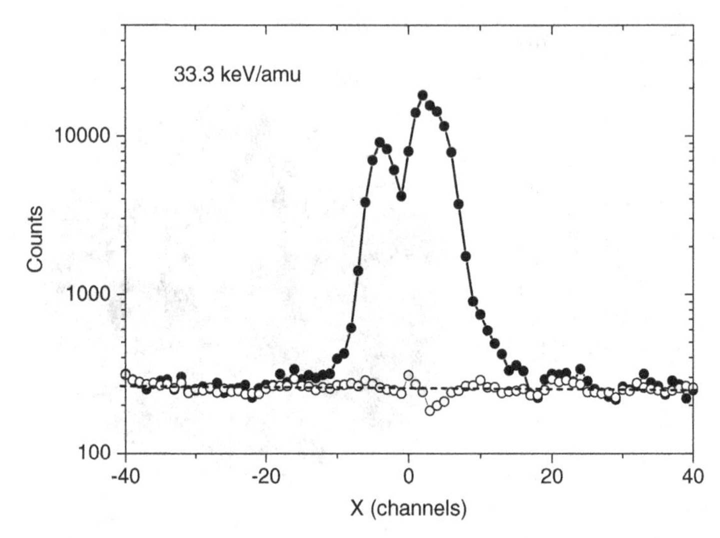

Fig. 3. Projection along the x-axis of the image from Fig. 2 for 33.3 keV/amu incident $^3He^{2+}$ projectiles on He (full circles). The depicted distribution is equivalent to a p_x or longitudinal momentum distribution of the recoil-ions. Also displayed is an independent measurement of the background gas distribution (open circles). A linear interpolation (dashed line) used for background subtraction and joining the tails of the x distribution is seen to represent well the background distribution.

and more recently by Zhu *et al.*[56,57] at a lower energy range from 5 to 10 keV/amu. We present measurements performed for this collision system in the range of impact energies from 13.3 to 100 keV/amu,[58] covering the gap between the previously reported ones. The present measurements were performed with the COLTRIMS setup described above and in Ref. 40. The process studied was the single charge capture for the $^3He^{2+} + He$ collision in the range of impact energies from 40 to 300 keV.

From the recoil-ion longitudinal momentum distribution, state-selective capture information can be obtained. Along the projectile beam axis, energy and momentum conservation lead to the following relation for a single charge exchange process (in atomic units)[25]:

$$p_l = -\frac{Q}{v_p} - \frac{v_p}{2}. \tag{4}$$

Here Q is the inelastic energy defect or change of electronic energy of the reaction (given as the difference between the initial and the final sum over the electronic binding energies), and v_p is the projectile velocity. The first term in equation (4) reflects the momentum change of the projectile, and the second the momentum change due to the electron transferred from the target to the projectile. Usually capture reactions are preferentially exoergic ($Q > 0$), this means that p_l takes negative values, i.e. the recoiling target ion is emitted backwards. In contrast to ionization processes, where the p_l distribution is given by a smooth structure starting in an abrupt rise corresponding to electron capture into the continuum of the projectile,[59] charge exchange processes lead to discrete values of p_l.

As a result, the obtained p_l distributions not only reflect the way in which the different final states of the projectile are populated by the captured electron, but are also determined by the electronic final state of the remaining target core. Since the target has two electrons in its initial state, once the first electron is captured either to the ground or an excited state, there still remains the possibility that the second electron could be found in an excited state of the target. We follow the notation used by Mergel *et al.*[36] to describe the final projectile-target electronic states. The case in which the target and the projectile end with one electron in the ground state each will be denoted $(1, 1)$. In this notation (n, n') the first element is given by the n value of the captured electron, while the second element is given by the n' value of the electron that remains bound to the target. For the present symmetric collision system, configurations where the projectile or target is in a given excited state cannot be distinguished based on p_l spectra due that the Q values are the same.

In Fig. 4 we show the measured p_l distributions for the different impact energies used. From equation (4), for the present collision system that involve only hydrogenic ions, all the Q values are known and their values help to identify the states involved in the collision. As can be seen, all distributions show a similar two-peak structure that we could identify as capture to different states. The peak on the left was identified as a single one, resulting from a charge exchange collision in which the target and projectile end up with one electron in the ground state each. The peak on the right consists of several contributions that we can identify as belonging to the capture (1, 2)&(2, 1), (1, $\geq$3)&($\geq$3, 1) and ($\geq$2, $\geq$2). The resolution of our apparatus, under the present working conditions, for the longitudinal momentum component was 0.8 ± 0.2 a.u. (full width at half maximum). The resolution was estimated to result from the combined contribution of target size and detector resolution (the latter was changed slightly, by changing the electric extraction field, at different collision energies), and allowed to resolve only, at the lower collision energies, the $(1, 1)$ channel from those involving excited states. The temperature of the supersonically

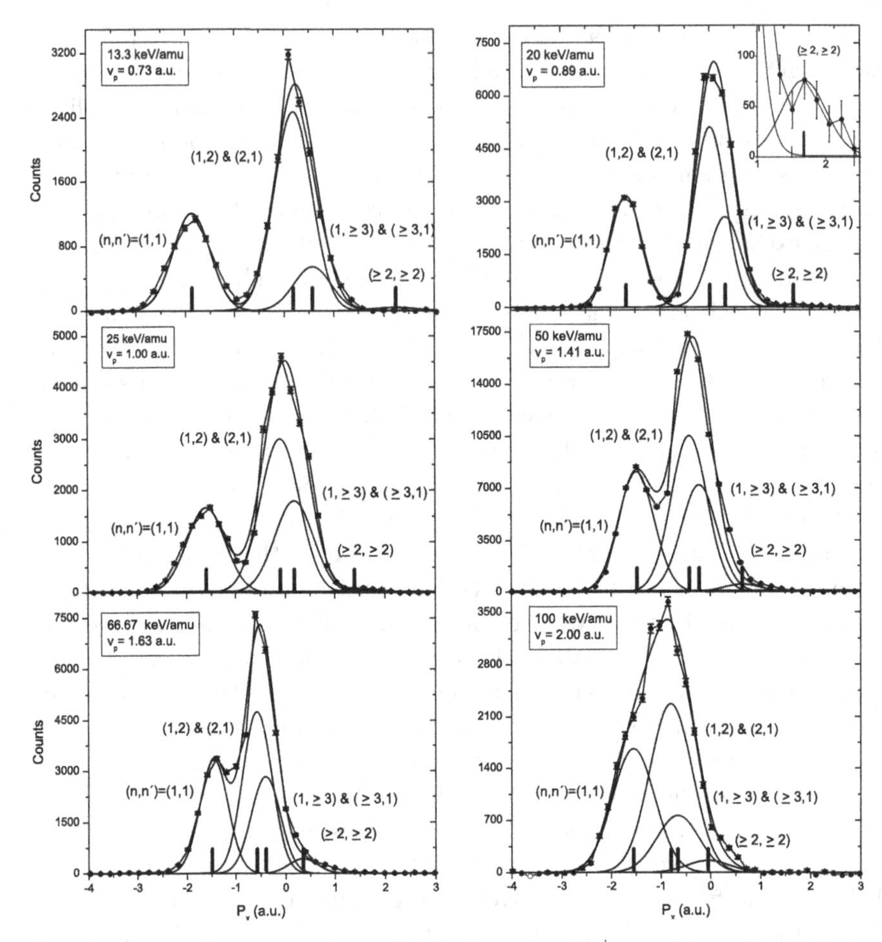

Fig. 4. Longitudinal momentum distributions for He^+ recoil-ions for 13.3 to 100 kev/amu $^3He^{2+} + He \rightarrow ^3 He^+ + He^+$ single capture reaction. The bars indicate the theoretical position of the peaks. Also displayed are the different state-selective capture components resulting from a fitting procedure.

expanded target gas does not have any effect on the longitudinal momentum resolution. The geometrical cooling contributes only with 0.08 a.u. and represents a small contribution.

The state-selective cross sections were obtained by fitting the longitudinal momentum distributions shown in the Fig. 4 by means of Gaussian functions under the condition of constant peak widths at each collision energy. The peak positions were fixed at the positions given by equation (4), except for two cases, where a fit would result better, this condition was relaxed. The parameters resulting from the fit are displayed in Tables 1 and 2. The cross sections have been determined, for the present measurement, by normalization to total cross sections published by Dubois[51,52] and are also listed in table 1 and displayed in Fig. 5. The state-selective cross sections obtained are shown in Fig. 6 as a function of the impact energy. Experimental data of Mergel *et al.*[36] and Zhu *et al.*[56,57] are included and seen to match the present data at the edges of our energy range within the associated error bars.

The p_l distributions shown in Fig. 4 were obtained after a background subtraction procedure. First, from the measured distribution of recoil-ions we chose those events that were in coincidence with detected $^3He^+$ projectiles and He^+ recoils. This last distribution is affected by background events resulting mainly from capture processes produced by the interaction of the projectiles with the uniformly distributed gas in the chamber located along their path. The background, which in some cases has been determined by an independent measurement, is displayed in Fig. 3. From Fig. 3 we see that a simple procedure to account for this background, when handling longitudinal momentum distributions, may be to fit the measured distribution at their ends with a linear interpolation which is later subtracted from the total momentum distributions. A more general background subtraction procedure has been described in Ref. 40 but was not necessary to apply in the present case. The errors resulting in the presented p_l distributions result from counting statistics in the total distribution and an estimated error assigned to the background counts. This last one is more of a systematic nature.

A possible source of error in our data could be the fact that the incident projectile beam has a contamination component of $^3He^+$ ions, produced by charge capture collisions along the transport line and inside the collision chamber before reaching the interaction region. If this contamination is present, detection of He^+ recoil-ions in coincidence with $^3He^+$ emerging projectiles would correspond to an ionization process. However, these ionization events would produce, as mentioned before, a continuous p_l distribution starting at a threshold and extending to positive p_l, due to ionization processes being endoergic ($Q < 0$). For He^+ ions the ionization threshold begins at a p_l where the charge transfer to the continuum of He^{2+} projectiles is located. For our case, taking into account the finite resolution in momentum, the ionization may start to contribute at the location of the $(1, \geq 3)\&(\geq 3, 1)$ peaks. We assume that the ionization, due to the low contamination rate we have and due to low cross section at the present

Table 1. Experimental state-selective cross sections for $^3He^{2+} + He$ single electron capture collisions (in units of $10^{-16} cm^2$).

Final State	Energy keV/amu						
(n, n')	13.3	20	25	33.3	50	66.67	100
(1, 1)	0.39 ± 0.16	0.69 ± 0.19	0.67 ± 0.18	0.71 ± 0.18	0.77 ± 0.21	0.66 ± 0.17	0.33 ± 0.09
(1, 2)&(2, 1)	0.79 ± 0.26	1.12 ± 0.33	1.21 ± 0.36	1.24 ± 0.34	1.01 ± 0.30	0.89 ± 0.27	0.45 ± 0.14
(1, $\geq$3)&(3, 1)	0.18 ± 0.08	0.56 ± 0.21	0.73 ± 0.34	0.80 ± 0.36	0.69 ± 0.33	0.56 ± 0.27	0.17 ± 0.09
($\geq$2,$\geq$2)	0.014 ± 0.008	0.02 ± 0.01	0.02 ± 0.01	0.03 ± 0.02	0.05 ± 0.03	0.05 ± 0.03	0.03 ± 0.02

Table 2. As table 1 but values are give in percent of the total capture.

Final State (n, n')	Energy keV/amu						
	13.3	20	25	33.3	50	66.67	100
(1, 1)	28 ± 12	29 ± 8	25 ± 7	26 ± 6	31 ± 8	31 ± 8	34 ± 9
(1, 2)&(2, 1)	58 ± 19	47 ± 14	46 ± 14	44 ± 12	40 ± 12	41 ± 12	46 ± 15
(1, $\geq$3)&(3, 1)	13 ± 6	23 ± 9	28 ± 13	29 ± 13	27 ± 13	26 ± 12	17 ± 9
($\geq$2, $\geq$2)	1.0 ± 0.6	0.8 ± 0.4	0.9 ± 0.5	1.2 ± 0.8	2 ± 1	2 ± 1	3 ± 2

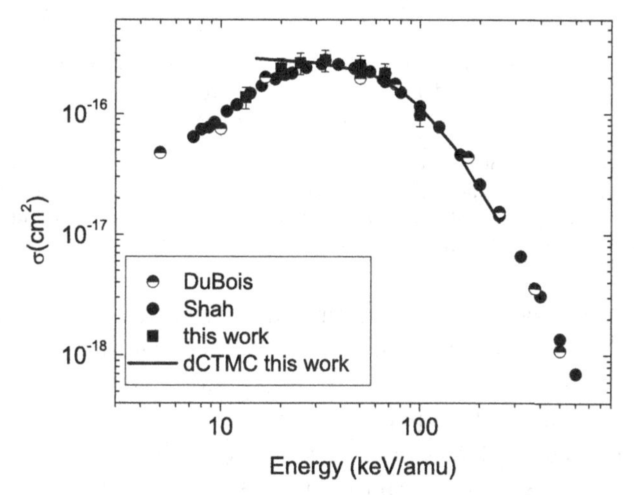

Fig. 5. Total cross sections for single charge transfer. Experimental data used for normalization: solid circles, Shah *et al.*[49,50]; partially filled circles, DuBois[51,52]; solid squares show the location used for normalization for this work. Solid line, present dCTMC calculation.

collision energies, may have a negligible effect on the (1, 3)&(3, 1) peak, but would probably affect the ($\geq$2, $\geq$2) peak. Therefore, these cross sections displayed in Fig. 4 may be affected by a systematic excess value. From control measurements we found that the contribution of He^{+} projectiles was never higher than 1% of the He^{2+} beam. From known total cross sections for the present reactions,[60] we made a rough estimate for the contribution of ionization events. We found that this eventual contamination may not be higher than 0.6% at 50 keV to 0.9% at 200 keV of the total capture events, and that it may be spread over a wide p_l region.

In Fig. 6 we see that all the partial cross section reach their maximum value in the present studied collision energy range. At the lower energies

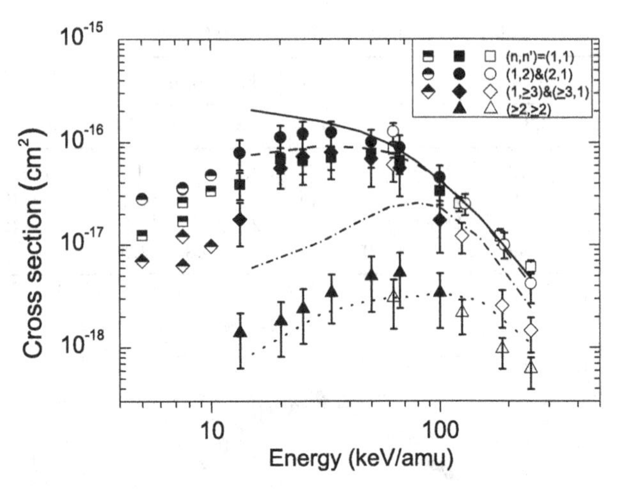

Fig. 6. State-selective capture cross sections as indicated in the figure. Solid symbols, this work; open symbols, data from Mergel *et al.*[36]; half-filled symbols, data from Zhu *et al.*[56,57] Lines, present dCTMC calculations: solid line: (1,1); dashed line: (1,2) & (2,1); dot-dashed line: (1,3) & (3,1); and dotted line: (>2,>2).

the capture proceeds mainly populating the excited state of the projectile (1, 2)&(2, 1). The same tendency is observed in the data presented by Zhu *et al.*[56,57] Around 200 keV/amu the situation is reverted and the ground state is more likely populated, as can be seen from the data of Mergel *et al.*[36] This difference in behavior may be an indication that two mechanisms are present in the capture process. At the lower energies, where the relative velocity of the colliding particles is smaller than the velocity of the target electron, the quasi-molecular picture of a collision process may be applied and capture to states that imply the lowest $|Q|$ may be expected to be favored. On the contrary, at higher collision energies, beyond the maximum of the cross section, the kinematic electron capture picture may provide a description for the population of the most strongly bounded projectile states.

Concerning the question whether the target or the projectile is in the excited state for this symmetric collision system, we cannot infer such a trend from the experimental data. In order to contrast the present data and complement our experimental understanding of the collision system under study, the present data are contrasted against classical trajectory Monte Carlo simulations in which the Hamilton equations for the classical 4-body system are numerically solved. Since the classical He atom autoionizes spontaneously, 4-body CTMC codes generally adopt one of the following strategies in order to circumvent the problem and incorporate as much information as it is possible concerning the electron-electron interaction: (a) the Bohr atom,[61] in which the explicit $1/r_{12}$ interaction is fully considered and the electrons are located in circular orbits with opposite momentum at equal distances from the nucleus, which lies right between the electrons. In some cases, stabilizing potentials or Heisenberg cores are also employed.[62] (b) The split-shell model,[63] in which the interelectonic interaction $1/r_{12}$ is neglected during the whole collision process, and the bound electrons are initialized with the corresponding sequential binding energies. (c) The dCTMC model,[64,65] which considers a dynamical screening for the electrons that depends on their binding energies. In this sense, the dCTMC model incorporates radial correlation among the initially bound electrons (without the risk of non-physical autoionization) at the expense of sacrificing the angular correlation.[64,65] Since capture + excitation provides a good fraction of the total single charge exchange events recorded, and electronic correlation is expected to play a role, we have chosen the dCTMC model to perform our theoretical analysis.

In short, each electron during the simulation is subject to an interaction potential which can be split in two terms: one corresponding to the electron-nucleus interaction, and one which contemplates a dynamical screening based on the other electron energy. The explicit form chosen is that suggested by Montemayor and Schiwietz[64] and later on used by Meng *et al.*[65]:

$$V_i(r) = -\frac{Z}{r_i} + \frac{1 - (1 + \lambda_j r_i)e^{-2\lambda_j r_i}}{r_i}, \tag{5}$$

with

$$\lambda_j = \begin{cases} 0 & \text{for } I_j > 0 \\ (\lambda_{ls}/I_{ls})\, I_j & \text{for } I_{ls} < I_j < 0 \\ \lambda_{ls} & \text{for } I_j < I_{ls}. \end{cases} \tag{6}$$

For the screening constant we have used the Slater screening constant for the $1s$ orbital of He $\lambda_{ls} = 1.6875$ and the ionization potential $I_{ls} = 24.6$. From equation (5) it can be easily inferred that the force acting on each electron depends not only on the radial position of that electron, but on the position and velocity of the other one. Montemayor and Schiwietz[64] have shown that this procedure improved the agreement with available data from several laboratories worldwide for the proton + He double ionization total cross section in comparison with the split-shell model.

As usual, the classical number n_c is obtained from the binding energy E_p of the electron relative to the projectile by:

$$E_p = -Z_{\rm p}^2/(2n_c^2), \tag{7}$$

where $Z_{\rm p}$ is the charge of the projectile. Then, n_c is related to the quantum number n of the final state by the Becker and McKellar condition[66]:

$$[(n-1)(n-1/2)n]^{1/3} \leq n_c \leq [n(n+1)(n+1/2)]^{1/3}. \tag{8}$$

Similar exit tests were performed for both electrons on those events corresponding to single charge exchange. This procedure allowed the identification of the n-values of each electron with respect either to the projectile or the remaining target core according to their final binding energies.

In Fig. 7 we display dCTMC state-selective capture cross sections for the process under study. It is seen that the one electron reaction of capture to the excited state in the projectile $(2, 1)$ is more probable than the two electron reaction of capture of one electron to the ground state of the projectile with simultaneous excitation of the second target electron into an excited target state. One electron processes are more probable than two electron processes. Furthermore, in Fig. 8 we show the scattering angle and impact parameter distributions of the captured events recorded with the dCTMC model at impact energy of 60 keV/amu. It can be observed that

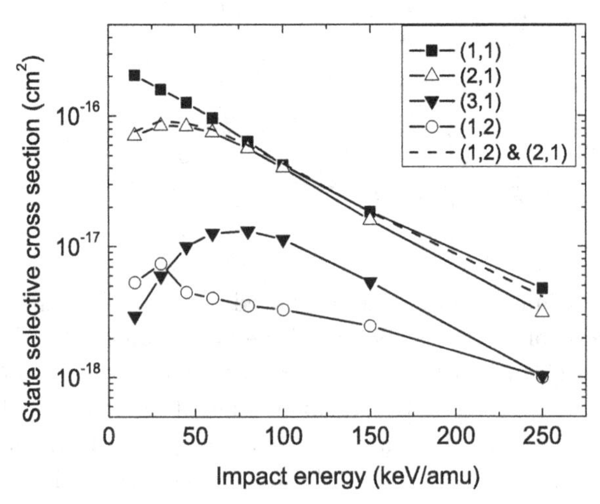

Fig. 7. State-selective capture cross sections as a function of impact energy from dCTMC calculations.

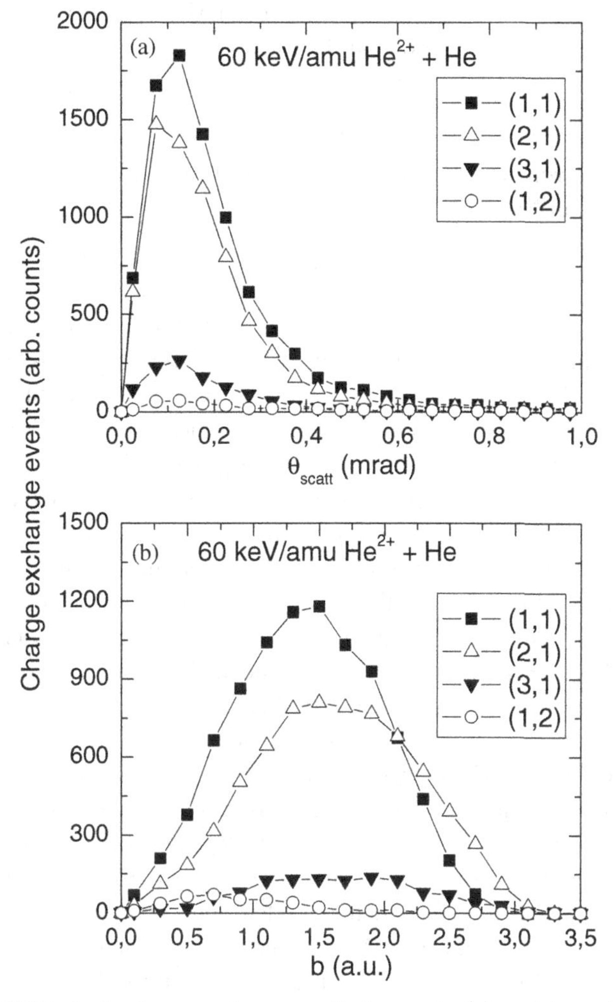

Fig. 8. dCTMC single charge exchange scattering angle (a) and impact parameter (b) distributions for 60 keV/amu $^3He^{2+}$ + He collisions.

the (1, 1), (2, 1) and (3, 1) channels mainly populate from the same range of impact parameters, but the (1, 2) channel clearly feeds from an inner range of impact parameters as could be expected. In terms of scattering angle, most of the capture events recorded are associated to angular deflections of the projectile which are inferior than about 0.4 mrad.

What has been considered until now was the p_l distribution obtained from the p_x momentum component measurements. We can also measure the p_y and p_z components. From these both the transverse momentum component, p_t, of the recoil-ion can be found by:

$$p_t = \sqrt{p_y^2 + p_z^2}. \tag{9}$$

For an electron capture collision it follows from energy and momentum conservation that[54]:

$$p_{\mathrm{t}} = -M_{\mathrm{p}} v_{\mathrm{p}} \theta. \tag{10}$$

Here M_{p} is the projectile mass and θ is the scattering angle of the projectile. Equation (10) is valid for small scattering angles and small changes in projectile energy in the collision; a condition fulfilled for ion atom collisions at intermediate collision energies. Equation (10) shows that measurement of the p_t of the recoil-ion gives information about the scattering angle of the projectile and hence about the impact parameter and approach distances of the colliding particles.

In our experiment at the present stage it was not possible to go into studies of the p_t component. In Figs. 9 and 10 we show a typical distribution for p_y and p_z, with background subtracted, for 60 keV incident $^3He^{2+}$ projectiles. From considerations about cylindrical symmetry of the collision process, it is expected that both distributions should display the same shape. This is not so; the p_y distributions for capture to the ground and excited states have nearly the same shape, while the p_z distributions are narrower and display different shapes for the ground and excited states. The p_y distributions show a width at half maximum of 2.2 a.u. that we explain as resulting dominantly from the longitudinal jet residual temperature after the adiabatic expansion, to which is added the detector resolution. The important contribution of the initial thermal momentum of the target atoms strongly shadows the contribution of the momentum resulting from the collision process. On the contrary, the p_z distributions obtained from the time-of-flight spectrum, have a better resolution as expected from the first order time focusing, the electronics timing and geometrical cooling for the present experimental setup. Hence, in this direction the distributions may reflect mainly the momentum transferred in the collision. The width of p_z distribution for the $(1, 1)$ state is 1.07 a.u. and for the excited states is 0.83 a.u. Although we cannot find the p_t distribution using equation (9), it is possible to draw some qualitative information from the p_z distributions. From the observation in Fig. 10 that the excited states have a narrower width, it follows that the scattering angles associated with this capture channel are

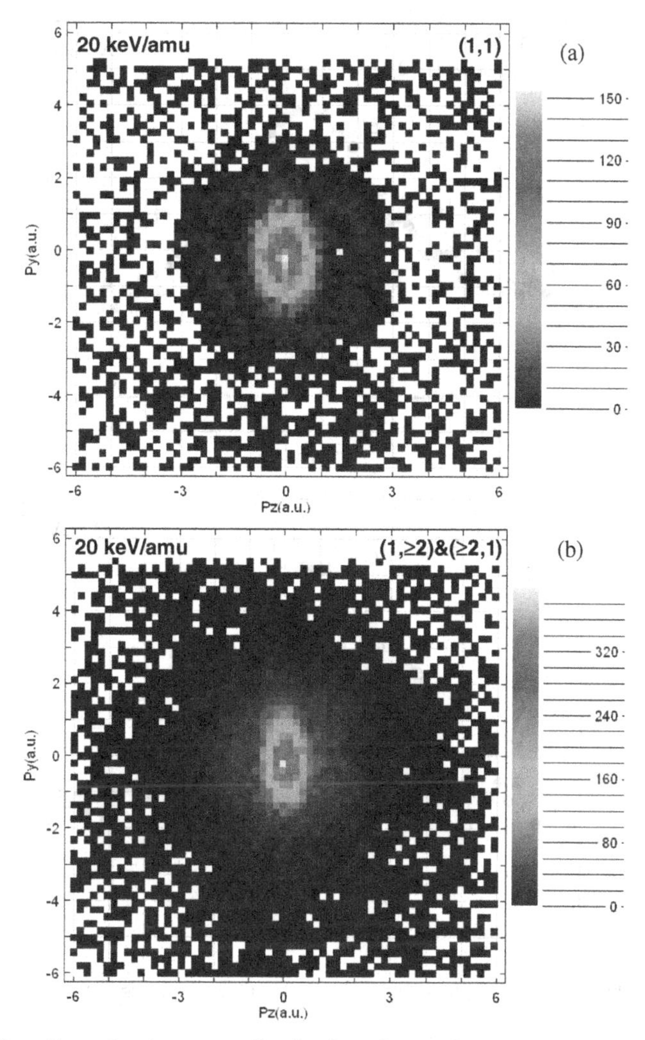

Fig. 9. Two dimensional $p_y - p_z$ distributions for single capture collision for 60 keV $^3He^{2+}$ projectiles. (a) For the (1, 1) capture to the ground state, and (b) for capture to excited states.

smaller than for the ground state, and therefore larger impact parameters are involved. A similar observation has been reported by Cassimi *et al.*[67] for single capture from He by highly ionized Ne and Ar projectiles, where they show that the higher excited states result from smaller p_t collisions. Another observation is that the p_z distribution corresponding to the (1, 1) capture shows oscillations with minima at 1.2 a.u. and 3 a.u. A similar oscillatory

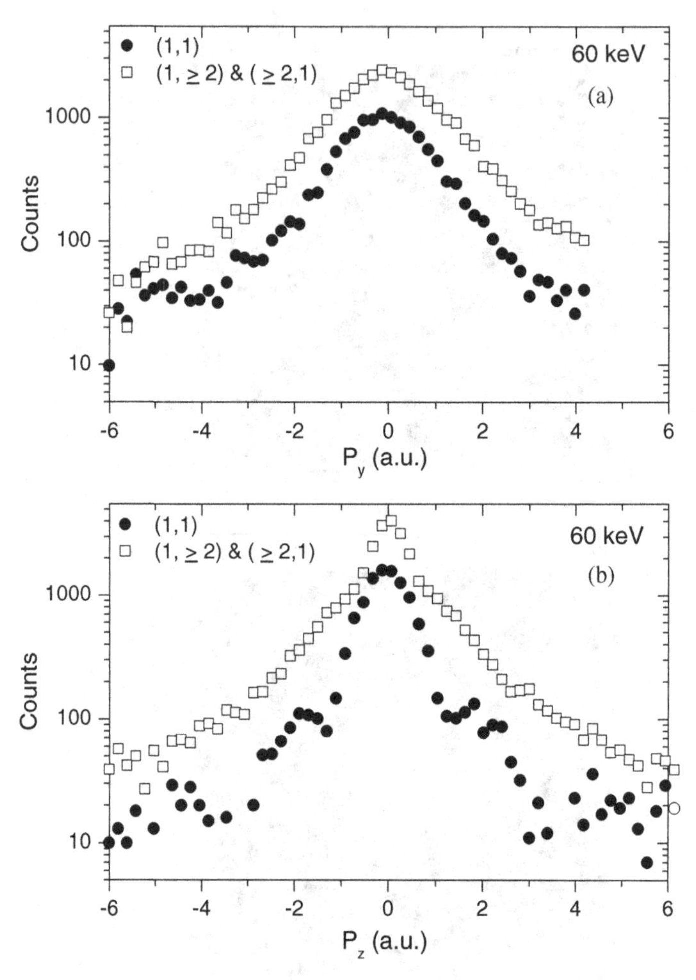

Fig. 10. p_z and p_y distributions for 60 keV $^3He^{2+}$ projectiles, resulting from the ones displayed in Fig. 9.

structure has been reported by Mergel *et al.*[36] and Schöffler *et al.*[54] for K capture in He^{2+} + He system at higher collision energies. The distribution corresponding to the excited states does not show such oscillation, except for a hump at 1 a.u. Figures 9 and 10 clearly show for the present studied collision system that the two dimensional $p_y - p_z$ distributions display in one of the axes a strong contribution of thermal random momenta. In order to obtain the transverse momentum distribution, it may be necessary to include a precooling facility to our target.

5. Summary

We presented a short survey of traditional experimental techniques used since the 1960s and 1970s to study collision processes, aimed at detecting all the reaction products. Among the different techniques the most successful were the coincidence measurements and translational spectroscopy. This latter technique, still successfully used, is a powerful spectroscopic technique for the measurement of the Q values of inelastic collisions. From the efforts and progress of experimentation, the detection of the recoil-ions evolved, establishing the powerful COLTRIMS technique and the development of the "reaction microscopes". This innovative technique provides spectroscopic studies of Q of reaction as a competitive alternative to the traditional translational spectroscopy, and also makes possible kinematically complete experiments for the study of collisional processes. A brief account was given of the concept behind the COLTRIMS technique. We described a COLTRIMS spectrometer assembled in our laboratory in a basic version that allows the detection of recoil-ion momenta.[40] This instrument was used for the study of single capture processes in the reaction He^{2+} + He.[58] The present study, performed in an intermediate range of collision energies, represents a continuation of previously reported measurements performed by Mergel *et al.*[36] at higher collision energies. We identified state-selective capture reactions from longitudinal momentum spectra. The results have been compared with dCTMC calculations, showing a good agreement in the energy range under study. Our presented data, at the edges of the accessible energy range, closely match published data from other laboratories,[36,56,57] giving confidence in the performance of our COLTRIMS apparatus. With regard to the measurement of the transverse momentum component for the present capture process, we note that the results obtained were limited. Only some qualitative observations could be made. The main limitation was the inadequate cooling of the He target-gas. In order to step forward towards a complete determination of the recoil-ion momenta with our COLTRIMS facility, efforts are under way to overcome this issue.

References

1. R. C. Isler, *Plasma Phys. Control. Fusion.*, **36**, 171–208 (1994).
2. F. P. Ziemba, G. L. Lockwood, G. H. Morgan, and E. Everhart, *Phys. Rev.*, **118**, 1552–1561 (1960).
3. V. V. Afrosimov, Yu. S. Gordev, M. N. Panov, and N. V. Fedorenko, *Sov. Phys.-Tech. Phys.*, **9**, 1248–1255, 1256–1264, 1265–1271 (1965).

4. V. V. Afrosimov, Yu. A. Mamaev, M. N. Panov, V. Uroshevich, and N. V. Fedorenko, *Sov. Phys.-Tech. Phys.*, **12**, 394–402 (1967).
5. W. Aberth and D. C. Lorents, *Phys. Rev.*, **144**, 109–115 (1966).
6. J. T. Park and F. T. Showengert, *Rev. Sci. Instr.*, **40**, 753–760 (1969).
7. Y. Y. Makhdis, K. Birkinshaw, and J. B. Hasted, *J. Phys. B: At. Mol. Phys.*, **9**, 111–121 (1976).
8. W. W. Siegel, Y. H. Chen, and J. W. Boring, *Phys. Rev. Lett.*, **28**, 465–468 (1972).
9. Q. C. Kessel, *Case Studies in Atomic Collision Physics*, **I**, E. W. McDaniel and M. R. C. McDowell (eds.), North Holland Publ. Comp. Amsterdam London, pp. 399–462 (1969).
10. G. H. Morgan and E. Everhart, *Phys. Rev.*, **128**, 667–676 (1962).
11. E. Everhart and Q. C. Kessel, *Phys. Rev. Lett.*, **14**, 247–249 (1965).
12. N. Stolterfoht, *Phys. Rep.*, **146**, 315–424 (1987).
13. J. T. Park, *Adv. At. Mol. Phys.*, **19**, 67–133 (1983).
14. J. T. Park, *Collision Spectroscopy*, R. G. Cooks (ed.) Plenum, New York, pp. 19–90 (1978).
15. R. K. Janev and H. Winter, *Phys. Rep.*, **117**, 265–387 (1985).
16. A. Lahmam-Bennani, *J. Phys. B: At. Mol. Opt. Phys.*, **24**, 2401–2442 (1991).
17. M. A. Coplan, J. H. Moore, and J. P. Doering, *Rev. Mod. Phys.*, **66**, 985–1014 (1994).
18. Q. C. Kessel, E. Pollack, and W. W. Smith, *Collision Spectroscopy*, R. G. Cooks (Ed.) Plenum New York, pp. 147–225 (1978).
19. J. C. Levin, R. T. Short, O. S. Elston, J. P. Gibbons, I. A. Sellin, and H. Schmidt-Böcking, *Phys. Rev. A*, **36**, 1649–1652 (1987).
20. J. Ullrich and H. Schmidt-Böcking, *Phys. Lett. A*, **125**, 193–196 (1987).
21. J. Ullrich, H. Schmidt-Böcking, and C. Kelbch, *Nucl. Instr. Meth. Phys. Res. A*, **268**, 216–224 (1988).
22. R. Dörner, J. Ullrich, O. Jagutzki, S. Lencinas, A. Gensmantel, and H. Schmidt-Böcking, *XVII Int. Conf. on the Physics of Electronic and Atomic Collisions*, Invited Papers, p. 351. W. R. McGillivray, I. E. McCarthy, and M. C. Standage (eds.), Adam Hilger, Bristol (1991).
23. J. Ullrich, R. Dörner, S. Lencinas, O. Jagutzki, H. Schmidt-Böcking, and U. Buck, *Nucl. Instr. Meth. Phys. Res. B*, **61**, 415–422 (1991).
24. J. Ullrich, R. Dörner, V. Mergel, O. Jagutzki, L. Spielberger, and H. Schmidt-Böcking, *Comm. At. Mol. Phys.*, **30**, 285–304 (1994).
25. J. Ullrich, R. Moshammer, R. Dörner, O. Jagutzki, V. Mergel, H. Schmidt-Böcking, and L. Spielberger, *J. Phys. B: At. Mol. Opt. Phys.*, **30**, 2917–2974 (1997).
26. R. Dörner, V. Mergel, O. Jagutzki, L. Spielberger, J. Ullrich, R. Moshammer, and H. Schmidt-Böcking, *Phys. Rep.*, **330**, 95–192 (2000).
27. J. Ullrich, R. Moshammer, A. Dorn, R. Dörner, L. Ph. H. Schmidt, and H. Schmidt-Böcking, *Rep. Prog. Phys.*, **66**, 1463–1545 (2003).
28. R. Ali, V. Frohne, C. L. Cocke, M. Stöckli, S. Cheng, and M. L. A. Raphaelian, *Phys. Rev. Lett.*, **69**, 2491–2494 (1992).
29. S. D. Kravis *et al. Phys. Rev. A*, **54**, 1394–1403 (1996).

30. R. Dörner, H. Khemliche, M. H. Prior, C. L. Cocke, J. A. Gary, R. E. Olson, V. Mergel, J. Ullrich, and H. Schmidt-Böcking, *Phys. Rev. Lett.*, **77**, 4520–4523 (1996).
31. R. Moshammer, M. Unverzagt, W. Schmitt, J. Ullrich, and H. Schmidt-Böcking, *Nucl. Instr. Meth. Phys. Res. B*, **108**, 425–445 (1996).
32. M. van der Poel, C. V. Nielsen, M. A. Gearba, and N. Andersen, *Phys. Rev. Lett.*, **87**, 123201-1–4 (2001).
33. J. M. Tukstra, R. Hoekstra, S. Knoop, D. Meyer, R. Morgenstern, and R. E. Olson, *Phys. Rev. Lett.*, **87**, 123202-1–4 (2001).
34. X. Flechard, H. Nguyen, E. Wells, I. Ben-Itzhak, and B. D. DePaola, *Phys. Rev. Lett.*, **87**, 123203-1–4 (2001).
35. D. R. Miller, *Atomic and Molecular Beam Methods*. Vol, **1**, pp. 14–53. G. Scoles *et al.* (eds.), Oxford Univ. Press, New York-Oxford (1988).
36. V. Mergel *et al.*, *Phys. Rev. Lett.*, **74**, 2200–2203 (1995).
37. W. C. Wiley and I. H. McLaren, *Rev. Sci. Instr.*, **26**, 1150–1157 (1955).
38. R. Dörner *et al.*, *Nucl. Instr. Meth. Phys. Res. B*, **124**, 225–231 (1997).
39. M. A. Abdallah, W. Wolff, H. E. Wolf, L. F. S. Coelho, C. L. Cocke, and M. Stöckli, *Phys. Rev. A*, **62**, 012711-1–10 (2000).
40. M. Alessi, D. Fregenal, and P. Focke, *Nucl. Instr. Meth. Phys. Res. B*, **269**, 484–491 (2011).
41. S. E. Sobottka and M. B. Williams, *IEEE Trans. Nucl. Sci.*, **35**, 348–351 (1988).
42. O. Jagutzki, V. Mergel, K. Ullmann-Pfleger, L. Spielberger, U. Meyer, R. Dörner, and H. Schmidt-Böcking, *Imaging Spectroscopy IV*, in M. R. Descour and S. S. Shen (eds.), *Proceedings of the International Symposium on Optical Science Engineering and Instrumentation, SPIE.*, **3438**, 322–333 (1998).
43. O. Jagutzki, V. Mergel, K. Ullmann-Pfleger, L. Spielberger, U. Spillmann, R. Dörner, and H. Schmidt-Böcking, *Nucl. Instr. Meth. Phys. Res. A*, **477**, 244–249 (2002).
44. J. P. Toennies and K. Winkelmann, *J. Chem. Phys.*, **66**, 3965–3979 (1977).
45. C. L. Cocke, *Phys. Scr.*, **T110**, 9–21 (2004).
46. M. A. Abdallah, A. Landers, M. Singh, W. Wolff, H. E. Wolf, E. Y. Kamber, M. Stöckli, and C. L. Cocke, *Nucl. Instr. Meth. Phys. Res. B*, **154**, 73–82 (1999).
47. M. Barat and P. Roncin, *J. Phys. B: At. Mol. Opt. Phys.*, **25**, 2205–2243 (1992).
48. Dž. Belkić, I. Mančev, and J. Hansen, *Rev. Mod. Phys.*, **80**, 249–314 (2008).
49. M. B. Shah and H. B. Gilbody, *J. Phys. B: At. Mol. Phys.*, **18**, 899–913 (1985).
50. M. B. Shah, P. McCallion, and H. B. Gilbody, *J. Phys. B: At. Mol. Opt. Phys.*, **22**, 3037–3045 (1989).
51. R. D. DuBois, *Phys. Rev. A*, **33**, 1595–1601 (1986).
52. R. D. DuBois, *Phys. Rev. A*, **36**, 2585–2593 (1987).
53. R. Dörner, V. Mergel, L. Spielberger, O. Jagutzki, J. Ullrich, and H. Schmidt-Böcking, *Phys. Rev. A*, **57**, 312–317 (1998).

54. M. S. Schöffler, J. Titze, L. Ph. Schmidt, T. Jahnke, N. Neumann, O. Jagutzki, H. Schmidt-Böcking, R. Dörner, and I. Mancev, *Phys. Rev. A*, **79**, 064701-1–4 (2009).
55. M. S. Schöffler, J. N. Titze, L. Ph. H. Schmidt, T. Jahnke, O. Jagutzki, H. Schmidt-Böcking, and R. Dörner, *Phys. Rev. A*, **80**, 042702-1–6 (2009).
56. X. L. Zhu *et al.*, *Chin. Phys. Lett.*, **23**, 587–590 (2006).
57. X. L. Zhu *et al.*, *J. Phys: Conf. Ser.*, **163**, 012064-1–4 (2009).
58. M. Alessi, S. Otranto, and P. Focke, *Phys. Rev. A*, **83**, 014701-1–4 (2011).
59. V. D. Rodriguez, Y. D. Wang, and C. D. Lin, *Phys. Rev. A*, **52**, R9–R12 (1995).
60. *Atomic Data for Controlled Fusion Research, ORNL 5206*, Vol, **I** (1977).
61. R. E. Olson, *Phys. Rev. A*, **36**, 1519–1521 (1987).
62. M. L. McKenzie, and R. E. Olson, *Phys. Rev. A*, **35**, 2863–2868 (1987).
63. A. E. Wetmore and R. E. Olson, *Phys. Rev. A*, **38**, 5563–5570 (1988).
64. V. J. Montemayor and G. Schiwietz, *Phys. Rev. A*, **40**, 6223–6230 (1989).
65. L. Meng, R. E. Olson, R. Dörner, J. Ullrich, and H. Schmidt-Böcking. *J. Phys. B: At. Mol. Opt. Phys.*, **26**, 3387–3401 (1993).
66. R. Becker and A. D. McKellar, *J. Phys. B: At. Mol. Phys.*, **17**, 3923–3942 (1984).
67. A. Cassimi, S. Duponchel, X. Flechard, P. Jardin, P. Sortais, D. Hennecart, and R. E. Olson, *Phys. Rev. Lett.*, **76**, 3679–3682 (1996).

Chapter 3

Recent Advances in the Theory and Modelling of Multiple Processes in Heavy-Particle Collisions

T. Kirchner*
Department of Physics and Astronomy, York University, Toronto, Ontario, Canada M3J 1P3, tomk@yorku.ca

M. Zapukhlyak
Institut für Theoretische Physik, Leibniz Universität Hannover, D-30167 Hannover, Germany

M. F. Ciappina
ICFO-The Institute of Photonic Sciences, 08860 Castelldefels (Barcelona), Spain

M. Schulz
Department of Physics and LAMOR, Missouri University of Science & Technology, Rolla, Missouri 65409, USA

Progress in experimental methods and data analysis techniques has enabled detailed and in-depth views on the few-body dynamics which govern heavy-particle atomic collisions. Still, theoretical and computational support is needed to gain a thorough understanding of the measured spectra. This is a challenging task, in particular when multiple-electron transitions are studied in regimes in which standard perturbative methods are not applicable.

This chapter explores two avenues that have been followed to meet this challenge. First, the coupled-channel basis generator method has been extended in order to calculate projectile angular differential cross

*The author to whom corrspondence should be addressed.

sections for processes involving electron transfer. Somewhat surprisingly, it is found that the independent particle model works well, in some cases even for double capture, which is commonly believed to be a correlated process. Second, first- and higher-order perturbative models have been used in conjunction with a Monte Carlo event generator to investigate double ionization of helium. The Monte Carlo event generator provides theoretical event files which are analogous to data obtained from kinematically complete experiments. As it turns out, these event files form a seminal starting point to shed new light on the roles of different double-ionization mechanisms.

1. Introduction

A large variety of multi-electron processes occur when a charged heavy particle collides with an atom or a molecule. This chapter concentrates on the prototype examples: double ionization (DI) and two-electron processes involving electron capture,[a] particularly double capture (DC), which have attracted the attention of both experimentalists and theorists over many years. One question that arises when one studies these processes is whether and to which extent electron-correlation effects play a role in them. Loosely speaking, we have learned that single-electron transitions do not involve significant electron correlations, but multiple transitions do. To deal with them in an appropriate way is a challenge for any theoretical approach to these problems.

An additional issue (and challenge) arises if the heavy projectile is a highly-charged ion (HCI): a HCI is associated with a strong Coulomb field, and this implies that the induced electron dynamics is typically of a non-perturbative nature. On top of this, experimental advances, most notably the cold target recoil ion momentum spectroscopy (COLTRIMS) technique, have enabled these possibly nonperturbative and possibly correlated few-body dynamics in collisions to be mapped in depth and in detail (see, e.g., Refs. 1, 2). They have produced a wealth of interesting and surprising results, whose description and explanation is what theorists are asked to give.

There is little doubt that, in principle, one consistent quantum theory should suffice to fulfil this request. In practice, however, we are pretty far away from having a theory at hand, which can describe e.g. both DC processes in relatively slow and DI processes in relatively fast collisions.

While the ultimate answer is still lacking, partial answers based on different methods for different processes have become available. This

[a]We will use the terms capture and transfer as synonyms in this chapter.

chapter is concerned with two such approaches and the partial answers they provide: first, the coupled-channel basis generator method has been extended in order to calculate projectile angular differential cross sections for processes involving electron transfer. Somewhat surprisingly, it was found that the independent electron model works well, in some cases even for DC, which is commonly believed to be a correlated process. These findings were discussed in detail in Refs. 3, 4–6. We summarize the approach in Sec. 2.1 and discuss sample results in Sec. 3.1.

Second, a Monte Carlo event generator (MCEG) technique has been developed that allows the generation of theoretical event files which are analogous to data obtained from kinematically complete experiments and, accordingly, can be analyzed in the same way.[7–9] First- and higher-order perturbative models have been implemented to investigate specifically DI of helium by ion[9–13] and very recently also by electron[14] impact. As it turns out, this approach sheds new light on the roles of different ionization mechanisms, some of which are well known, while at least one has not been discussed before. The perturbative methods and the MCEG technique are described in Sec. 2.2, while sample results are discussed in Sec. 3.2. A few conclusions are offered in Sec. 4.

Atomic units, characterized by $\hbar = e^2 = m_e = 1$ are used throughout this chapter unless stated otherwise.

2. Theory

An appropriate starting point for a full quantum-mechanical description of an ion-atom collision system is the (nonrelativistic) Hamiltonian

$$\hat{H} = \hat{K} + \hat{H}_e \tag{1}$$

that consists of the kinetic energy $\hat{K}$ associated with the heavy-particle motion and the electronic Hamiltonian $\hat{H}_\mathrm{e}$. We denote by $\mathbf{x}_j$ and $\mathbf{s}_j$ the position vectors of the jth electron with respect to the target and the projectile nuclei respectively, by $r_{ij} = |\mathbf{x}_i - \mathbf{x}_j| = |\mathbf{s}_i - \mathbf{s}_j|$ the distance between the ith and the jth electron, and by $\mathbf{R}$ the relative vector between the heavy particles such that $\mathbf{s}_j = \mathbf{x}_j - \mathbf{R}$. The electronic Hamiltonian then reads

$$\hat{H}_\mathrm{e} = \sum_{j=1}^{N} \left(-\frac{1}{2}\nabla_j^2 - \frac{Z_\mathrm{t}}{x_j} - \frac{Z_\mathrm{P}}{s_j} \right) + \sum_{i<j}^{N} \frac{1}{r_{ij}} + V_\mathrm{nn} \tag{2}$$

$$V_\mathrm{nn} = \frac{Z_\mathrm{t} Z_\mathrm{P}}{R} \tag{3}$$

with the charges Z_t and Z_p of the target and the projectile nuclei, respectively. Note that for the helium target atom under consideration in this chapter we have $N = Z_t = 2$.

A standard approach to this collision problem can be formulated as follows: depending on the process of interest one decomposes $\hat{H}_e$ into a part $\hat{H}_0$ that defines the asymptotic channels and a perturbation $\hat{V}$. Then one sets up the T-matrix, whose evaluation is the central task. The usual first step in doing so is to separate heavy-particle and electronic motions by representing the former in terms of plane waves. The T-matrix can then be written in the form

$$T_{if} = \frac{1}{(2\pi)^3}\langle e^{i\mathbf{K}_f\cdot\mathbf{R}}\Phi_f|\hat{V}|e^{i\mathbf{K}_i\cdot\mathbf{R}}\Psi_i\rangle = \frac{1}{(2\pi)^3}\int\langle\Phi_f|\hat{V}|\Psi_i\rangle e^{i\mathbf{q}\cdot\mathbf{R}}d^3R \tag{4}$$

with the initial and final projectile momenta $\mathbf{K}_i$ and $\mathbf{K}_f$ and the momentum transfer $\mathbf{q} = \mathbf{K}_i - \mathbf{K}_f$. The electronic wave functions Ψ_i and Φ_f solve stationary Schrödinger equations with appropriate boundary conditions for the Hamiltonians $\hat{H}_e$ and $\hat{H}_0 = \hat{H}_e - \hat{V}$ and characterize the full scattering state and a given final state, respectively.

There are two options to proceed further. The first one is to tackle the T-matrix element (4) directly. This is usually done in the context of perturbation theory. We follow this avenue in our treatment of DI which is described in Sec. 2.2. The second option is to rely on the impact parameter method and assume that the wave function Ψ_i solves the *time-dependent* Schrödinger equation (TDSE) for $\hat{H}_e$, in which $\mathbf{R}$ is treated as a classical coordinate, normally a straight-line trajectory $\mathbf{R}(t) = \mathbf{b} + \mathbf{v}_0 t$ characterized by the impact parameter $\mathbf{b}$ and the constant projectile velocity $\mathbf{v}_0$ with $\mathbf{b}\cdot\mathbf{v}_0 = 0$. Exploiting the specifics of ion-atom collisions, namely small scattering angles and the large mass ratio of the heavy particles to the electrons, one can cast the T-matrix element (4) into the form

$$T_{if} = \frac{iv_0}{(2\pi)^3}\int e^{i\mathbf{q}\cdot\mathbf{b}}A_{if}(\mathbf{b})d^2b, \tag{5}$$

where the impact-parameter-dependent transition amplitude

$$A_{if}(\mathbf{b}) = \lim_{t\to\infty}\langle\Phi_f|\Psi_i\rangle \tag{6}$$

is obtained from the solution of the TDSE.[15] If the TDSE is solved in perturbation theory, this treatment is equivalent to applying perturbation theory to the T-matrix (4) itself.[16] The advantage of the impact parameter method is that *nonperturbative* solutions of the TDSE are feasible — in particular if one is ready to compromise on the treatment of the electron-electron (*ee*) interaction in $\hat{H}_e$. A related benefit is that one can account for the nucleus-nucleus (*nn*) interaction (3) to all orders without much pain: since it is a merely time-dependent term in the TDSE for $\hat{H}_e$ the amplitude (6) can be rewritten as

$$A_{if}(\mathbf{b}) = a_{if}(\mathbf{b}) \exp[-2i \int_0^\infty V_{\mathrm{nn}}(R(t))\mathrm{d}t], \tag{7}$$

where $a_{if}(\mathbf{b})$ is the corresponding transition amplitude for the TDSE with the Hamiltonian $\hat{H}_e - V_{\mathrm{nn}}$. We follow this avenue in our account on electron-capture processes which is described in the next subsection.

2.1. *Independent Electron Approximation for Capture Processes*

The *ee* interaction and the correlation effects it might cause make the full solution of the TDSE for $\hat{H}_e$ (or for $\hat{H}_e - V_{\mathrm{nn}}$) a challenging task. Nevertheless, this challenge has been accepted by a number of researchers when considering the two-electron helium target. Roughly, one can distinguish two groups of nonperturbative calculations which take electron-correlation effects into account: the first one addresses collisions at relatively low impact energies E_p, for which only a few states have to be considered and the influence of continuum channels on bound-state transitions is deemed to be small.[17] The second one is concerned with ionization in situations in which the coupling to electron-transfer channels can be neglected; i.e., with high E_p[18,19] or antiproton collisions in which electron transfer is absent.[20]

However, if these conditions are not met, convergence is hard to achieve and simplifications are in order. The most popular one is to replace the Hamiltonian $\hat{H}_e - V_{\mathrm{nn}}$ by a one-body operator

$$\hat{H}_e - V_{\mathrm{nn}} \to \sum_{j=1}^{N} \hat{h}_j. \tag{8}$$

The TDSE then separates into a set of single-particle equations

$$i\partial_t \psi_j(\mathbf{r}, t) = \hat{h}\psi_j(\mathbf{r}, t) \quad j = 1, \ldots, N \tag{9}$$

for the Hamiltonian

$$\hat{h} = -\frac{1}{2}\nabla^2 + V_{\text{eff}}(x) - \frac{Z_{\text{P}}}{s} \tag{10}$$

that contains the kinetic energy and (effective) target and projectile potentials. Different choices can be made for V_{eff}. We use a helium ground-state potential obtained from the optimized potential method (OPM) of density functional theory, in which screening and exchange are treated exactly, but correlation effects are neglected.[21] For the spin-singlet helium atom the ground state is realized by populating the K shell with two equivalent electrons. As a consequence, only one (initial $1s$) orbital has to be propagated. Note also that the OPM potential is equivalent to the Hartree-Fock ground-state potential in this case.

For the propagation we use the two-center (TC) version of the nonperturbative basis generator method (BGM). The BGM was introduced as a general means to solve the TDSE in terms of a finite, system-adapted basis set.[22] The underlying idea is that completeness of a basis is not a necessary condition for obtaining a converged result. Rather, the basis must be able to represent the time-evolving solution of the TDSE, which for a given initial condition defines nothing more than a one-dimensional subspace of the Hilbert space. A finite basis can do this job if it has the flexibility to adapt to the dynamics of the system. In Ref. 22 a basis with this property was found: it consists of a set of target eigenfunctions and a set of pseudostates, which are constructed by a repeated application of the (regularized) Coulombic projectile potential onto these eigenstates. Subsequently, such BGM basis sets were applied to a number of collision systems with considerable success.[23] The only issue with this method was its somewhat limited ability to describe electron-capture processes. This was overcome by extending the BGM to its two-center formulation, the TC-BGM, in which bound projectile states are also included.[24]

To be more specific, the TC-BGM basis sets used in the studies described in this chapter consist of finite sets of U_t target and $U - U_t$ projectile states. Galilean invariance is ensured by attaching electron translation factors to these states:

$$\phi_u^0(\mathbf{r}) = \begin{cases} \phi_u(\mathbf{x})\mathrm{e}^{i\mathbf{v}_t\cdot\mathbf{r}} & u \le U_t \\ \phi_u(\mathbf{s})\mathrm{e}^{i\mathbf{v}_p\cdot\mathbf{r}} & \text{else.} \end{cases} \tag{11}$$

Here, $\mathbf{v}_t$ and $\mathbf{v}_p$ denote the constant velocities of the atomic target and projectile frames and $\mathbf{r}$ the position vector of the electron in the center-of-mass (c.m.) frame. Equation (11) defines a standard two-center atomic orbital expansion. It is augmented by a set of pseudostates

$$\chi_u^\mu(\mathbf{r},t) = [W_p]^\mu \phi_u^0(\mathbf{r}) \quad \mu = 1,\ldots,M, \;\; u = 1,\ldots,U_t \tag{12}$$

$$W_p = \frac{1}{s}(1 - \mathrm{e}^{-s}), \tag{13}$$

which depend on time in any given reference frame, since they depend on both $\mathbf{x}$ and $\mathbf{s}$. When orthogonalized to the generating two-center atomic orbital basis (11) the pseudostates (12) account for ionization channels and for quasimolecular effects at low collision velocity, i.e., they give the TC-BGM basis the desired flexibility for the representation of the single-particle wave function:

$$\psi_{j=1s}(\mathbf{r},t) = \sum_{\mu=0}^{M(u)} \sum_{u=1}^{U} c_{\mu u}(t)\chi_u^\mu(\mathbf{r},t) \quad M(u) = \begin{cases} M & \text{if } u \le U_t \\ 0 & \text{else.} \end{cases} \tag{14}$$

If inserted in Equation (9) one obtains a set of ordinary differential equations for the coefficients $c_{\mu u}$ that can be solved by standard methods. The single-particle transition amplitudes for excitation and capture channels are directly represented by the coefficients for $\mu = 0$ at an asymptotic time $t = t_f$ when the propagation is stopped. Total ionization can be obtained from summing up the populations of the pseudostates, but we will not look into ionization processes in this section. Rather, we are interested in obtaining differential cross sections (DCSs) and total cross sections (TCSs) for two-electron transitions to bound states such as DC.

To this end, we need the corresponding two-electron amplitudes. If we want to be consistent with the one-body Hamiltonian (8) we have to adhere to the independent electron model (IEM) and calculate these amplitudes as products of one-electron amplitudes. Taking into account that both electrons start from the same $1s$ initial state, the two-electron amplitudes a_{if} for a transition of one electron to $\phi_{f_1}^0$ and one electron to $\phi_{f_2}^0$ is

$$\begin{aligned} a_{if} &= \sqrt{2} c_{0f_1} c_{0f_2} \quad \text{for } f_1 \ne f_2 \\ &= c_{0f_1}^2 \quad \text{else,} \end{aligned} \tag{15}$$

where the factor $\sqrt{2}$ takes the indistinguishability of the electrons into account.

The amplitude (15) is multiplied by the nn phase factor, the so-called eikonal phase, according to equation (7) and inserted in the T-matrix element (5). The azimuthal integration over the impact parameter can be carried out and the two-dimensional Fourier integral reduces to a one-dimensional Bessel integral that can be performed numerically or by expansions which are better adapted to deal with the eikonal phase. For heavy-particle collisions the magnitude of the momentum transfer can be related to the polar scattering angle θ according to

$$q \approx 2K_i \sin\frac{\theta}{2} \approx K_i\theta \qquad (16)$$

and the square of the T-matrix element is directly proportional to the projectile angular DCS in the c.m. system:

$$\left(\frac{\mathrm{d}\sigma_{if}}{\mathrm{d}\Omega}\right)_{c.m.}(\theta) = (2\pi)^4 \mu_{pt}^2 |T_{if}(\theta)|^2, \qquad (17)$$

where $\mu_{pt} = M_\mathrm{p} M_\mathrm{t}/(M_\mathrm{p} + M_\mathrm{t})$ is the reduced mass of the heavy particles. The procedure described in the last paragraph is of course nothing but the well-established eikonal approximation.[25] The TCS can be calculated either by integrating equation (17) over the projectile solid angle, or by using Parseval's identity and integrating weighted transition probabilities $b|a_{if}(b)|^2$ over the impact parameter b.[15] The latter procedure shows immediately that TCSs are not sensitive to the nn interaction, since $|a_{if}(b)|^2 = |A_{if}(b)|^2$.

2.2. *Perturbative Models of Double-Ionization Processes*

Let us come back to the form (4) of the T-matrix element. For the case of DI the natural decomposition of $\hat{H}_e$ is such that $\hat{H}_0$ is the helium Hamiltonian

$$\hat{H}_0 = \sum_{j=1}^{2}\left(-\frac{1}{2}\nabla_j^2 - \frac{Z_\mathrm{t}}{x_j}\right) + \frac{1}{r_{12}} \qquad (18)$$

and the perturbation $\hat{V}$ is the sum of the Coulomb interactions of the projectile with all constituents of the target

$$\hat{V} = -\sum_{j=1}^{2}\frac{Z_\mathrm{p}}{s_j} + V_\mathrm{nn}. \qquad (19)$$

In the first-order Born approximation (FBA) one replaces the scattering state Ψ_i by an unperturbed eigenstate Φ_i of $\hat{H}_0$. Using the Bethe integral the T-matrix can then be cast into the form[26]

$$T_{if} = \frac{Z_{\mathrm{p}}}{2\pi^2 q^2}\left[Z_{\mathrm{t}} M_0 - M_1 - M_2\right] \tag{20}$$

with

$$M_0 = \int \mathrm{d}\mathbf{x}_1 \int \mathrm{d}\mathbf{x}_2 \Phi^{(-)*}_{\mathbf{k}_1,\mathbf{k}_2}(\mathbf{x}_1,\mathbf{x}_2)\Phi_i(\mathbf{x}_1,\mathbf{x}_2) \tag{21}$$

$$M_1 = \int \mathrm{d}\mathbf{x}_1 \int \mathrm{d}\mathbf{x}_2 \Phi^{(-)*}_{\mathbf{k}_1,\mathbf{k}_2}(\mathbf{x}_1,\mathbf{x}_2)\mathrm{e}^{i\mathbf{q}\cdot\mathbf{x}_1}\Phi_i(\mathbf{x}_1,\mathbf{x}_2) \tag{22}$$

$$M_2 = \int \mathrm{d}\mathbf{x}_1 \int \mathrm{d}\mathbf{x}_2 \Phi^{(-)*}_{\mathbf{k}_1,\mathbf{k}_2}(\mathbf{x}_1,\mathbf{x}_2)\mathrm{e}^{i\mathbf{q}\cdot\mathbf{x}_2}\Phi_i(\mathbf{x}_1,\mathbf{x}_2), \tag{23}$$

where we use the more explicit notation $\Phi^{(-)}_{\mathbf{k}_1,\mathbf{k}_2}(\mathbf{x}_1,\mathbf{x}_2)$ for the coordinate representation of a final state (with incoming boundary conditions) that corresponds to detecting the two electrons with the momenta $\mathbf{k}_1$ and $\mathbf{k}_2$. Note that the overlap integral (21) would vanish if the initial and final states were exact eigenstates of $\hat{H}_0$. In practice, they are not and M_0 is nonzero. We found, however, that for the approximate wave functions which we use, the nonorthogonality is small and of no significance for the results presented in Sec. 3.2.

The matrix elements M_1 and M_2 would be zero if the initial and final states were uncorrelated. This is to say that DI occurs in first-order only as a consequence of electron-correlation effects. This corresponds to what is often dubbed the two-step-one (TS-1) projectile-electron interaction and the shake-off (SO) mechanism: the projectile interacts with only one electron, while the second electron is ejected through correlation with the first one. In the case of TS-1 this correlation is thought to come about as an *ee* collision, while in the SO mechanism the rearrangement in the target after the ejection of the first electron is blamed for the removal of the second one. In an approach in which the *ee* interaction is part of the undisturbed Hamiltonian H_0 both processes cannot be distinguished rigorously. Hence, we use the label TS-1 for both of them — or more accurately, for DI due to a single interaction of the projectile as it is described by the FBA.

Let us now specify the initial and final two-electron states we use in equations (21) to (23). The helium ground-state wave function is taken as a symmetrized product of hydrogenlike $1s$ orbitals

$$\Phi_i(\mathbf{x}_1,\mathbf{x}_2) = N_i(\mathrm{e}^{-Z_{\mathrm{a}}x_1}\mathrm{e}^{-Z_{\mathrm{b}}x_2} + \mathrm{e}^{-Z_{\mathrm{b}}x_1}\mathrm{e}^{-Z_{\mathrm{a}}x_2}), \tag{24}$$

where N_i is a normalization factor. Using two different effective charges corresponds to a model in which an "inner" electron screens the atomic nucleus such that the "outer" electron experiences a weaker attractive force. The charges Z_a and Z_b are determined variationally and have the values 2.183171 and 1.188530. This model yields the ground-state energy $\varepsilon_i = -2.8757$ a.u.,[27] which is below the Hartree Fock result $\varepsilon_i = -2.8617$ a.u. for the $1s^2$ configuration. In this sense (radial) correlations are included to some extent in equation (24).

For the two-electron continuum states we tested three relatively simple models.[10] The first and simplest one is a symmetrized product of one-electron scattering eigenstates $\phi_{\mathbf{k}}^{(-)}(\mathbf{x})$ of the bare helium nucleus (with incoming boundary conditions):

$$\phi_{\mathbf{k}_1,\mathbf{k}_2}^{(-),2C}(\mathbf{x}_1,\mathbf{x}_2) = \frac{1}{\sqrt{2}}[\phi_{\mathbf{k}_1}^{(-)}(\mathbf{x}_1)\phi_{\mathbf{k}_2}^{(-)}(\mathbf{x}_2) + \phi_{\mathbf{k}_2}^{(-)}(\mathbf{x}_1)\phi_{\mathbf{k}_1}^{(-)}(\mathbf{x}_2)]. \tag{25}$$

This wave function is known as the 2-Coulomb (2C) model and was used extensively in studies of electron impact ionization.[28] It describes the two one-electron-nucleus subsystems exactly, but neglects the interaction between the electrons completely. As a consequence, nothing prevents them from having the same or very similar momenta — an unphysical feature that was pointed out by several authors (see, e.g., Ref. 29 and references therein).

A more realistic final-state model has to take the asymptotic boundary conditions of the three-body Coulomb problem of the two ionized electrons and the residual target ion into account. One way to achieve this is to follow the so-called 3-Coulomb (3C) ansatz,[31,32] in which a third Coulomb distortion factor is added to equation (25) in order to model the *ee* interaction. To keep the numerical and computational burden manageable, we content ourselves with simplified versions of the 3C wave function, in which the relative Coulomb wave of the two-electron subsystem is replaced by its value at zero spatial distance. The explicit expression for this wave function is[33]

$$\phi_{\mathbf{k}_1,\mathbf{k}_2}^{(-),2C+}(\mathbf{x}_1,\mathbf{x}_2) = \frac{1}{\sqrt{2}}[\phi_{\mathbf{k}_1}^{(-)}(\mathbf{x}_1)\phi_{\mathbf{k}_2}^{(-)}(\mathbf{x}_2) + \phi_{\mathbf{k}_2}^{(-)}(\mathbf{x}_1)\phi_{\mathbf{k}_1}^{(-)}(\mathbf{x}_2)]\alpha(k_{12}), \tag{26}$$

with

$$\alpha(k_{12}) = \mathrm{e}^{-\pi\xi_{12}}\Gamma(1 - i\xi_{12}), \quad \xi_{12} = \frac{Z_{12}}{k_{12}} \tag{27}$$

and $k_{12} = |\mathbf{k}_1 - \mathbf{k}_2|$ being the relative momentum between the two electrons. $Z_{12} = 1$ represents the case of *static screening*. Compared to the 2C model the quantity

$$|\alpha(k_{12})|^2 = \frac{2\pi\xi_{12}}{e^{2\pi\xi_{12}} - 1} \tag{28}$$

known as the Gamow factor appears as a prefactor in the transition probabilities and cross sections. The Gamow factor suppresses exponentially the probability to find both ionized electrons with close momenta, but lets electrons with very different momenta move independently.

It was found that $Z_{12} = 1$ yields too strong a repulsion between electrons with low emission energies, and consequently strongly underestimated cross sections near threshold. To remedy this flaw *dynamic screening* models with effective charges that depend on the electron momenta were introduced.[34–37] We use a somewhat simplified version proposed to describe $(e, 3e)$ reactions,[38] and also applied to model the correlation function in DI by ion impact.[39] It consists of using the effective charge

$$Z_{12} = 1 - \frac{k_{12}^2}{(k_1 + k_2)^2} \tag{29}$$

in equations (27) and (28). Not surprisingly, we found that among our three final-state models, the most advanced one based on equation (29) gives the most convincing results. In Sec. 3.2 we only show spectra obtained from using equation (29).

We also found that the FBA is not sufficient to explain the experimental DI data — not even for 6 MeV p-He collisions for which the perturbation parameter $\eta = Z_\mathrm{p}/v_0$ is rather small. This signals the importance of higher-order effects. Performing second- or even higher-order Born calculations is, however, a daunting task, which we did not pursue. Rather, we made use of the MCEG technique mentioned in the Introduction (and described in more detail in Ref. 9) and simulated higher-order DI mechansims in terms of combinations of single-ionization events.

Working with theoretical event files that are analogous to event files obtained from a kinematically complete experiment has several advantages besides the possibility of simulating higher-order processes. The first one is that the theoretical data can be manipulated and analyzed in the same way as the experimental data, e.g., one can easily account for experimental uncertainties and phase space limitations. Furthermore, one can generate any type of cross sections, in particular those that are numerically too intensive to compute with conventional methods, simply by sorting histograms in the same way as the experimental data. A good example is the four-particle Dalitz (4-D) plot,[40] which provides a comprehensive yet detailed picture of the momentum balance of the four outgoing particles after DI. 4-D plots are explained and discussed in Sec. 3.2.

Another advantage of the MCEG technique is that one can add actual physics to the theoretical event file, which was not taken into account in the evaluation of the T-matrix. For example, effects of the nn interaction (3) are neglected in the FBA if one disregards the insignificant but nonzero contribution of the overlap integral (21). It is not impossible to deal with V_{nn} in a quite rigorous way by Fourier transforming the T-matrix to the impact parameter picture, multiplying the impact-parameter-dependent transition amplitude by the eikonal phase and transforming the result back to momentum space (cf. equations (5) and (7)), but this procedure is computationally costly. Using the MCEG instead and accounting for V_{nn} by convoluting the original FBA results with *classical* elastic scattering is less demanding and, interestingly, not necessarily inferior to the Fourier method when it comes to the comparison of the results with experimental data.[41] For all the results presented in Sec. 3.2 these convolutions have been performed using the method described in Ref. 8.

The simulation of higher-order processes is based on similar convolutions. The most prominent higher-order process is the two-step-two (TS-2) projectile-electron interaction mechanism, in which both electrons are ejected through independent interactions with the projectile. In the Born expansion framework, TS-2 is a part of the second-order amplitude which would have to be added to the first-order (TS-1) amplitude and might interfere with it. This aspect is ignored in our simulation — as are possible phase effects in the multiplication of two projectile-electron interactions in the TS-2 term itself. We simply convolute two single-ionization events, i.e., TS-2 is treated completely incoherently. The first ionization step corresponds to the single ionization of the neutral helium atom by the incoming projectile, while the second step is the ionization of the He^+ ion. Each ionization step can be calculated either in the FBA or in a higher-order theory such as the continuum-distorted wave with eikonal initial state (CDW-EIS) approach.[42] We have done both, based on similar undisturbed initial- and final-state wave functions: for the description of the initial state we employ a semi-analytical Hartree-Fock-Roothaan wave function for the neutral helium atom[43] and a hydrogenlike $1s$ state with

$Z_t = 2$ for the He^+ ion. The final states of the ejected electrons are Coulomb waves with effective charges, namely $Z_t = 1.69$ for the first ionization step in order to account for the partial screening of the nuclear charge and $Z_t = 2$ for the second step.

Another higher-order mechanism that can be simulated by using the MCEG technique is a hybrid between TS-1 and TS-2 which we introduced recently under the label TS-1-EL.[11] This is described and discussed along with the results obtained from it in Sec. 3.2.

3. Results

3.1. *Processes Involving Electron Capture*

Multiple-electron processes involving capture have been studied over many years for quite a few ion-atom collision systems. Earlier studies concentrated on TCSs and their impact-energy dependencies. Fully differential measurements over a broad range of impact energies E_p have become feasible only more recently, mainly thanks to the development of COLTRIMS and related experimental techniques. Since the two-electron helium atom has been the primary target under investigation, the question about the role of electron-correlation effects has been a rather popular one.[30]

It appears natural to assume that electron correlations play an important role in two-electron processes, such as electron transfer with simultaneous target excitation[44] or DC.[45] Over the course of time, however, it has become clear also that the footprints of the nn interaction are seen in cross sections which are differential in the projectile deflection. The question arises whether nn and ee interaction effects can be disentangled.

This question has been a major motivation for our own work in this area. Our strategy has been to adhere to the IEM, i.e., to neglect electron-correlation effects, but to include all other aspects as accurately as possible. As described in Sec. 2.1 these aspects are the nonperturbative nature of the electron dynamics (treated with the TC-BGM) and the nn interaction (dealt with in the eikonal approximation). Discrepancies in calculated and measured data can then be blamed on electron-correlation effects. If, on the other hand, agreement is found — and, in fact, it has been found in quite a few cases — these correlations are unimportant.

We can also obtain more explicit information on the role and the nature of the nn interaction with our approach. In an early account[44] we calculated

DCSs for various inelastic processes by using

$$\frac{d\sigma_{if}}{d\Omega}(\theta) = \frac{d\sigma_{\rm class}}{d\Omega}(\theta)\,|a_{if}(b(\theta))|^2, \tag{30}$$

where $d\sigma_{\rm class}/d\Omega$ is the classical cross section for elastic scattering and the impact parameter b is classically related to the scattering angle θ. Specifically, we looked at a double ratio of DCSs, for which a peak structure was found in the experimental data at 50 keV impact energy as a function of θ.[44] The processes involved in this double ratio are transfer excitation, single transfer, double excitation and single excitation in p-He collisions. Our IEM calculations based on equation (30) did not result in any structure, leaving the question open as to whether it is caused by *ee* correlations or by *nn* effects beyond classical scattering. Subsequent IEM calculations based on the eikonal approximation did result in a peak.[4] Hence, we were able to conclude that its appearance is indeed caused by *nn* interaction effects, or more precisely by quantum-mechanical heavy particle - electron couplings, which are taken into account in the eikonal approximation.

The calculated and measured double ratios are plotted in Fig. 1. Two calculations based on the eikonal approximation are shown: the full curve is obtained from the standard IEM according to equation (15), while the dashed curve corresponds to a slightly modified analysis in which single transfer and single excitation are evaluated in terms of a one-active-electron model.[4] The difference among both results might indicate that the peak height is sensitive to *ee* interaction effects, which are treated differently in both analyses. Even though we cannot prove this conjecture at this point, the example shows that our approach is indeed capable of disentangling *ee* and *nn* effects.

As mentioned above, we are especially interested in DC processes in this chapter. For the p-He system DC is not an impossible, but a weak and delicate process.[43,46] We do not consider it here, but address He^{2+} and highly-charged Ar ion impacts. To get a first idea about the possible role of electron-correlation effects, we take a look at the TCS for single and double capture in He^{2+}–He collisions. Single capture can be a pure one-electron process or can occur together with target excitation — the TCS shown in Fig. 2 does not distinguish between these situations, i.e., it is inclusive in target excitations.

Obviously, some of the experimental data sets are in conflict with each other at low E_p, but they clearly show that DC is stronger than single capture in this region. This is a consequence of the resonance character of

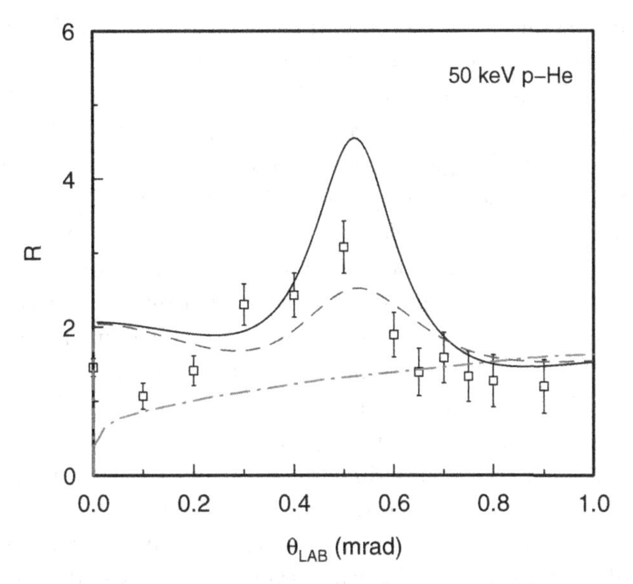

Fig. 1. Ratio of transfer excitation to single transfer versus double to single excitation cross sections as a function of laboratory scattering angle for 50 keV p–He collisions. Theory: solid curve, TC-BGM calculation within IEM for all amplitudes; dashed curve, TC-BGM calculation within one-active-electron model for single transfer and single excitation amplitudes; dash-dotted curves, TC-BGM calculation based on equation (30).[44] Experiments: (□).[44]

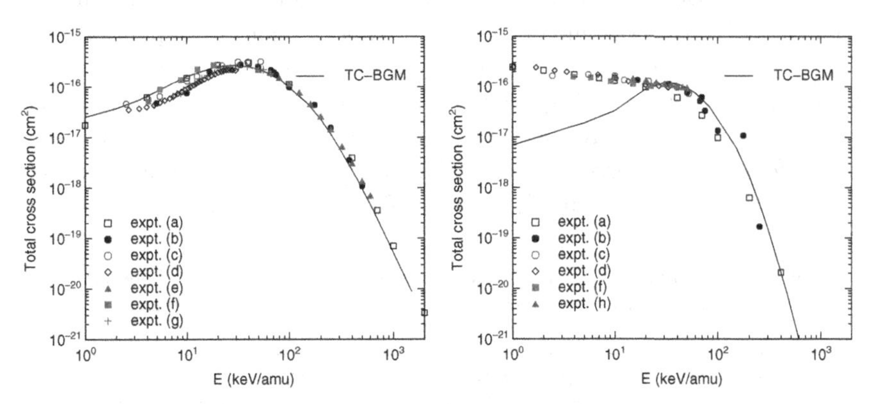

Fig. 2. Total single (left panel) and double (right panel) electron-capture cross sections as functions of impact energy for He^{2+}–He collisions. Theory: present TC-BGM calculation within IEM; Experiment: (a),[47] (b),[48,49] (c),[50] (d),[51,52] (e),[53] (f),[54] (g),[55] (h).[56]

the He^{2+}–He system and is reproduced by two-electron calculations which include electron-correlation effects.[57–59] For the sake of conciseness, those theoretical results are not displayed in Fig. 2. Rather, only our TC-BGM results within the IEM are compared to the experimental data. The IEM gives a good account of single capture (even though the situation at low E_p is somewhat unclear due to the spread of the experimental data), but fails to describe the resonant DC below $E_p \approx 20\,\text{keV/amu}$. However, at higher E_p the agreement is good, which implies that electron-correlation effects are either unimportant in this region or concealed by the integration over the impact parameter (or the scattering angle) involved in the calculation of the TCS. Which of both alternatives is true can only be decided by investigating *differential* cross sections.

This is what we do next. We compare DCSs calculated in the eikonal approximation with recent experimental data obtained by the Frankfurt COLTRIMS group.[60–62] By determining the Q-value, i.e. the difference between the electronic binding energies before and after the collision, the experimentalists were able to identify several single- and double-capture processes, some of which are listed in Table 1. The processes dubbed SC2 and SC3 are sums of pure transfer to excited states and transfer-excitation processes, which correspond to the same Q-values and could not be distinguished. In our calculations we summed up all state-to-state DCSs which correspond to the channels specified in Table 1.

In Fig. 3 DCSs for different single-capture processes are shown. It should be noted that the impact energies are not the same in all panels. The SC1 channel (top left figure) is the only pure single-electron process: one electron is transferred to the ground state of the projectile, while the other electron remains in the K shell of the target ion. At low impact energies, an interference structure is present in the data at small scattering angles. It disappears with increasing E_p, and the TC-BGM calculations reproduce

Table 1. Q-values for relevant final states and abbreviations for different one- and two-electron processes in He^{2+}–He collisions as introduced in Ref. 60.

Abbrev.	Final state	Note	Q-value [a.u.]
SC1	$He_p^+(1s) + He_t^+(1s)$		1.1
SC2	$He_p^+(nl) + He_t^+(1s)$; $He_p^+(1s) + He_t^+(nl)$	$n = 2$	−0.4
SC3	$He_p^+(nl) + He_t^+(1s)$; $He_p^+(1s) + He_t^+(nl)$	$n \geq 3$	−0.68
DC1	$He_p(1s^2) + He_t^{2+}$		0.0
DC2	$He_p(1s, nl) + He_t^{2+}$	$n \geq 2$	−0.73

this behavior quite precisely. At larger scattering angles oscillations are seen in the data in the energy range of $E_p = 40$–$60\,\text{keV/amu}$, which are missing in the theoretical cross sections. However, the averages of the experimental DCSs are well reproduced. Also shown are results of a four-body Born distorted wave (BDW-4B) calculation, which were published along with the measurements. They exhibit pronounced dips in the small angle region, which were attributed to a mutual cancellation of terms in the perturbation potential and considered unphysical.[61]

Results for SC2 and SC3 are shown in the other panels of Fig. 3. In the case of SC2, excellent agreement with the experimental data is obtained. For SC3, the agreement is also good. Only at $E_p = 40\,\text{keV/amu}$ are the experimental DCSs somewhat larger than the theoretical ones. At the

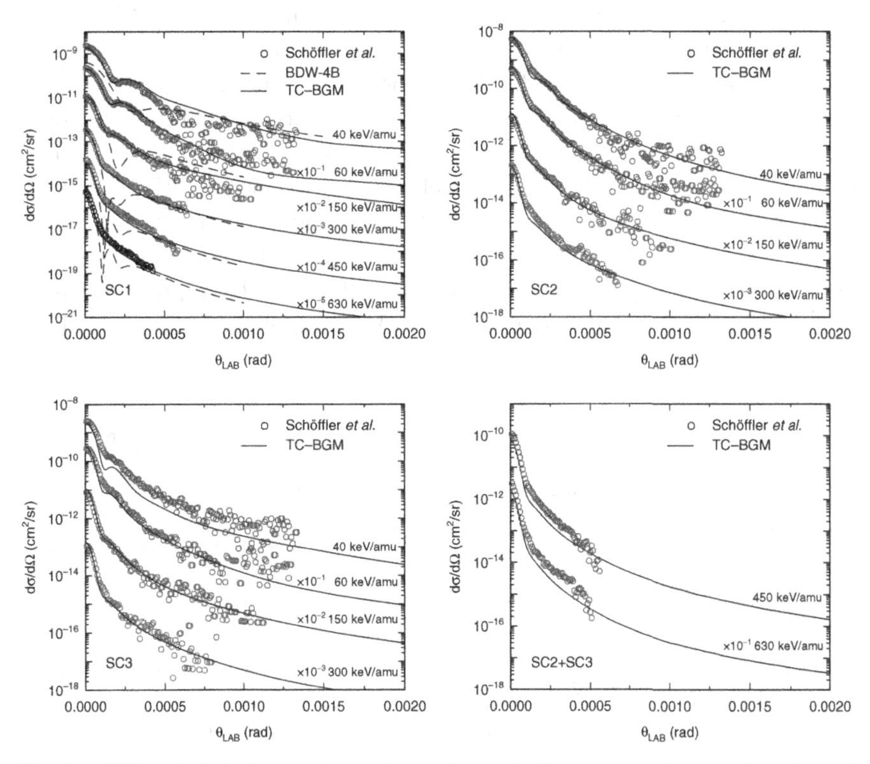

Fig. 3. Differential single-transfer and transfer-excitation cross sections as functions of laboratory scattering angle for $E_p = 40$–$630\,\text{keV/amu}$ $^3He^{2+}$–^4He collisions. Theory: present TC-BGM calculations within IEM for processes corresponding to SC1 (top left), SC2 (top right), SC3 (bottom left), and SC2 + SC3 (bottom right) in Table 1; BDW-4B calculations for SC1.[61] Experiment: Schöffler *et al.*[60,61]

higher energies $E_p = 450$–$630\,\mathrm{keV/amu}$ only the sum of SC2 and SC3 are resolved experimentally. Also in these cases are the data somewhat underestimated by the calculations except for larger scattering angles, where the experimental DCS drops off rapidly.

The favorable comparison of our calculations and the data for SC1 is pleasing, but not too surprising: we can take it as a proof that we have indeed managed to describe all aspects accurately except electron correlations which one would not expect to be important for this one-electron process. For the SC2 and SC3 channels, the overall good agreement between measurements and calculations is somewhat more remarkable, since two-electron processes contribute to them. But the single-electron processes dominate and might mask electron-correlation effects.

DC1 and DC2, in contrast, are pure two-electron processes. They are displayed in Fig. 4. Surprisingly, the level of agreement of our IEM results with the experimental data is only marginally reduced compared to the single-transfer processes. As the figure shows, it is even better than for the BDW-4B calculation for DC1 reported in Ref. 61, in which correlated initial and final states were used. This might indicate that even on the differential level DC could be an uncorrelated process for He^{2+}–He collisions in this particular energy range. In fact, we found good agreement with experimental data which is inclusive in all DC channels also at the slightly higher impact energy $E_p = 375\,\mathrm{keV/amu}$.[5] For the DC2 process some discrepancies between our calculations and the experimental data are visible

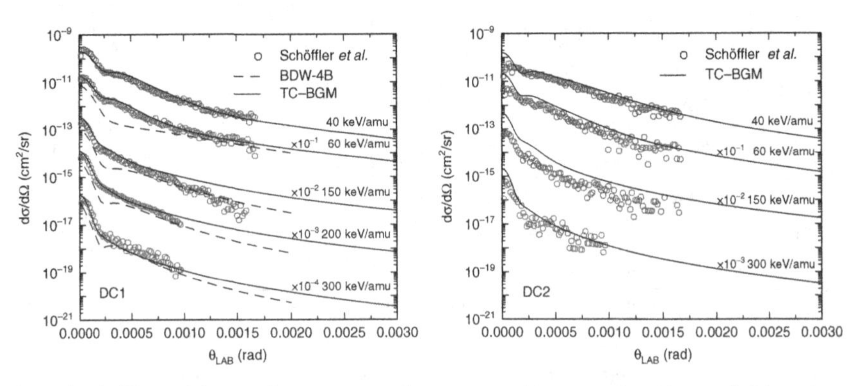

Fig. 4. Differential two-electron transfer cross sections as functions of laboratory scattering angle for $E_p = 40$–$300\,\mathrm{keV/amu}$ $^3He^{2+}$–^4He collisions. Theory: present TC-BGM calculations within IEM for DC1 (left panel) and DC2 (right panel) of Table 1; BDW-4B calculations for DC1.[61] Experiment: Schöffler *et al.*[60,61]

in Fig. 4, but they do not negate the conclusion about the minor role of electron-correlation effects.

Q-value-resolved capture has also been measured for HCI impact. In Ref. 3 the COLTRIMS technique was used to determine Q-value- and momentum-transfer-differential data for single-capture processes in keV $Ar^{15+,\ldots,18+}$–He collisions. Preliminary data were also taken for DC, but have remained unpublished. These systems are challenging for theoretical descriptions since capture takes place in high n-shells, which implies that a large number of final states must be taken into account. However, present-day computational power makes calculations on the same level as those discussed for He^{2+} impact feasible. We have used a large set of projectile orbitals in our TC-BGM basis including all states from $n = 3$ to $n = 10$ and carried out total and differential cross section calculations for shell-specific capture in the $Ar^{15+,\ldots,18+}$–He collision systems considered in Ref. 3.

Figure 5 shows relative TCSs for single capture into the dominating shells. Three sets of theoretical results, labeled as TC-BGM(I), TC-BGM(II), and TC-BGM(III) are included: TC-BGM(I) refers to a standard TCS calculation, for which IEM probabilities were integrated over the impact parameter. For TC-BGM(II) calculated DCSs were integrated over the scattering angle, but only up to the experimental maximum angle $\theta_{\max}$, which is determined by the detector acceptance. We checked that both models give very similar results if a larger upper bound is used in the integration over θ. Finally, TC-BGM(III) denotes an impact-parameter-based calculation like TC-BGM(I), but for probabilities that correspond to the one-active-electron model instead of the IEM. The one-active-electron probability for capture into the nth shell is obtained by multiplying the corresponding single-particle probability by 2 to account for the fact that helium has two electrons. In the IEM, in contrast, the single-particle capture probability is multiplied by a probability that one electron does not undergo any transition. The latter corresponds to what is measured and indeed gives better agreement with the experimental data. This should be taken as a warning that the one-active-electron model should not be used when the final states of both electrons are known.

TC-BGM(I) and TC-BGM(II) results also differ indicating that contributions from $\theta > \theta_{\max}$ are, in general, not negligible. This is most clearly seen in the relatively weakly populated $n = 6$ channel for $Ar^{16+,\ldots,18+}$. Here, the data agree well with TC-BGM(II), while the relative cross sections obtained in the impact parameter picture are significantly larger. The situation is less clear for the strong channels. For Ar^{15+} and Ar^{18+}

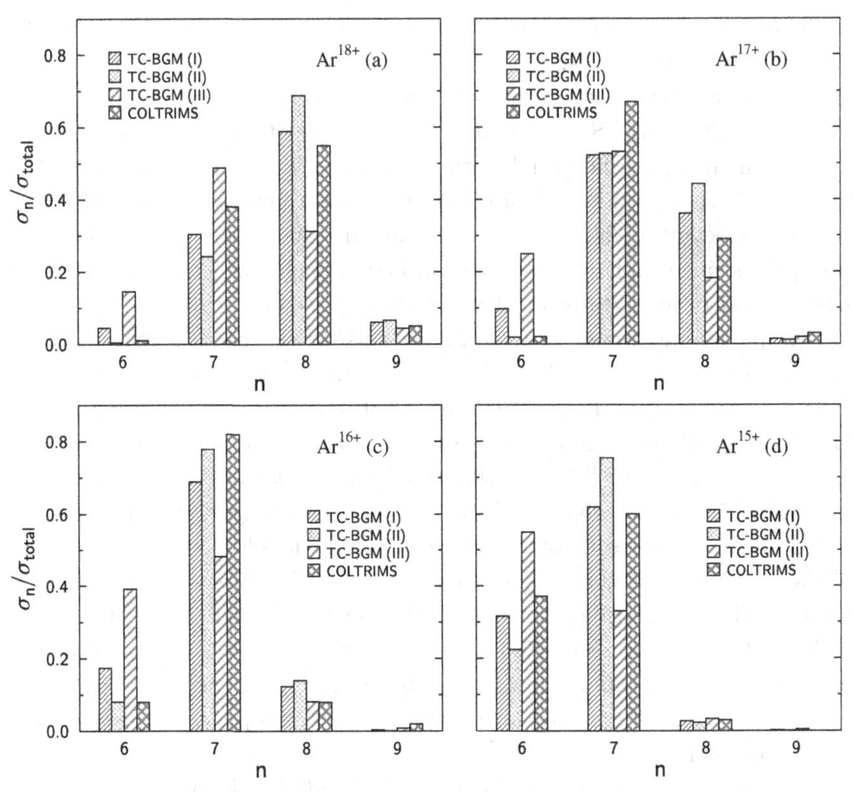

Fig. 5. Partial cross sections for single capture into the nth shell, comparing COLTRIMS data[3] with TC-BGM(I), TC-BGM(II), and TC-BGM(III) calculations, which are explained in the text. (a) 6.30 keV/amu Ar^{18+}–He; (b) 5.95 keV/amu Ar^{17+}–He; (c) 5.60 keV/amu Ar^{16+}–He; (d) 5.25 keV/amu Ar^{15+}–He.

TC-BGM(I) fares better, while for Ar^{16+} the dominant $n = 7$ channel is described more adequately by TC-BGM(II). For Ar^{17+} both models give very similar relative cross sections for $n = 7$, but differ considerably from the experimental result. However, given the complexity of experiments and calculations for such HCI collisions, one can state that the overall agreement between the COLTRIMS data, the TC-BGM results, and also previous data from Refs. 63, 64 which were shown in Ref. 3, but are not included in Fig. 5, is satisfactory. This is also true on the level of the DCSs, which were shown and discussed in Ref. 3.

Instead of repeating this discussion, we end this section with a look at theoretical DC cross sections. Double capture into slow HCIs and the role of electron correlations has attracted considerable attention (see, e.g.,

Ref. 64 and references therein). The situation is complicated by the fact that DC into high n-shells is often followed by autoionization, such that the process actually contributes to transfer ionization. A COLTRIMS experiment, however, detects the situation before the decay, since it is the Q-value which is measured and which is of course not changed by the subsequent decay process. Autoionizing and non-autoionizing DC can be distinguished by measuring the charge states of the projectile and target ions in coincidence.[64]

For the $Ar^{15+,...,18+}$–He collision systems discussed above, DC data were taken, but not fully analyzed. Our TC-BGM calculations include DC automatically, and we show the results for DC into (n, n') for Ar^{17+} and Ar^{18+} projectiles in Fig. 6. Not surprisingly, the standard IEM favors

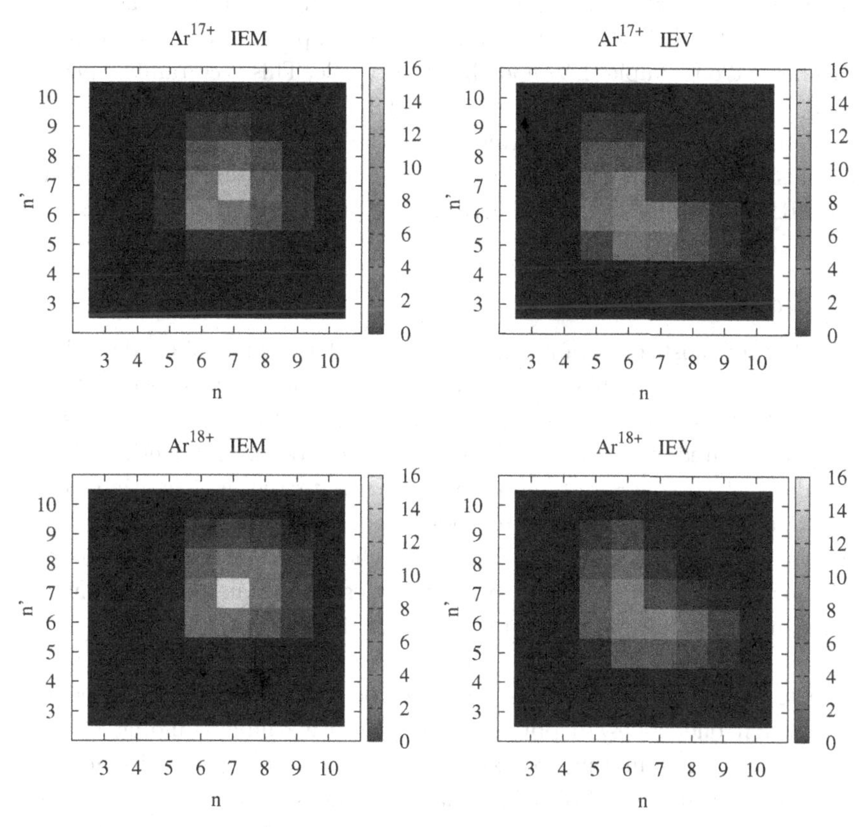

Fig. 6. Partial cross sections for DC into (n, n') shells. Shown are IEM and IEV results for 6.30 keV/amu Ar^{18+}–He and 5.95 keV/amu Ar^{17+}–He collisions.[65]

symmetric distributions, where both electrons are captured into the same shell. However, previous experiments for similar collision systems indicate that asymmetric distributions (which typically do not autoionize) should contribute more strongly.[64] This can qualitatively be reproduced by a calculation based on a two-step (so-called independent event (IEV)) model for DC. Here it is assumed that DC consists of single capture from neutral He and single capture from He^+. We have performed a TC-BGM calculation for the hydrogenlike He^+ target ion and have combined the final single-capture probabilities accordingly.[65] These IEV results are also included in Fig. 6.

As expected, more asymmetric final-state distributions are observed. Also, the absolute values of the TCSs are reduced compared to the IEM, since it is less likley to capture an electron from the more compact He^+ ion than from neutral He. Still, we have indications that the TCS for DC is too high even in the IEV model. If true this would imply that electron correlations beyond our simple IEV modelling play a role. This is certainly a possibility — and in fact one that was advocated in a number of publications (see the discussion and references in Ref. 64). In the light of our results for He^{2+}–He collisions, however, it is not obvious that this explanation is correct. New COLTRIMS data for DC would certainly help to clarify the situation.

3.2. *Double Ionization*

Probably the most important advantage of the MCEG technique described in Sec. 2.2 is the capability of extracting any type of cross section from the event file that can be obtained from experimental data, regardless of how complicated it is to be calculated by "conventional" methods and regardless of the degree of symmetry inherent to the cross section. The MCEG approach thus offers tremendous flexibility of testing theoretical models. On the other hand, apart from the choice of the theoretical model, the choice of the cross section to be investigated is equally important in optimizing the information output obtained from a comparison between theory and experiment.

The optimal choice of the cross section depends on the type of information one hopes to obtain. If a qualitative understanding of the reaction dynamics in the process of interest already exists, the goal is usually to extract as detailed information as possible from the measured and calculated data. To this end multiple, ideally fully differential cross sections (FDCSs) are most suitable, as they offer the most sensitive tests of the theoretical model. If, on the other hand, such a qualitative understanding is still incomplete or lacking, FDCSs suffer from some detriments: first, the advantage of sensitively testing theory can turn into a disadvantage. For example, a model which makes basically correct assumptions about the underlying reaction mechanism may still yield poor agreement with experiment exactly because the calculation may be very sensitive to the details of incorporating the description of the reaction dynamics. The investigator could easily be led to incorrect conclusions by such discrepancies. Second, by their very nature FDCSs only cover a tiny fraction of the TCS. Features that are pronounced in the FDCS could thus be insignificant in the TCS or vice versa, again potentially leading the investigator to incorrect conclusions.

In the case of DI of atoms by charged particle impact, our understanding of the underlying reaction dynamics must be viewed as rather incomplete, even on a qualitative level. For example, the role of higher-order contributions at relatively large projectile energies is still controversially debated.[66–71] It is therefore advantageous to first study spectra which unambiguously provide information about the underlying dominant reaction mechanism. Ideally, such spectra should fulfill several conditions: (1) they should be sensitive enough to exhibit differences between various models depending on which process is assumed to be dominant, but not too sensitive to the details of the theoretical implementations of the underlying physics (see above). (2) The integral over the entire spectrum should reflect the TCS or at least a significant fraction thereof in order to provide a comprehensive picture of the collision dynamics. (3) The spectra should be differential in parameters of all involved particles in order to visualize the correlations between all particles simultaneously in a single plot.

One plot satisfying all of the above conditions, the 4-D plot mentioned in Sec. 2.2, was recently developed[40] and later applied to DI.[10–12] It is an extension of "conventional" Dalitz plots originally devised to analyze three-body decay processes in particle physics[72] and later applied to atomic fragmentation processes (see, e.g., Refs. 73, 74). A 4-D plot is based on a tetrahedral coordinate system, where each tetrahedron plane represents one particle. For a given data point, which can only occur in the interior of the tetrahedron, the distances to the four planes represent the relative squared momenta $\pi_i = p_i^2/\sum_j p_j^2$, where p_j is the momentum of the jth particle.[b]

[b]In the case of the scattered projectile the momentum transfer to the target atom q is used instead of the total final-state projectile momentum.

It is straightforward to generate a 4-D plot in Cartesian coordinates using the following transformations:

$$x = \pi_1 \tag{31}$$

$$y = \frac{1}{\sqrt{8}}(3\pi_2 + \pi_1) \tag{32}$$

$$z = \sqrt{\frac{3}{2}}\,(\pi_3 + 0.5\pi_2 + 0.5\pi_1)\,. \tag{33}$$

It should be noted that these transformations do not contain π_4 because it is already fixed by the π_j of the other three particles by $\pi_4 = 1 - \pi_1 - \pi_2 - \pi_3$.

In Fig. 7a an experimental 4-D plot is shown for DI of helium by 6 MeV proton impact.[10,11] Here, the front and bottom planes represent the ejected electrons, the right plane the projectile, and the back plane the helium nucleus. Only the transverse momentum components in the scattering plane, spanned by the initial and final projectile momenta, were used in the computation of the π_j. As a first step, it is useful to analyze the 4-D plots in terms of events where the momenta of two particles are negligible compared to the momenta of the other two particles. In such events two particles appear to be essentially passive and we thus associate such events with binary interactions, although this is, in a strict sense, somewhat misleading. In 4-D plots, binary interactions occur at the intersection lines between two tetrahedron planes, where the particles represented by the intersecting planes have zero momenta. For example, at the line labeled 1 in Fig. 7a the planes representing the projectile and the recoil ion intersect so that data points along this line would represent binary *ee* interactions.

The dominant feature in the experimental data of Fig. 7a is a peak structure at intersection line no 6, representing binary *nn* interactions. This illustrates the overwhelming role played by elastic scattering between the heavy particles in the momentum balance of fast doubly ionizing ion–atom collisions. Furthermore, significant contributions from binary interactions between the projectile and one electron are seen at intersection lines 4 and 5. All other binary interactions (intersection lines 1–3) are rather unimportant.[c] Finally, there is considerable intensity in the lower left corner of the tetrahedron, which represents events in which the momentum transferred by the projectile is shared approximately equally between the

[c] It should be noted that these spectra are plotted with an offset. The cross section near intersection lines 1–3 is very small, but not zero as suggested by Fig. 7a.

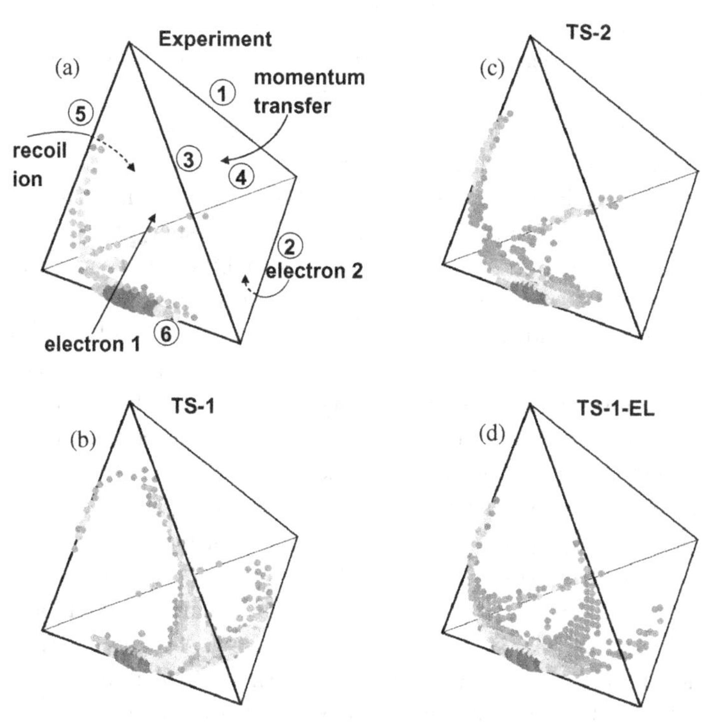

Fig. 7. DI in 6 MeV p-He collisions: experimental and theoretical 4-D plots for the transverse momentum components in the scattering plane. (a) Experimental data, (b) TS-1 calculation, (c) TS-2 calculation based on the FBA, (d) TS-1-EL calculation.

two electrons and the target nucleus. Overall, these various contributions give rise to a shape of the data to which we refer as the "three-finger structure".

The very small contributions from the binary *ee* interaction could be misinterpreted as an indicator that TS-1 is relatively weak, which would be quite surprising for such a large projectile speed. However, it should be noted that TS-1 is a two-step process and a large fraction of the momentum exchange may occur in the first step involving the projectile and one electron. This, in turn, could lead to the opposite interpretation, namely that the contributions seen in the data at intersection lines 4 and 5 are a signature of TS-1, assuming that the momentum exchange between the electrons in the second step is small. Our TS-1 calculation, shown in Fig. 7b, demonstrates that this is not the case, either. Although some enhancement of the cross section near these lines is indeed seen, it is nevertheless significantly weaker than in the data. More importantly, apart from the nn interaction, the most prominent features in the calculation are maxima at intersection lines 2 and 3 reflecting binary electron–target nucleus interactions, which are thus the signature of TS-1.

Surprisingly, the TS-1 calculation is in rather poor agreement with the data. Even more surprising, the TS-2 calculation, plotted in Fig. 7c, is in very nice qualitative agreement with the data, i.e. the "three-finger structure" is well reproduced. This comparison thus suggests that even at this large projectile speed, there is a large probability that the projectile interacts with both electrons directly. This observation seems to be in conflict with earlier studies, which suggest a dominance of TS-1 at projectile speeds larger than approximately 10 a.u. (see, e.g., Ref. 75). In fact, the same data set investigated here also shows features, if analyzed in terms of four-fold differential cross sections (4DCSs), which suggest a dominance of TS-1. These cross sections we analyze next.

In Fig. 8a measured 4DCSs for two electrons of equal energy (within 2.5 eV) both ejected into the scattering plane and for a momentum transfer of $q = 1.1 \pm 0.3$ a.u. are plotted as a function of the two polar ejection angles.[76] Zero degrees is the initial projectile beam direction and the symmetry between positive and negative angles is broken by $\mathbf{q}$, which points (depending on the sum energy of both electrons) in a direction between 62° and 83°. Two main maxima are observed at angle combinations of about (0, 120°) and (120°, 0) and additional, although rather weak, maxima at angle combinations of about (−60°, 160°) and (160°, −60°). Two particularly important regions for the following discussions are indicated by the dashed lines. Here, the angle between the two ejected electrons is 180°. Such back-to-back emission of electrons of equal energy originating from a ^1S state by an electric dipole (E1) transition is prohibited by selection rules.[77] Indeed, a suppression of the cross sections along these lines is found in the data, while without the E1 selection rules, one would expect strong maxima in these regions because of the Coulomb repulsion between the electrons in the continuum.

Since TS-1 is a first-order process in the projectile–target atom interaction, the cross sections for this mechanism are expected to be dominated by E1 transitions. Indeed, our TS-1 calculation, plotted in Fig. 8b yields pronounced minima along the E1-forbidden lines. In contrast, TS-2 is a second-order process which basically can be viewed as two subsequent single-ionization steps, each of which is dominated by E1 transitions as well. Because of angular momentum conservation and helicity considerations, one would then expect that TS-2 leads mostly to final

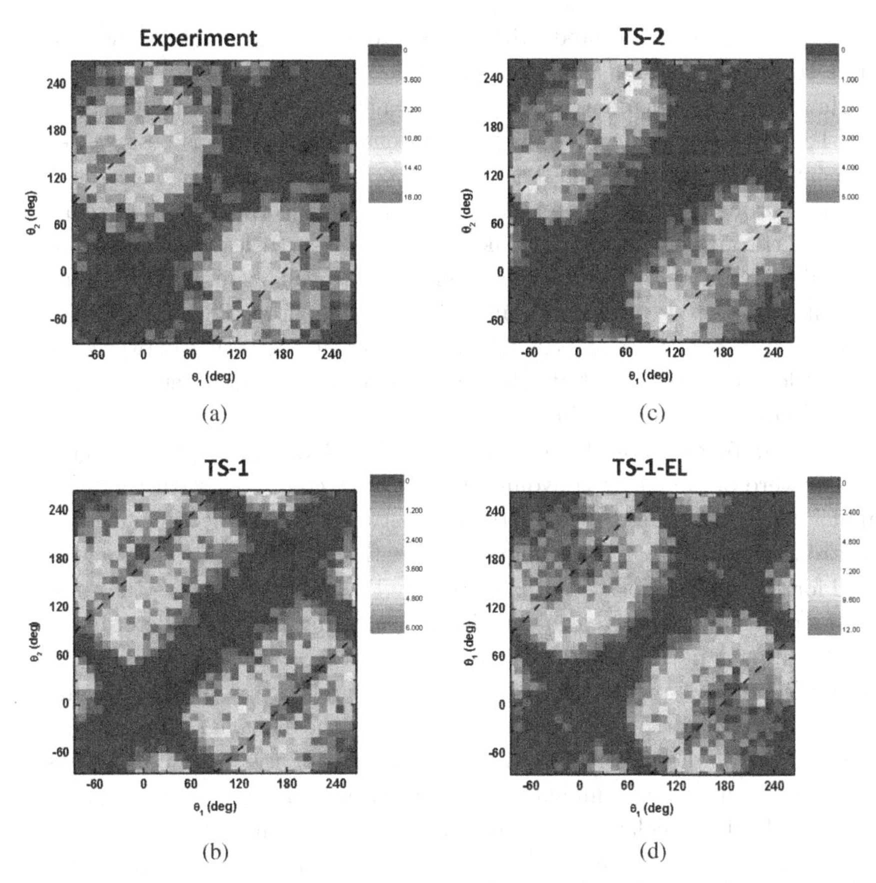

Fig. 8. DI in 6 MeV p-He collisions: 4DCS for electrons of equal energy (within 2.5 eV) ejected into the scattering plane for a momentum transfer of $q = 1.1 \pm 0.3$ a.u. as a function of the two polar ejection angles. (a) Experimental data, (b) TS-1 calculation, (c) TS-2 calculation based on the FBA, (d) TS-1-EL calculation.

two-electron states with an angular momentum of 0 or 2, i.e., the E1 selection rules should not hold for TS-2. As mentioned above, one would then expect pronounced maxima for back-to-back emission due to the Coulomb repulsion. Again, this expectation is confirmed by our TS-2 calculation presented in Fig. 8c, which shows four maxima on top of the two E1-forbidden lines. The TS-1 calculation is in much better agreement with the data than the TS-2 calculation, in sharp contrast to the 4-D plots. Most importantly, the small cross sections along the E1-forbidden lines seen in the experimental plots clearly demonstrate the importance of E1 transitions and are inconsistent with significant contributions from TS-2.

We are now confronted with an apparent conflict: while the 4-D plots seem to suggest a dominance of TS-2, the 4DCS clearly favor TS-1. However, this dilemma is resolved if a new DI mechanism, which we call TS-1-EL, is considered. This latter mechanism is perhaps best viewed as a hybrid process between TS-1 and TS-2.[11] Here, only one electron is ejected through a direct interaction with the projectile and the second electron through *ee* correlation, like in TS-1. Nevertheless, the projectile does directly interact with the second electron as well, as in TS-2, but only after it was ejected to the continuum already. Therefore, TS-1-EL can also be described as a TS-1 process followed by elastic scattering between the projectile and one of the electrons. Like TS-2 it is a higher-order process in the projectile–target atom interaction.

In Fig. 7d our TS-1-EL calculation of the 4-D plot is shown. These results were obtained by convoluting the FDCS computed with the TS-1 model with classical elastic scattering between the projectile and one electron using the MCEG technique. The method is essentially the same as used for the convolution with *nn* scattering. The only difference is that now the momentum transferred in the elastic scattering process is determined by the impact parameter relative to the electron, which is given by the vector difference between the impact parameter relative to the nucleus and the position vector of the electron relative to the nucleus. The position distribution of the electron in the ground state of the target atom was simulated using random numbers as described in Ref. 11.

The TS-1-EL calculation leads to dramatically improved agreement with the experimental data compared to the TS-1 model. The "three-finger structure" is now reproduced and the large contributions at intersection lines 2 and 3, seen in the TS-1 calculation, but not in the data, are strongly suppressed. Nevertheless, some discrepancies along these lines remain, which are probably due to small (but nonzero) contributions from TS-2. At the same time, the TS-1-EL calculation of the 4DCS (Fig. 8d) yields similarly good qualitative agreement with the experimental data as the TS-1 model. Therefore, the above mentioned conflict arising from the comparison of the experimental 4-D plots and 4DCSs with the TS-1 and TS-2 calculation can be resolved if TS-1-EL is the dominant DI mechanism.

As mentioned above, the analysis of the 4-D plots suggest that TS-2 is not negligible. In order to determine the relative importance between TS-1-EL and TS-2, we have to compare the experimental data to both models also on an absolute scale. The measured cross sections were normalized to the TCS. A value of $1.2 \times 10^{-20}\,\text{cm}^2$ was obtained by multiplying the measured double to single ionization ratio reported by Andersen *et al.*[78] with the recommended cross section for single ionization.[47] Our calculated TCS for TS-2 is $3.09 \times 10^{-21}\,\text{cm}^2$, which brings the ratio to the single-ionization cross section in excellent agreement with values reported by McGuire[79] and Díaz *et al.*[80] Our absolute total TS-1-EL cross section is more problematic because the initial state contains only radial, but no angular correlation. As pointed out by Byron and Joachain, the absence of angular correlation can lead to a reduction of the TCS by as much as an order of magnitude.[81] Indeed, our value of $3.8 \times 10^{-21}\,\text{cm}^2$ appears to be approximately a factor 2 to 3 too small.[d] Relative to the experimental TCS, our TS-2 value corresponds to an approximate 25% contribution to DI.

With increasing perturbation η (projectile charge to speed ratio) TS-2 becomes increasingly important in DI. In the following, we therefore analyze data which were recently obtained for 0.8 MeV/amu Au^{33+}–He collisions.[12] These highly charged projectiles correspond to an extreme perturbation of $\eta = 5.8$, for which everything but a clear dominance of TS-2 would be very surprising. This collision system thus offers an ideal test case for our TS-2 model. However, here another type of higher-order effect is known to be very important: not only does the projectile directly interact with both electrons (and the target nucleus), but each of the single-ionization steps in TS-2 is strongly affected by multiple projectile-electron interactions (see, e.g., Ref. 82). A well known manifestation of such higher-order effects is the post-collision interaction (PCI) between the outgoing projectile and an electron already ejected in a preceding interaction with the projectile (see, e.g., Ref. 83). This PCI is not accounted for in the TS-2 model in which the FDCS for single ionization were calculated within the FBA. We therefore use the TS-2 model, in which single ionization is treated within the CDW-EIS model, and we refer to it as TS-2-CDW, to analyze DI for this very large η.

In Fig. 9a the same type of 4-D plot as shown in Fig. 7a for 6 MeV p-He is plotted for 0.8 MeV/amu Au^{33+}–He. Surprisingly, hardly any differences at all between the data for small and large η are noticeable. This may look disappointing at first since it seems to suggest that these 4-D plots are not sufficiently sensitive to distinguish the various DI mechanisms. However, the comparison to our theoretical models shows that this is not

[d]This TCS is obtained from using the uncorrelated 2C final-state wave function (25). The 3C-type final-state models reduce this value.

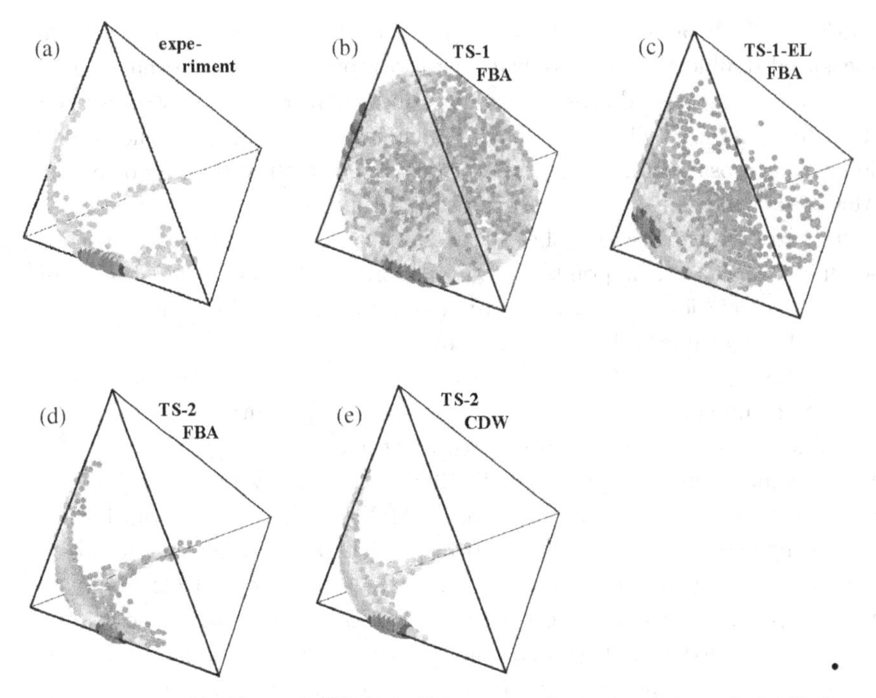

Fig. 9. DI in 0.8 MeV/amu Au^{33+}–He collisions: experimental and theoretical 4-D plots for the transverse momentum components in the scattering plane. (a) Experimental data, (b) TS-1 calculation, (c) TS-1-EL calculation, (d) TS-2 calculation based on the FBA, (e) TS-2 calculation based on the CDW-EIS approximation.

the case. The TS-1 calculation (Fig. 9b) does not even remotely resemble the experimental data. Some improved agreement is achieved with the TS-1-EL model (Fig. 9c), but nevertheless major qualitative discrepancies remain. On the other hand, once again very nice agreement is achieved with our TS-2 model. Surprisingly, it makes very little difference whether single ionization is treated within the FBA (TS-2-FBA, (Fig. 9d)) or within the CDW-EIS approach (Fig. 9e). We therefore conclude that at this very large η the 4-D plots for the transverse momentum components are quite sensitive to the DI mechanism, but rather insensitive to PCI.

A very different picture emerges if the 4-D plots are computed from the longitudinal momentum components of each particle, which are plotted in Fig. 10a to Fig. 10e for the experimental data and the various calculations in the same order as in Fig. 9a to Fig. 9e. This time all calculations based on the FBA, regardless of DI mechanism, are very similar to each other and

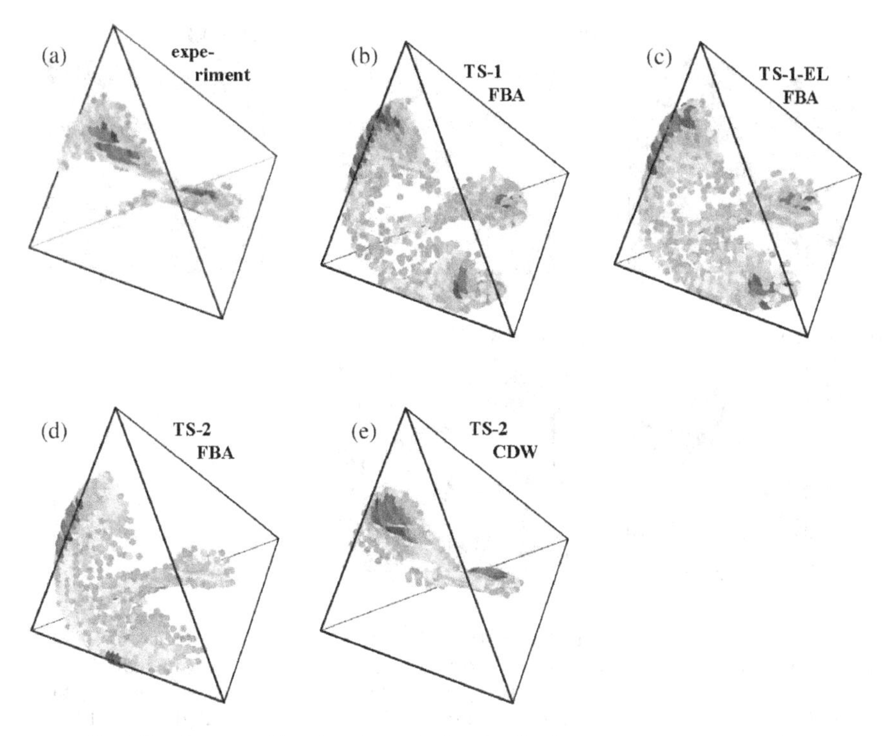

Fig. 10. Same as Fig. 9, but for the longitudinal momentum components.

in poor agreement with the experimental data. However, the plot calculated with the TS-2-CDW model looks very different from the other theoretical results and very well reproduces the measured spectrum. The longitudinal 4-D plot thus shows the exact opposite behavior as the transverse plot at very large η in that it is very sensitive to PCI, but insensitive to the DI mechanism.

Finally, in Fig. 11 we present the experimental and theoretical 4DCSs, again in the same order as in Figs. 10. Here, none of the calculations is in staggering agreement with the data, to say the least. Nevertheless, the TS-2-CDW model fares best with the TS-1-EL approach being a close second. This comparison reinforces a point we made at the beginning of this section: the nearly fully differential 4DCSs are too sensitive to the technical details of incorporating the underlying physics in the various models to identify the dominant DI process. The 4-D plots demonstrate convincingly that at this very large η DI is dominated by TS-2 accompanied by strong PCI. But yet, the TS-2-CDW model reproduces the measured 4DCS only

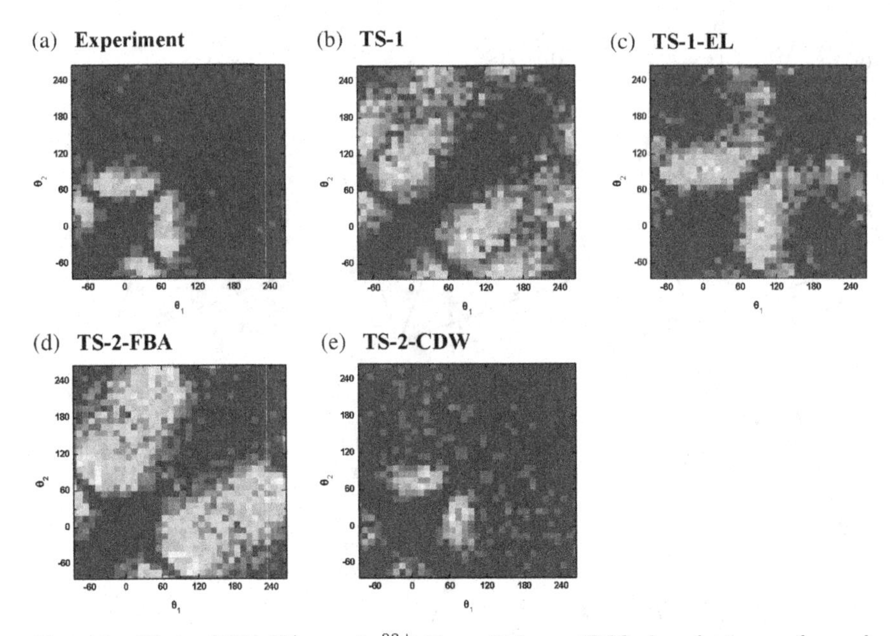

Fig. 11. DI in 0.8 MeV/amu Au^{33+}–He collisions: 4DCS for electrons of equal energy (within 2.5 eV) ejected into the scattering plane for a momentum transfer of $q = 1.1 \pm 0.3$ a.u. as a function of the two polar ejection angles. (a) Experimental data, (b) TS-1 calculation, (c) TS-1-EL calculation, (d) TS-2 calculation based on the FBA, (e) TS-2 calculation based on the CDW-EIS approximation.

marginally better than the TS-1-EL calculation, although the latter makes an incorrect assumption about the dominant DI mechanism and does not even include PCI.

Overall the measured 4-D plots can be well reproduced by the TS-1-EL model at small η and by the TS-2-CDW model at large η. The agreement achieved in the 4DCSs is significantly less impressive. We therefore conclude that 4-D plots are much better suited to identify the underlying DI process than 4DCSs (and probably FDCSs). Especially at large η it seems important to analyze the 4-D plots for the transverse and longitudinal momentum components simultaneously. As a next step, more sophisticated models need to be developed which go beyond merely qualitatively identifying the dominant process and which should be tested with measured 4DCSs and FDCSs. These models should include all processes coherently so that they can also be applied to intermediate η, where contributions from 2 or more processes could be comparable. Such efforts are currently in progress and experimental data for such η exist already.[12]

4. Conclusions

Considerable theoretical efforts have been devoted to the problem of multi-electron transitions in ion–atom collisions over the years. This chapter has been concerned with two approaches to specifically deal with two-electron processes involving electron transfer and with double ionization. The first one relies on the impact parameter picture and addresses the time-dependent Schrödinger equation in the independent electron appoximation. The nonperturbative two-center basis generator method is used to propagate the orbitals and the eikonal approximation is applied to calculate capture cross sections which are differential in the projectile scattering angle. The second approach directly addresses the T-matrix for ionization processes in first- and higher-order approximations and makes use of a Monte Carlo event generator technique to produce a theoretical event file that can be analyzed in the same way as data obtained from kinematically complete experiments. Electron-correlation effects are taken into account to some extent.

The comparison with measurements shows that in general both approaches do a good job in describing and explaining the data. In the case of the capture processes, we have found that the independent electron model works surprisingly well for intermediate-energy He^{2+}–He collisions. For relatively slow highly-charged ion impact, correlation effects might be more important, but ultimate conclusions are hampered by the lack of detailed experimental data — at least for the collision systems we have studied so far.

On the other hand, double ionization of helium at higher impact energies cannot be described without taking electron-correlation effects into account. Relatively simple perturbative models seem to be sufficient to answer the question as to which ionization mechanisms dominate. However, these perturbative models show limitations when it comes to comparisons with data on the level of multiple-differential cross sections. It will be interesting to see whether more sophisticated models that allow for a coherent treatment of first- and higher-order processes will lead to improved agreement.

Notwithstanding these successes and the planned refinements of the models, we are still quite far away from a complete qualitative and quantitative understanding of the few-body dynamics in such collisions. At this point, different methods for the description of different processes are available, but a simultaneous treatment of all processes (and for a broad range of collision systems) by one single quantum-mechanical theory is lacking. One might wonder: how should such a theory look?

If one accepts the separation of the heavy-particle and the electronic motions, the impact parameter picture is an appropriate framework. The task is then to develop a method for the full solution of the many-electron time-dependent Schrödinger equation. The ideas of the eikonal approximation can be used to translate the impact-parameter-dependent amplitudes into momentum-transfer-differential quantities. Depending on the process and cross section of interest, this step should be coupled with the Monte Carlo event generator to generate a theoretical event file for the further analysis. Some steps in this direction have been taken in recent years. However, the simultaneous description of capture and ionization processes on a differential level is still an open problem. Moreover, there are strong indications that the impact parameter framework is not sufficient to explain all of the available experimental data, i.e., the assumption that the full scattering state is a product of an electronic state and a plane wave that represents the projectile motion might not always be justified. This signals that ultimately one has to address the full T-matrix of the problem. In any case, there is still a lot to be done in this field.

Acknowledgements

T.K. is supported by the Natural Sciences and Engineering Research Council of Canada, and M.S. by the National Science Foundation under Grant No. PHY-0969299. M.F.C acknowledges the MINCIN projects (FIS2008-00784 TOQATA and Consolider Ingenio 2010 QOIT) for financial support.

References

1. J. Ullrich, R. Moshammer, A. Dorn, R. Dörner, L. Ph. H. Schmidt, and H. Schmidt-Böcking, *Rep. Prog. Phys.*, **66**, 1463–1545 (2003).
2. M. Schulz, R. Moshammer, D. Fischer, M. Dürr, J. Ullrich, A. Hasan, M. F. Ciappina, and T. Kirchner, *Nucl. Instrum. Methods Phys. Res. B*, **267**, 187–191 (2009).
3. S. Knoop, D. Fischer, Y. Xue, M. Zapukhlyak, C. J. Osborne, Th. Ergler, T. Ferger, J. Braun, G. Brenner, H. Bruhns, C. Dimopoulou, S. W. Epp, A. J. González Martínez, G. Sikler, R. Soria Orts, H. Tawara, T. Kirchner, J. R. Crespo López-Urrutia, R. Moshammer, J. Ullrich, and R. Hoekstra, *J. Phys. B*, **41**, 195203(1–6) (2008).
4. M. Zapukhlyak, T. Kirchner, A. Hasan, B. Tooke, and M. Schulz, *Phys. Rev. A*, **77**, 012720(1–9) (2008).
5. M. Zapukhlyak and T. Kirchner, *Phys. Rev. A*, **80**, 062705(1–7) (2009).
6. M. Zapukhlyak, N. Henkel, and T. Kirchner, *J. Phys. Conf. Series*, **212**, 012030(1–6) (2010).

7. M. Dürr, B. Najjari, M. Schulz, A. Dorn, R. Moshammer, A. B. Voitkiv, and J. Ullrich, *Phys. Rev. A*, **75**, 062708(1–14) (2007).
8. M. Schulz, M. Dürr, B. Najjari, R. Moshammer, and J. Ullrich, *Phys. Rev. A*, **76**, 032712(1–8) (2007).
9. M. F. Ciappina, T. Kirchner, and M. Schulz, *Comput. Phys. Commun.*, **181**, 813–820 (2010).
10. M. F. Ciappina, M. Schulz, T. Kirchner, D. Fischer, R. Moshammer, and J. Ullrich, *Phys. Rev. A*, **77**, 062706(1–12) (2008).
11. M. Schulz, M. F. Ciappina, T. Kirchner, D. Fischer, R. Moshammer, and J. Ullrich, *Phys. Rev. A*, **79**, 042708(1–7) (2009).
12. D. Fischer, M. Schulz, K. Schneider, M. F. Ciappina, T. Kirchner, A. Kelkar, S. Hagmann, M. Grieser, K. -U. Kühnel, R. Moshammer, and J. Ullrich, *Phys. Rev. A*, **80**, 062703(1–8) (2009).
13. M. F. Ciappina, T. Kirchner, M. Schulz, D. Fischer, R. Moshammer, and J. Ullrich, *J. At. Mol. Opt. Phys.*, **2010** 231329(1-7) (2010).
14. M. F. Ciappina, M. Schulz, and T. Kirchner, *Phys. Rev. A*, **82**, 062701(1–8) (2010).
15. M. R. C. McDowell and J. P. Coleman, *Introduction to the Theory of Ion-Atom Collisions*, North-Holland Publishing Company, Amsterdam, (1970).
16. M. McGovern, D. Assafrao, J. R. Mohallem, C. T. Whelan, and H. R. J. Walters, *Phys. Rev. A*, **79**, 042707(1–16) (2009).
17. W. Fritsch and C. D. Lin, *Phys. Rep.*, **202**, 1–97 (1991).
18. X. Guan and K. Bartschat, *Phys. Rev. Lett.*, **103**, 213201(1–4) (2009).
19. M. S. Pindzola, F. Robicheaux, and J. Colgan, *Phys. Rev. A*, **82**, 042719(1–8) (2010).
20. T. Kirchner and H. Knudsen, *J. Phys. B*, **44**, 122001(1–49) (2011).
21. E. Engel and S. H. Vosko, *Phys. Rev. A*, **47**, 2800–2811 (1993).
22. O. J. Kroneisen, H. J. Lüdde, T. Kirchner, and R. M. Dreizler, *J. Phys. A*, **32**, 2141–2156 (1999).
23. T. Kirchner, H. J. Lüdde, and M. Horbatsch, *Recent Res. Dev. Phys.*, **5**, 433–461 (2004).
24. M. Zapukhlyak, T. Kirchner, H. J. Lüdde, S. Knoop, R. Morgenstern, and R. Hoekstra, *J. Phys. B*, **38**, 2353–2369 (2005).
25. R. McCarroll and A. Salin, *J. Phys. B*, **1**, 163–171 (1968).
26. S. Keller, B. Bapat, R. Moshammer, J. Ullrich, and R. M. Dreizler, *J. Phys. B*, **33**, 1447–1461 (2000).
27. J. N. Silverman, O. Platas, and F. A. Matsen, *J. Chem. Phys.*, **32**, 1402–1406 (1960).
28. C. DalCappello and H. LeRouzo, *Phys. Rev. A*, **43**, 1395–1404 (1991).
29. B. Bapat, R. Moshammer, S. Keller, W. Schmitt, A. Cassimi, L. Adoui, H. Kollmus, R Dörner, Th. Weber, K. Khayyat, R. Mann, J. P. Grandin, and J. Ullrich, *J. Phys. B*, **32**, 1859–1872 (1999).
30. Dž. Belkić, I. Mančev, and J. Hanssen, *Rev. Mod. Phys.*, **80**, 249–314 (2008).
31. A. L. Godunov, Sh. D. Kunikeev, V. N. Mileev, and V. S. Senashenko, *Zh. Tekh. Fiz.*, **53**, 436–443 (1983).
32. M. Brauner, J. S. Briggs, and H. Klar, *J. Phys. B*, **22**, 2265–2287 (1989).

33. F. Maulbetsch and J. S. Briggs, *J. Phys. B*, **26**, 1679–1696 (1993).
34. J. Berakdar and J. S. Briggs, *Phys. Rev. Lett.*, **72**, 3799–3802 (1994).
35. J. Berakdar, *Phys. Rev. A*, **53**, 2314–2326 (1996).
36. J. Berakdar, *Phys. Rev. A*, **54**, 1480–1486 (1996).
37. S. Zhang, *J. Phys. B*, **33**, 3545–3553 (2000).
38. J. R. Götz, M. Walter, and J. S. Briggs, *J. Phys. B*, **38**, 1569–1579 (2005).
39. L. Gulyás, A. Igarashi, and T. Kirchner, *Phys. Rev. A*, **74**, 032713(1–9) (2006).
40. M. Schulz, D. Fischer, T. Ferger, R. Moshammer, and J. Ullrich, *J. Phys. B*, **40**, 3091–3099 (2007).
41. A. C. Laforge, K. N. Egodapitiya, J. S. Alexander, A. Hasan, M. F. Ciappina, M. A. Khakoo, and M. Schulz, *Phys. Rev. Lett*, **103**, 053201(1–4) (2009).
42. D. S. F. Crothers and J. F. McCann, *J. Phys. B*, **16**, 3229–3242 (1983).
43. E. Clementi and C. Roetti, *At Data Nucl. Data Tables*, **14**, 177–478 (1974).
44. A. Hasan, B. Tooke, M. Zapukhlyak, T. Kirchner, and M. Schulz, *Phys. Rev. A*, **74**, 032703(1–5) (2006).
45. R. Schuch, E. Justiniano, H. Vogt, G. Deco, and N. Grün, *J. Phys. B*, **24**, L133–L138 (1991).
46. M. Schulz, T. Vajnai, and J. A. Brand, *Phys. Rev. A*, **75**, 022717(1–6) (2007).
47. C. F. Barnett, *Atomic Data for Fusion*, Oak Ridge National Laboratory, Tennessee, **1**, (1990). url=http://www-cfadc.phy.ornl.gov/redbooks/one/1.html,
48. R. D. DuBois, *Phys. Rev. A*, **33**, 1595–1601 (1986).
49. R. D. DuBois, *Phys. Rev. A*, **35**, 2585–2590 (1987).
50. K. H. Berkner, R. V. Pyle, J. W. Stearns, and J. C. Warren, *Phys. Rev.*, **166**, 44–46 (1968).
51. V. V. Afrosimov, G. A. Leiko, Y. A. Mamaev, and M. N. Panov, *Sov. Phys. JETP*, **40**, 661–666 (1975).
52. V. V. Afrosimov, A. A. Basalaev, G. A. Leiko, and M. N. Panov, *Sov. Phys. JETP*, **47**, 837–842 (1978).
53. M. B. Shah and H. B. Gilbody, *J. Phys. B*, **18**, 899–913 (1985).
54. J. E. Bayfield and G. A. Khayrallah, *Phys. Rev. A*, **11**, 920–929 (1975).
55. M. B. Shah, P. McCallion, and H. B. Gilbody, *J. Phys. B*, **22**, 3037–3045 (1989).
56. M. B. Shah and H. B. Gilbody, *J. Phys. B*, **7**, 256–268 (1974).
57. C. Harel and A. Salin, *J. Phys. B*, **13**, 785–798 (1980).
58. M. Kimura, *J. Phys. B*, **21**, L19–L24 (1988).
59. K. Gramlich, N. Grün, and W. Scheid, *J. Phys. B*, **22**, 2567–2579 (1989).
60. M. S. Schöffler, *Grundzustandskorrelationen und dynamische Prozesse untersucht in Ion-Helium-Stößen*, Goethe-Universität Frankfurt a.M., (2006). url= http://publikationen.ub.uni-frankfurt.de/volltexte/2006/3536.
61. M. S. Schöffler, J. Titze, L. Ph. H. Schmidt, T. Jahnke, N. Neumann, O. Jagutzki, H. Schmidt-Böcking, R. Dörner, and I. Mančev, *Phys. Rev. A*, **79**, 064701(1–4) (2009).
62. M. S. Schöffler, J. N. Titze, L. Ph. H. Schmidt, T. Jahnke, O. Jagutzki, H. Schmidt-Böcking, and R. Dörner, *Phys. Rev. A*, **80**, 042702(1–6) (2009).

63. A. Cassimi, S. Duponchel, X. Flechard, P. Jardin, P. Sortais, D. Hennecart, and R. E. Olson, *Phys. Rev. Lett.*, **76**, 3679–3682 (1996).
64. M. A. Abdallah, W. Wolff, H. E. Wolf, E. Y. Kamber, M. Stöckli, and C. L. Cocke, *Phys. Rev. A*, **58**, 2911–2919 (1998).
65. M. Zapukhlyak, *Einblicke in die atomare Vielteilchendynamik von Streuprozessen durch ab-initio Rechnungen*, Technische Universität Clausthal, (2008).
66. A. Dorn, A. Kheifets, C. D. Schröter, C. Höhr, G. Sakhelashvili, R. Moshammer, J. Lower, and J. Ullrich, *Phys. Rev. A*, **68**, 012715(1–4) (2003).
67. A. Lahmam-Bennani, F. Catoire, A. Duguet, and C. Dal Cappello, *Phys. Rev. A*, **71**, 026701(1–3) (2005).
68. A. Dorn, A. Kheifets, C. D. Schröter, C. Höhr, G. Sakhelashvili, R. Moshammer, J. Lower, and J. Ullrich, *Phys. Rev. A*, **71**, 026702(1–3) (2005).
69. A. Lahmam-Bennani, A. Duguet, C. Dal Cappello, H. Nebdi, and B. Piraux, *Phys. Rev. A*, **67**, 010701(1–4) (2003).
70. J. R. Götz, M. Walter, and J. S. Briggs, *J. Phys. B*, **36**, L77–L83 (2003).
71. A. Lahmam-Bennani, E. M. Staicu Casagrande, A. Naja, C. Dal Cappello, and P. Bolognesi, *J. Phys. B*, **43**, 105201(1–7) (2010).
72. R. H. Dalitz, *Phil. Mag.*, **44**, 1068–1080 (1953).
73. L. M. Wiese, O. Yenen, B. Thaden, and D. H. Jaecks, *Phys. Rev. Lett.*, **79**, 4982–4985 (1997).
74. M. Schulz, R. Moshammer, W. Schmitt, H. Kollmus, R. Mann, S. Hagmann, R. E. Olson, and J. Ullrich, *Phys. Rev. A*, **61**, 022703(1–9) (2000).
75. L. H. Andersen, P. Hvelplund, H. Knudsen, S. P. Møller, K. Elsener, K. G. Rensfelt, and E. Uggerhøj, *Phys. Rev. Lett.*, **57**, 2147–2150 (1986).
76. D. Fischer, R. Moshammer, A. Dorn, J. R. Crespo López-Urrutia, B. Feuerstein, C. Höhr, C. D. Schröter, S. Hagmann, H. Kollmus, R. Mann, B. Bapat, and J. Ullrich, *Phys. Rev. Lett.*, **90**, 243201(1–4) (2003).
77. F. Maulbetsch and J. S. Briggs, *J. Phys. B*, **28**, 551–564 (1995).
78. L. H. Andersen, P. Hvelplund, H. Knudsen, S. P. Møller, A. H. Sørensen, K. Elsener, K. G. Rensfelt, and E. Uggerhøj, *Phys. Rev. A*, **36**, 3612–3629 (1987).
79. J. H. McGuire, *Phys. Rev. Lett.*, **49**, 1153–1157 (1982).
80. C. Díaz, F. Martín, and A. Salin, *J. Phys. B*, **33**, 4373–4388 (2000).
81. F. W. Byron and C. J. Joachain, *Phys. Rev. Lett.*, **16**, 1139–1142 (1966).
82. M. Schulz, R. Moshammer, A. N. Perumal, and J. Ullrich, *J. Phys. B*, **35**, L161–L166 (2002).
83. G. B. Crooks and M. E. Rudd, *Phys. Rev. Lett.*, **25**, 1599–1601 (1970).

Chapter 4

A 4-Body Model for Charge Transfer Collisions

A. L. Harris[†], J. L. Peacher and D. H. Madison[*]
Missouri University of Science and Technology,
1315 N Pine St, Rolla, Missouri 65409, USA
[*]*madison@mst.edu*

Although 4-body scattering problems have been studied for several decades, almost all of this work has been for total cross sections. It is only recently that experiment has advanced to the point where fully differential cross sections (FDCS) can be measured. The fundamental T-matrix for a 4-body process contains a 9-dimensional (9D) integral. For the work that has previously been done, a scattering model was developed such that at least part of the T-matrix could be integrated analytically. This reduced the dimensionality of any required numerical integration, making the calculation more tractable. To avoid model-dependent integrals, we have developed a computer code to directly evaluate the 4-body 9D integral numerically. The advantage of this approach is that we can calculate cross sections for any arbitrary model. The disadvantage is the enormous amount of required computer time. We have tested this approach by calculating very recently measured FDCS for proton–helium charge exchange collisions using a few first order models. The two processes examined are electron capture with target excitation and double electron capture.

1. Introduction

Charge transfer collisions involve a positively charged projectile colliding with a target atom, or ion. During the collision, at least one of the electrons

[*]The author to whom corrspondence should be addressed.
[†]Current address: Henderson State University, 1100 Henderson St, Arkadelphia, Arkansas 71999, USA.

from the target is captured by the projectile. The simplest case of a charge transfer collision is the 3-body process of single capture (SC), where the projectile captures one electron, and all the remaining electrons remain in their initial states. If the target atom or ion contains at least two electrons, more complex 4-body charge transfer collisions can occur.

To illustrate some of these 4-body charge transfer collisions, consider a proton colliding with a helium atom. If one electron is captured from the helium atom into the ground state of the outgoing hydrogen atom, a number of things can happen to the second electron. It can be left in the ground state of the helium ion, which results in SC. The second electron can also be left in an excited state of the helium ion, a process known as transfer with target excitation (TTE). This second electron could also be ionized into the continuum, which is known as transfer-ionization (TI). If the captured electron is captured into an excited state of the outgoing hydrogen atom, the process is known as transfer-excitation (TE). Another possible 4-body charge transfer process occurs when both electrons from the helium atom are captured by the incident projectile. This process is referred to as double capture (DC).

For proton + helium collisions, the reactions for SC, TTE, TI, TE, and DC are shown below:

$$\text{SC} \quad \mathrm{p + He(1s^2) \rightarrow H(1s) + He^+(1s)} \tag{1a}$$

$$\text{TTE} \quad \mathrm{p + He(1s^2) \rightarrow H(1s) + He^+(nl)} \tag{1b}$$

$$\text{TI} \quad \mathrm{p + He(1s^2) \rightarrow H(1s) + e^- + He^{++}} \tag{1c}$$

$$\text{TE} \quad \mathrm{p + He(1s^2) \rightarrow H(nl) + He^+(1s)} \tag{1d}$$

$$\text{DC} \quad \mathrm{p + He(1s^2) \rightarrow H^-(1s^2) + He^{++}}. \tag{1e}$$

In this chapter, a 4-body model that can be applied to any 4-body charge transfer process is discussed, and results are presented for collisions (1a), (1b), and (1e).

2. General Theoretical Approach

The general theory of multichannel scattering can be found in Ref. 1. A summary of the results will be presented here. Let H be the Hamiltonian of a general collision system. For a given energy E, the wave function Ψ for the collision system satisfies

$$(H - E)\Psi = 0. \tag{2}$$

The Hamiltonian H can be split in different ways corresponding to each final possible reaction or arrangement. For instance, for a given arrangement c, the Hamiltonian would be split as

$$H = H_c + V_{c,} \tag{3}$$

where H_c is the unperturbed Hamiltonian and V_c is the interaction potential acting between the reaction products in the final state arrangement c. The unperturbed wave functions Φ_n for arrangement c satisfy

$$(H_c - E)\Phi_n = 0. \tag{4}$$

There are many possible eigenfunctions for arrangement c. The center of mass system is used so that the total energy in each channel is the same and the total momentum in each channel is equal to zero.

For example, the reactions (1a), (1b), and (1d) all belong to the same arrangement, say c. Thus the reaction (1a), corresponding to the final state H(1s) + He^+(1s), is one of the many possible final state channels in that arrangement. If it is labeled as channel n, the unperturbed wave function can be expressed as

$$\Phi_n(c, \vec{x}_c, \vec{R}_c) = X_n(\vec{x}_c)\beta_{\vec{k}_n}(\vec{R}_c), \tag{5}$$

where $X_n(\vec{x}_c)$ is the product of the normalized $H(1s)$ atom and the normalized $He^+(1s)$ ion wave functions. The n and $\vec{x}_c$ represent the collective quantum numbers and internal coordinates of the $H(1s)$ atom and the $He^+(1s)$ ion in arrangement c. Thus, the n labels a particular channel in arrangement c. The relative motion in the center of mass system is represented by

$$\beta_{\vec{k}_n}(\vec{R}_c) = \frac{e^{i\vec{k}_n \cdot \vec{R}_c}}{(2\pi)^{3/2}}, \tag{6}$$

where a delta function normalization has been used for the plane wave. The total energy in the final channel n of arrangement c is given by

$$E_n = \varepsilon_n + \frac{k_n^2}{2\mu_n}, \tag{7}$$

where ε_n is the sum of the binding energies of the $H(1s)$ atom and the $He^+(1s)$ ion and $\frac{k_n^2}{2\mu_n}$ is the relative kinetic energy in the center of mass system. The wave vector $\vec{k}_n$ describes the relative motion of the H atom and the He^+ ion in channel n, and μ_n is the reduced mass of the H atom and the He^+ ion in channel n.

2.1. *Transition Matrix and Differential Cross Section*

The time-dependent scattering formalism is used in order to derive the transition rate from an initial state channel i to a particular final state channel f. The result for the transition rate from an initial state channel i to a particular final state channel f, w_{fi}, is determined by the equation

$$w_{fi} = 2\pi\delta(E_i - E_f)|T_{fi}|^2. \tag{8}$$

The transition matrix element T_{fi} can be written in either the prior or post form as

$$T_{fi} = \langle\Psi^{(-)}_{\vec{k}_f}(E_f)|V_i|\Phi_{\vec{k}_i}(E_i)\rangle = \langle\Phi_{\vec{k}_f}(E_f)|V_f|\Psi^{(+)}_{\vec{k}_i}(E_i)\rangle. \tag{9}$$

The energy delta function in equation (8) ensures that the total energy in each channel is the same. The Hamiltonian has been split two ways corresponding to the initial and final channels; that is

$$H = H_i + V_i = H_f + V_f. \tag{10}$$

Thus, the unperturbed wave functions $\Phi_{\vec{k}_i}$ and $\Phi_{\vec{k}_f}$ satisfy

$$(H_i - E_i)\Phi_{\vec{k}_i} = 0 \tag{11}$$

and

$$(H_f - E_f)\Phi_{\vec{k}_f} = 0. \tag{12}$$

Likewise, $\Psi^{(+)}_{\vec{k}_i}$ and $\Psi^{(-)}_{\vec{k}_f}$ are solutions of $(H - E)\Psi = 0$ in the initial channel and final channels. Asymptotically, the solution $\Psi^{(+)}_{\vec{k}_i}$ has an outgoing spherical wave part, and the solution $\Psi^{(-)}_{\vec{k}_f}$ has an incoming spherical wave part.

The cross section σ_{fi} for the transition $i \to f$ is defined as the sum of the transition rate w_{fi} over the possible final states divided by the incident flux F_i in the initial channel. If the final reaction consists of atoms and ions in bound states, then the only continuum state is the relative motion in the final channel. Thus, σ_{fi} is defined as

$$\sigma_{fi} = 2\pi \int \frac{1}{F_i}\delta(E_i - E_f)|T_{fi}|^2 \mathrm{d}^3 k_f. \tag{13}$$

The incident flux, F_i, is defined as the probability current density in the direction of the relative motion in the center of mass system in the initial channel, which is chosen as the $\hat{z}$ the direction. Thus, the incident flux is given by

$$F_i = \frac{1}{2i\mu_i}\left[\beta^*_{\vec{k}_i}\vec{\nabla}\beta_{\vec{k}_i} - \beta_{\vec{k}_i}\vec{\nabla}\beta^*_{\vec{k}_i}\right]\cdot\hat{z}, \tag{14}$$

where μ_i is the reduced mass in the initial channel. Since $\beta_{\vec{k}_i}$ is a plane wave given by equation (6), one obtains

$$F_i = \frac{k_i}{(2\pi)^{3/2}\mu_i}. \tag{15}$$

The total energy in the final channel is

$$E_f = \frac{k_f^2}{2\mu_f} + \varepsilon_f, \tag{16}$$

so that $\mathrm{d}E_f = \frac{k_f}{\mu_f}\mathrm{d}k_f$ and $\mathrm{d}^3 k_f = k_f^2 \mathrm{d}k_f \mathrm{d}\Omega_{k_f} = \mu_f k_f \mathrm{d}E_f \mathrm{d}\Omega_f$. The energy delta function ensures that the total energy in the center of mass system is conserved so that

$$\frac{k_f^2}{2\mu_f} = \frac{k_i^2}{2\mu_i} + \varepsilon_i - \varepsilon_f. \tag{17}$$

This gives

$$\sigma_{fi} = (2\pi)^4 \frac{\mu_i \mu_f k_f}{k_i} \int |T_{fi}|^2 \mathrm{d}\Omega_f. \tag{18}$$

Therefore, the differential cross section in the center of mass frame is given by

$$\frac{\mathrm{d}\sigma_{fi}}{\mathrm{d}\Omega_f} = (2\pi)^4 \frac{\mu_i \mu_f k_f}{k_i} |T_{fi}|^2. \tag{19}$$

In order to compare the calculated differential cross sections with experiment, it is necessary to convert the center of mass differential cross section to the lab frame. The center of mass and lab frame differential cross sections are related by

$$\frac{d\sigma_{fi}}{d\Omega_f}^{LAB} = \frac{1+\gamma_v^2+2\gamma_v\cos\theta_{CM}}{1+\gamma_v\cos\theta_{CM}}\frac{d\sigma_{fi}}{d\Omega_f}^{CM} \tag{20}$$

where $\gamma_v = \frac{v_{CM}}{v_f}$. The quantity v_{CM} is the speed of the center of mass, and v_f is the speed of the outgoing hydrogen atom (or ion) in the center of mass frame. The angle θ_{CM} is the center of mass scattering angle.

2.2. *Two Potential Formulation*

The T-matrix for a 4-body process represents a 9-dimensional (9D) integral. Traditionally, analytic approximations for the wave function have been made in order to perform a portion of the 9D integral analytically. If one is going to evaluate the full 9D integral numerically, it is not necessary to use analytic wave functions, making it possible to use more accurate numerical solutions of the Schrödinger equation. Numerical solutions of the Schrödinger equation are normally called distorted waves. In the distorted wave approach, the interaction potential is separated into a part that can be treated numerically plus a part that cannot. This approach is typically called the two potential formulation. In this formulation, the initial or final potential (V_i or V_f) is written as the sum of two potentials such that

$$V_i = U_i + W_i \tag{21}$$

and

$$V_f = U_f + W_f, \tag{22}$$

where $U_i(U_f)$ is the part of the potential that will be included in the numerical solution of the Schrödinger equation. The distorting potentials U_i and U_f are central potentials, which are functions of the coordinates R_i and R_f joining the center of masses in the initial and final channels. The term $W_i(W_f)$ is the perturbation. The final state distorted wave Hamiltonian is $\overline{H}_f = H_f + U_f$. The distorted waves are continuum waves that have been distorted by the potential U_f, and are solutions of the Lippmann-Schwinger equation[2]

$$|\eta_{\vec{k}_f}^{(-)}\rangle = |\Phi_{\vec{k}_f}\rangle + \frac{1}{E_f - \bar{H}_f - i\varepsilon}U_f\,|\Phi_{\vec{k}_f}\rangle. \tag{23}$$

In terms of the two potentials, the T-matrix of equation (9) becomes

$$T_{fi} = \langle \Psi^{(-)}_{\vec{k}_f} \mid U_i + W_i \mid \Phi_{\vec{k}_i} \rangle = \langle \Phi_{\vec{k}_f} \mid U_f + W_f \mid \Psi^{(+)}_{\vec{k}_i} \rangle. \qquad (24)$$

Focusing on the post form, using equation (23) $\langle \Phi_{\vec{k}_f} \mid$ can be written in terms of distorted waves, $\langle \eta^{(-)}_{\vec{k}_f} \mid$

$$\langle \Phi_{\vec{k}_f} \mid = \langle \eta^{(-)}_{\vec{k}_f} \mid - \langle \Phi_{\vec{k}_f} \mid U_f \frac{1}{E_f - \bar{H}_f + i\varepsilon}. \qquad (25)$$

It can be shown using equation (25) that the post form of the T-matrix can be expressed as[2]

$$T^{post}_{fi} = \langle \eta^{(-)}_{\vec{k}_f} \mid V_i - W_f \mid \Phi_{\vec{k}_i} \rangle + \langle \eta^{(-)}_{\vec{k}_f} \mid W_f \mid \Psi^{(+)}_{\vec{k}_i} \rangle. \qquad (26)$$

Similarly the prior form of the T-Matrix is given by

$$T^{prior}_{fi} = \langle \Phi_{\vec{k}_f} \mid V_f - W_i \mid \eta^{(+)}_{\vec{k}_i} \rangle + \langle \Psi^{(-)}_{\vec{k}_f} \mid W_i \mid \eta^{(+)}_{\vec{k}_i} \rangle, \qquad (27)$$

where $\mid \eta^{(+)}_{\vec{k}_i} \rangle$ are now distorted waves for the potential U_i.

3. Four-Body Transfer with Target Excitation (4BTTE) Model

The 4BTTE model[3] can be used to calculate both SC and TTE. The only difference between these two processes is the final state of the residual ion. For simplicity, the specific example of proton + helium collisions will be used. However, the model can easily be generalized to other systems.

The post form of the T-Matrix (26) is used for the 4BTTE model, and can be rewritten as

$$T_{fi} = \langle \eta^{(-)}_{\vec{k}_f} \mid V_i \mid \Phi_{\vec{k}_i} \rangle + \langle \eta^{(-)}_{\vec{k}_f} \mid W_f \mid (\Psi^{(+)}_{\vec{k}_i} - \Phi_{\vec{k}_i}) \rangle, \qquad (28)$$

where the first term represents the contribution from first order perturbation theory, and the second term represents contributions from all higher order terms. If the exact initial state wave function $\Psi^{(+)}_{\vec{k}_i}$ is approximated as the unperturbed wave function $\Phi_{\vec{k}_i}$ (a plane wave times the helium atom wave function), then equation (28) becomes the first order Born

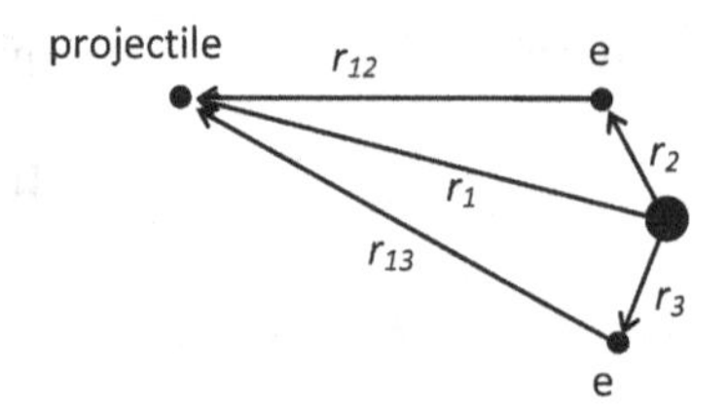

Fig. 1. Coordinate system for the projectile-helium atom system, in which r1, r12, and r13 are the magnitudes of the relative coordinates of the projectile to the helium nucleus and atomic electrons respectively.

approximation

$$T_{fi} = \langle \eta^{(-)}_{\vec{k}_f} \mid V_i \mid \Phi_{\vec{k}_i} \rangle. \tag{29}$$

The initial state projectile-atom interaction V_i is given by

$$V_i = \frac{Z_{\mathrm{p}} Z_{\mathrm{nuc}}}{r_1} + \frac{Z_{\mathrm{p}} Z_{\mathrm{e}}}{r_{12}} + \frac{Z_{\mathrm{p}} Z_{\mathrm{e}}}{r_{13}}, \tag{30}$$

where Z_{p}, Z_{nuc}, and Z_{e} are the charges of the projectile, nucleus, and electron respectively, and r_1, r_{12}, and r_{13} are the magnitudes of the relative coordinates for the projectile-target nucleus and projectile-target electrons (shown in Fig. 1).

If the initial state projectile wave function is approximated as something more appropriate than a plane wave, the second term of equation (28) does not vanish, and parts of all higher order terms remain in the perturbation series. For either case, the exact initial state wave function $\Psi^{(+)}_{\vec{k}_i}$ can be approximated as a product of a projectile wave function times an atomic wave function

$$\Psi^{(+)}_{\vec{k}_i} = \chi^{(+)}_{\vec{k}_i}(\vec{R}_i)\xi_{\mathrm{He}}(\vec{r}_2, \vec{r}_3), \tag{31}$$

where $\chi_{\vec{k}_i}(\vec{R}_i)$ is the incident projectile wave function and $\xi_{\mathrm{He}}(\vec{r}_2, \vec{r}_3)$ is the ground-state helium atom wave function. The final state distorted wave function $\eta^{(-)}_{\vec{k}_f}$ is also approximated as a product of wave functions for the final state particles

$$\eta^{(-)}_{\vec{k}_f} = \chi^{(-)}_{\vec{k}_f}(\vec{R}_f)\psi_{\mathrm{He}^+}(\vec{r}_3)\phi_H(\vec{r}_{12}), \tag{32}$$

where $\chi^{(-)}_{\vec{k}_f}(\vec{R}_f)$ is the scattered hydrogen wave function, $\psi_{\mathrm{He}^+}(\vec{r}_3)$ is the final state He^+ wave function, and $\phi_H(\vec{r}_{12})$ is the captured electron wave function. Both $\phi_H(\vec{r}_{12})$ and $\psi_{\mathrm{He}^+}(\vec{r}_3)$ are simply hydrogenic wave functions, and thus known exactly. Note that equation (32) is not properly symmetrized in the electron coordinates, and the properly symmetrized expression for the final state wave function is given by

$$\eta_{\vec{k}_f}^{(-)} = \frac{1}{\sqrt{2}}[\chi_{\vec{k}_f}^{(-)}(\vec{R}_f)\psi_{\mathrm{He}^+}(\vec{r}_3)\phi_H(\vec{r}_{12}) + \chi_{\vec{k}_f}^{(-)}(\vec{R}_f)\psi_{\mathrm{He}^+}(\vec{r}_2)\phi_H(\vec{r}_{13})]. \tag{33}$$

The calculations are performed in the center of mass frame, using the Jacobi coordinates[4] shown in Figs. 2 and 3. In this coordinate system, $\vec{R}_i$ is the relative vector between the projectile and the center of mass of the helium atom, and $\vec{R}_f$ is the relative vector between the center of mass of the hydrogen atom and the center of mass of the He^+ ion. They are given by

$$\vec{R}_i = \vec{r}_1 - \frac{m_e}{m_\alpha + 2m_e}(\vec{r}_2 + \vec{r}_3) \tag{34}$$

and

$$\vec{R}_f = \frac{m_e\vec{r}_2 + m_p\vec{r}_1}{m_p + m_e} - \frac{m_e}{m_e + m_\alpha}\vec{r}_3, \tag{35}$$

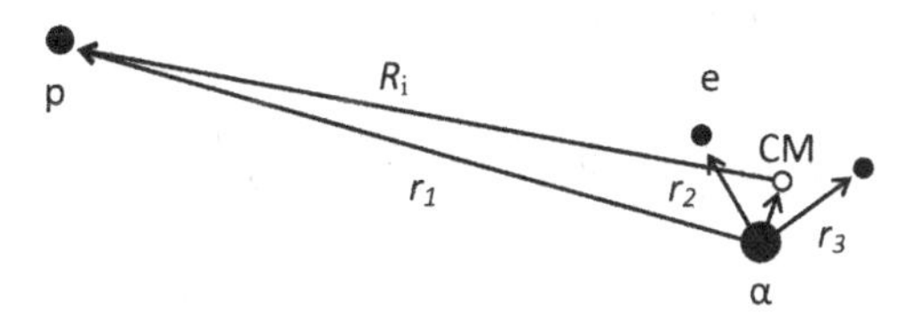

Fig. 2. Jacobi coordinate system for the projectile-helium atom system.

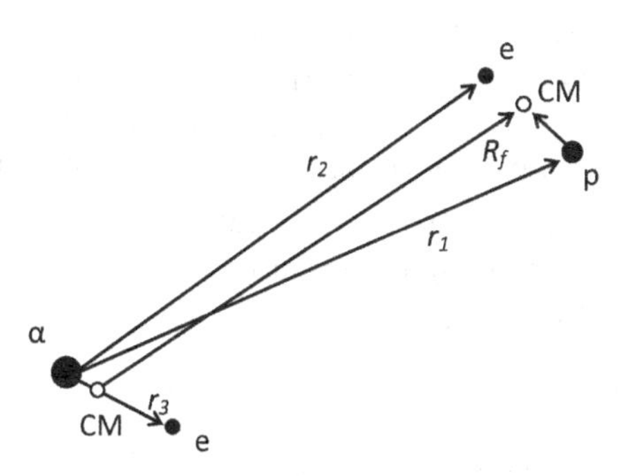

Fig. 3. Jacobi coordinate system for the hydrogen-helium ion system.

where m_e, m_α, and m_p are the masses of the electron, alpha particle, and projectile respectively. Because the 9D integral of the T-matrix is calculated numerically, any wave function can be used for the incident projectile, ground-state helium atom, or scattered projectile.

The incident projectile is treated as either a plane wave or an eikonal wave function.[5] The eikonal wave function has been very successful for treating three-body heavy-particle collisions. The plane wave approximation is obtained by replacing $\chi^{(+)}_{\vec{k}_i}(\vec{R}_i)$ with a plane wave. Thus,

$$\chi^{(+)}_{\vec{k}_i}(\vec{R}_i) = \beta_{\vec{k}_i}(\vec{R}_i) = \frac{e^{i\vec{k}_i\cdot\vec{R}_i}}{(2\pi)^{3/2}}, \tag{36}$$

where $\vec{k}_i = \mu_{pa}\vec{v}_i$, with $\vec{v}_i$ being the center of mass velocity and μ_{pa} the reduced mass of the projectile and target atom.

The eikonal wave function for a projectile incident on a one active electron atom is given by[5]

$$\psi^{1active} = \frac{e^{i\vec{k}_i\cdot\vec{r_1}}}{(2\pi)^{3/2}}\exp\left[i\frac{Z_\mathrm{p}}{v_i}\ln\left(\frac{(v_i r_1 - \vec{v_i}\cdot\vec{r_1})}{(v_i r_{12} - \vec{v_i}\cdot\vec{r_{12}})}\right)\right], \tag{37}$$

where Z_p is the charge of the projectile and $\vec{v}_i$ is the center of mass velocity of the incident projectile. Generalizing this to two active electrons gives

$$\chi^{(+)}_{\vec{k}_i}(\vec{R}_i) = \frac{e^{i\vec{k}_i\cdot\vec{R}_i}}{(2\pi)^{3/2}}\exp\left[i\frac{Z_\mathrm{p}}{v_i}\ln\left(\frac{(v_i r_1 - \vec{v}_i\cdot\vec{r}_1)^{Z_\mathrm{nuc}}}{(v_i r_{12} - \vec{v}_i\cdot\vec{r}_{12})(v_i r_{13} - \vec{v}_i\cdot\vec{r}_{13})}\right)\right]. \tag{38}$$

Note that asymptotically, $\vec{r_1} = \vec{r_{12}} = \vec{r_{13}}$ (see Fig. 2), so that the eikonal wave function becomes the plane wave of equation (36).

For the ground-state helium atom, any wave function can be used, allowing for the effects of electron-electron correlation to be studied. Here, either an analytic Hartree-Fock[6] wave function or a 20-term Hylleraas[7] wave function is used. The Hartree-Fock wave function has some radial correlation, but no angular correlation, while the Hylleraas wave function contains both radial and angular correlation.

For the scattered projectile, $\chi^{(-)}_{\vec{k}_f}(\vec{R}_f)$ is assumed to be either a plane wave given by

$$\chi^{(-)}_{\vec{k}_f}(\vec{R}_f) = \beta_{\vec{k}_f}(\vec{R}_f) = \frac{e^{i\vec{k}_f\cdot\vec{R}_f}}{(2\pi)^{3/2}} \tag{39}$$

or a Coulomb wave given by

$$\chi^{(-)}_{\vec{k}_f}(\vec{R}_f) = \frac{e^{i\vec{k}_f\cdot\vec{R}_f}}{(2\pi)^{3/2}} e^{-\pi\gamma/2}\Gamma(1-i\gamma)\,{}_1F_1(i\gamma,1;-i(k_fR_f+\vec{k}_f\cdot\vec{R}_f)), \quad (40)$$

where $\gamma = \frac{Z_{\rm P}Z_{\rm He^+}}{v_H}$ is the Sommerfeld parameter and ${}_1F_1(a,c;z)$ is a confluent hypergeometric function. The quantities $Z_{\rm p}$ and $Z_{\rm He^+}$ are the electric charges of the projectile and He^+ ion; v_H is the center of mass speed of the hydrogen atom.

In order to evaluate the post form of the T-matrix, one would need to know the perturbation that corresponds to the final state wave function. This final state perturbation satisfies

$$(H-E)\Psi_f^{(-)} = W_f\Psi_f^{(-)}, \quad (41)$$

where H is the full Hamiltonian

$$H = -\frac{1}{2\mu_{pa}}\nabla^2_{r_1} - \frac{1}{2}\nabla^2_{r_2} - \frac{1}{2}\nabla^2_{r_3} + V, \quad (42)$$

and E is the total center of mass energy

$$E = \frac{k_f^2}{2\mu_{pa}} - B_H - B_{\rm He^+}. \quad (43)$$

The quantities $B_{\rm H}$ and $B_{\rm He^+}$ are the binding energies of the hydrogen atom and He^+ ion respectively. The total interaction potential V for a proton + helium collision is given by

$$V = \frac{2}{r_1} - \frac{2}{r_2} - \frac{2}{r_3} - \frac{1}{r_{12}} - \frac{1}{r_{13}} + \frac{1}{r_{23}}.$$

When the exact final state wave function $\Psi^{(-)}_{\vec{k}_f}$ is approximated by $\eta^{(-)}_{\vec{k}_f}$, the perturbation W_f can then be obtained from

$$W_f = \frac{1}{\eta^{(-)}_{\vec{k}_f}}(H-E)\eta^{(-)}_{\vec{k}_f}. \quad (44)$$

The final state wave function $\eta^{(-)}_{\vec{k}_f}$ for the case of SC or TTE is approximated as a product of the wave functions for the three particles according to equation (32). To simplify the notation, three generic wave functions are defined such that

$$\eta^{(-)}_{\vec{k}_f} = A(\vec{r}_1)B(\vec{r}_3)C(\vec{r}_{12}), \tag{45}$$

where $A(\vec{r}_1)$ is the scattered projectile wave function $\chi^{(-)}_{\vec{k}_f}(\vec{R}_f)$, $B(\vec{r}_3)$ is the He^+ bound state wave function, and $C(\vec{r}_{12})$ is the hydrogen wave function. For the calculation of the perturbation, it is assumed that the center of mass coordinate $\vec{R}_f \approx \vec{r}_1$, which is a very good approximation for the present case. The He^+ bound state wave function and hydrogen wave functions are given by

$$B(\vec{r}_3) = \psi_{\mathrm{He}^+}(\vec{r}_3) \tag{46}$$

and

$$\begin{aligned} C(\vec{r}_{12}) &= \phi_{\mathrm{H}}(\vec{r}_{12}) \\ &= \frac{\mathrm{e}^{-r_{12}}}{\sqrt{\pi}}. \end{aligned} \tag{47}$$

The scattered projectile wave function is approximated as either a plane wave given by equation (39)

$$A(\vec{r}_1) = \beta_{\vec{k}_f}(\vec{r}_1) = \frac{\mathrm{e}^{i\vec{k}_f\cdot\vec{r}_1}}{(2\pi)^{3/2}} \tag{48}$$

or a Coulomb wave given by

$$A(\vec{r}_1) = \frac{N\mathrm{e}^{i\vec{k}_f\cdot\vec{r}_1}}{(2\pi)^{3/2}}\,{}_1F_1(i\gamma, 1; -i(k_f r_1 + \vec{k_f}\cdot\vec{r_1})), \tag{49}$$

with $N = \mathrm{e}^{-\pi\gamma/2}\Gamma(1 - i\gamma)$ and $\gamma = \frac{\mu_{pt} Z_{\mathrm{He}^+} Z_{\mathrm{P}}}{k_1}$. The left hand side of equation (41) can then be written as

$$\begin{aligned} (H - E)\Psi_f = &-\frac{B}{2\mu_{pa}}\nabla^2_{r_1}(AC) - \frac{AB}{2}\nabla^2_{r_2}C - \frac{AC}{2}\nabla^2_{r_3}B \\ &+ \left(\frac{2}{r_1} - \frac{2}{r_2} - \frac{2}{r_3} - \frac{1}{r_{12}} - \frac{1}{r_{13}} + \frac{1}{r_{23}}\right)ABC \\ &- \frac{k_f^2}{2\mu_{pa}}ABC - B_H ABC - B_{\mathrm{He}^+}ABC. \end{aligned} \tag{50}$$

Note that

$$-\frac{1}{2}\nabla^2_{r_3}B - \frac{2}{r_3}B = -B_{\mathrm{He}^+}B \tag{51}$$

and

$$-\frac{1}{2}\nabla^2_{r_{12}}C - \frac{2}{r_{12}}C = -B_H C. \tag{52}$$

Also, because $C(\vec{r}_{12})$ is only a function of $\vec{r}_{12}$,

$$\nabla^2_{r_{12}}C = \nabla^2_{r_1}C = \nabla^2_{r_2}C. \tag{53}$$

Then, equation (50) can be written as

$$(H-E)\Psi_f = -\frac{B}{2\mu_{pa}}\nabla^2_{r_1}(AC) + \left(\frac{2}{r_1} - \frac{2}{r_2} - \frac{1}{r_{13}} + \frac{1}{r_{23}}\right)ABC - \frac{k_f^2}{2\mu_{pa}}ABC. \tag{54}$$

Using the operator identity $\nabla^2_{r_1}(AC) = C\nabla^2_{r_1}A + A\nabla^2_{r_1}C + 2\nabla_{r_1}A \cdot \nabla_{r_1}C$ gives

$$(H-E)\Psi_f = -\frac{B}{2\mu_{pa}}[C\nabla^2_{r_1}A + A\nabla^2_{r_1}C + 2\nabla_{r_1}A \cdot \nabla_{r_1}C] + \left(\frac{2}{r_1} - \frac{2}{r_2} - \frac{1}{r_{13}} + \frac{1}{r_{23}}\right)ABC - \frac{k_f^2}{2\mu_{pa}}ABC. \tag{55}$$

Then, the perturbation W_f can be written as

$$W_f = -\frac{1}{2\mu_{pa}A}\nabla^2_{r_1}A - \frac{1}{2\mu_{pa}C}\nabla^2_{r_1}C - \frac{1}{\mu_{pa}AC}\nabla_{r_1}A \cdot \nabla_{r_1}C + \left(\frac{2}{r_1} - \frac{2}{r_2} - \frac{1}{r_{13}} + \frac{1}{r_{23}}\right) - \frac{k_f^2}{2\mu_{pa}}. \tag{56}$$

Using a plane wave for the scattered projectile gives

$$\nabla^2_{r_1}e^{i\vec{k}_f\cdot\vec{r}_1} = -k_f^2 e^{i\vec{k}_f\cdot\vec{r}_1} \tag{57}$$

and

$$\nabla_{r_1}e^{i\vec{k}_f\cdot\vec{r}_1} = i\vec{k}_f e^{i\vec{k}_f\cdot\vec{r}_1}. \tag{58}$$

Also, for SC or TTE,

$$\nabla^2_{r_1}e^{-r_{12}} = \frac{r_{12}-2}{r_{12}}e^{-r_{12}}. \tag{59}$$

Plugging these into equation (56) gives the final state perturbation when treating the scattered projectile as a plane wave

$$W_f^{PW} = \frac{2 - r_{12}}{2\mu_{pa} r_{12}} + i\frac{\vec{k_f} \cdot \vec{r_{12}}}{\mu_{pa} r_{12}} + \left(\frac{2}{r_1} - \frac{2}{r_2} - \frac{1}{r_{13}} + \frac{1}{r_{23}}\right). \tag{60}$$

If a Coulomb wave is used for the scattered projectile,

$$\begin{aligned}
&\nabla^2_{r_1}[{}_1F_1(i\gamma, 1; -ik_f r_1 - i\vec{k}_f \cdot \vec{r}_1)\mathrm{e}^{i\vec{k}_f \cdot \vec{r}_1}] \\
&= \mathrm{e}^{i\vec{k}_f \cdot \vec{r}_1}\left[\frac{2\gamma k_f}{r_1} {}_1F_1(1 + i\gamma, 1; -ik_f r_1 - i\vec{k}_f \cdot \vec{r}_1)\right. \\
&\quad + 2i\gamma k_f {}_1F_1(1 + i\gamma, 2; -ik_f r_1 - i\vec{k}_f \cdot \vec{r}_1)(\widehat{k}_f + \widehat{r}_1) \cdot \vec{k}_f \\
&\quad \left. - k_f^2 {}_1F_1(i\gamma, 1; -ik_f r_1 - i\vec{k}_f \cdot \vec{r}_1)\right]
\end{aligned} \tag{61}$$

and

$$\begin{aligned}
&\nabla_{r_1}[{}_1F_1(i\gamma, 1; -ik_f r_1 - i\vec{k}_f \cdot \vec{r}_1)\mathrm{e}^{i\vec{k}_f \cdot \vec{r}_1}] \\
&= \mathrm{e}^{i\vec{k}_f \cdot \vec{r}_1}[\gamma k_f {}_1F_1(1 + i\gamma, 2; -ik_f r_1 - i\vec{k}_f \cdot \vec{r}_1)(\widehat{k}_f + \widehat{r}_1) \\
&\quad + i {}_1F_1(1 + i\gamma, 1; -ik_f r_1 - i\vec{k}_f \cdot \vec{r}_1)\vec{k}_f].
\end{aligned} \tag{62}$$

Thus, the final state perturbation when treating the scattered projectile as a Coulomb wave is given by

$$\begin{aligned}
W_f^{CW} &= -\frac{\gamma k_f}{\mu_{pa} r_1} \frac{{}_1F_1(1 + i\gamma, 1; -ik_f r_1 - i\vec{k}_f \cdot \vec{r}_1)}{{}_1F_1(i\gamma, 1; -ik_f r_1 - i\vec{k}_f \cdot \vec{r}_1)} \\
&\quad + \frac{{}_1F_1(1 + i\gamma, 2; -ik_f r_1 - i\vec{k}_f \cdot \vec{r}_1)}{{}_1F_1(i\gamma, 1; -ik_f r_1 - i\vec{k}_f \cdot \vec{r}_1)} \\
&\quad \times \left[\frac{\gamma k_f}{\mu_{pa}}(\hat{k}_f + \hat{r}_1) \cdot \frac{\vec{r}_{12}}{r_{12}} - i\frac{\gamma k_f}{\mu_{pa}}\vec{k}_f \cdot (\hat{k}_f + \hat{r}_1)\right] \\
&\quad + \frac{2 - r_{12}}{r_{12}} + \left(\frac{2}{r_1} - \frac{2}{r_2} - \frac{1}{r_{13}} + \frac{1}{r_{23}}\right) + i\frac{\vec{k}_f \cdot \vec{r}_{12}}{\mu_{pa} r_{12}}.
\end{aligned} \tag{63}$$

3.1. *Single Charge Transfer without Target Excitation*

Single charge transfer without target excitation, or single electron capture, can be treated as a 3-body process, and there has been a significant amount of work done on this problem.[a] In a 3-body treatment, it is assumed that the bound state wave function for the passive electron is the same initially and finally and, as a result, the passive electron does not participate in the collision. The first 3-body SC calculations were reported by Oppenheimer,[17] Brinkman, and Kramers.[18] Their model is known as the OBK approximation. In the OBK approximation, the projectile-nuclear interaction in the perturbation is neglected, and only the interaction between the projectile and the active atomic electron is included. The projectile-nuclear interaction is neglected based on the assumption that the initial and final state wave functions should be orthogonal (even though they are not), and this term would not contribute for orthogonal wave functions.[17] The OBK approximation also treats both the incident and scattered projectile as plane waves. Calculations based on the OBK approximation led to total cross section results that overestimated experiment, and over a decade passed before Jackson and Schiff[19] (hereafter referred to as JS) performed a calculation that included both the projectile-electron and projectile-nuclear terms in the perturbation.

Total cross section results based on the JS model correctly predicted the magnitude of the cross sections and agreed with experiment much better than the OBK results.[19,20] Since the OBK and JS models were introduced, there has been a significant amount of discussion regarding the inclusion of the projectile-nuclear term. Belkić and Salin[21] have shown that inclusion of the projectile-nuclear term in the perturbation improves agreement with experiment for differential cross sections, particularly at large scattering angles. When the projectile-nuclear interaction is ignored, the differential cross section drops off much more rapidly than when it is included. This discrepancy between the JS and OBK models is attributed to the projectile scattering from the nucleus. Classically, large angle scattering results from small impact parameters, which occurs when the projectile penetrates the electron cloud and scatters elastically from the nucleus. It is now accepted that the projectile-nuclear term in the perturbation needs to be included to accurately predict the magnitude of the charge transfer cross section.

Most of the work done for charge transfer processes has focused on integrated cross sections, but the study of cross sections that are differential in projectile scattering angle is a much more stringent test of theory. In general, the differential cross sections for single charge transfer decrease

[a]See references 8–16 for further information.

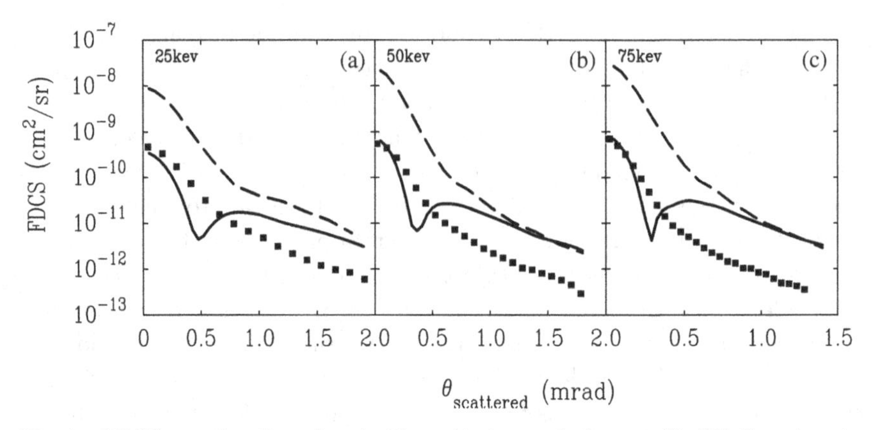

Fig. 4. FDCS as a function of projectile scattering angle for p + He SC. Experiment: ■ results of Schulz *et al.*[22] for the incident projectile energies shown in the figure. Both theoretical curves are from the 4BTTE[3] model with a plane wave for the incident projectile, Hylleraas wave function for the helium atom, and a Coulomb wave for the scattered projectile. Theoretical results: — all three terms in the perturbation; - - - no projectile-nuclear term in the perturbation.

rapidly as the scattering angle increases, changing as much as 3 orders of magnitude over 2 mrad. Typically, there is a change in slope in the differential cross section that has been attributed to the boundary between small and large angle scattering. At small angles, the projectile scattering is dominated by scattering from the atomic electrons, whereas for large angles, the scattering is a result of scattering from the nucleus, as mentioned above.[21]

Figure 4 shows the effect of the projectile-nuclear term in the perturbation on the fully differential cross sections (FDCS) for SC.[3] The calculation that contains all three terms in the perturbation potential uses the same perturbation as JS, while the calculation without the projectile-nuclear term uses the perturbation of OBK. However, these calculations are not simply JS and OBK calculations. In both the JS and OBK model, the projectile is treated as a plane wave in both the initial and final states, screening due to the passive electron on the initial ground state active electron wave function is neglected, and the passive electron wave function is assumed to be the same initially and finally. Thus, the passive electron wave function does not contribute to the integral. The calculations shown in Fig. 4 are full 4-body calculations because the passive electron is included in the calculation, and because the Coulomb interaction between the projectile and the passive electron is retained in V_i. This Coulomb interaction is neglected in 3-body calculations. For the SC results shown in Fig. 4, a Hylleraas wave function was used for the ground state of the helium atom, and the final passive electron wave function was the He^+ wave function. The scattered projectile was treated as a Coulomb wave that satisfies asymptotic boundary conditions, which has been shown to be very important.[23] It is quite evident that the calculation using the OBK perturbation overestimates the experiment by more than one order of magnitude, while the calculation using the JS perturbation correctly predicts the magnitude of experiment. The change in slope discussed above is also apparent in the calculation with the OBK perturbation.

An unphysical minimum can be seen in the calculation with the JS perturbation, which becomes deeper and shifts to smaller angles as the projectile energy increases. This was originally observed by Chen and Kramer,[24] and then again by Band[25] and Sil *et al.*[26] The minimum in the calculation that uses the JS perturbation is typically attributed to a cancellation of the terms in the perturbation.[27] Excluding the projectile-nuclear term (as is done in the OBK approximation) in the perturbation results in the elimination of this minimum, something that has also been seen by Belkić and Salin.[21]

3.2. *Charge Transfer with Target Excitation*

Charge transfer with target excitation is inherently a 4-body process, but there has been very little work done on this process. It is much more common to study charge transfer to an excited state of the projectile. For this TTE process, only two sets of experimental results and two theoretical models have been reported for the FDCS, all for proton + helium collisions. On the theoretical side, Kirchner[28,29] has presented a semi-classical, non-perturbative, impact parameter model that employs the independent electron approximation. Also, Harris *et al.*[3] have presented the 4BTTE model that will be discussed here. The reported experimental results are those of Hasan *et al.*[28] and Schöffler.[30] Both sets of measured differential cross sections show little structure as a function of scattering angle.

For the experiments that have been performed so far, the experimental results are absolute, and it is known that the outgoing hydrogen atom is in the ground state and the residual helium ion is in an excited state. However, the exact excited state of the helium ion could not be determined.

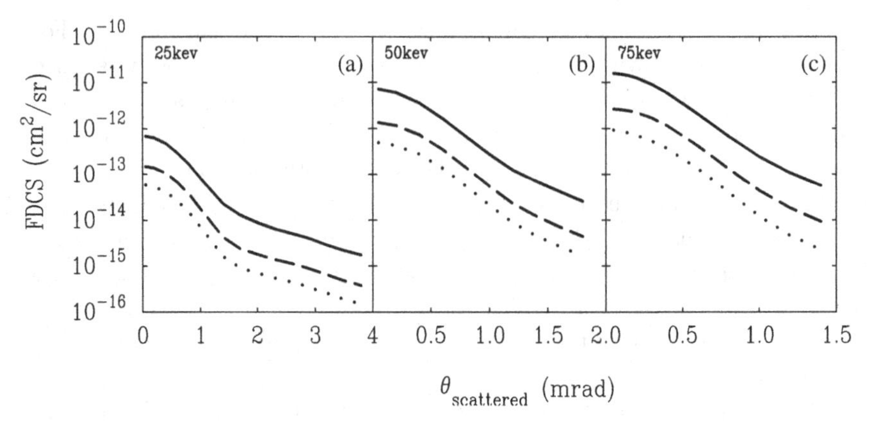

Fig. 5. FDCS as a function of projectile scattering angle for 25 keV, 50 keV and 75 keV p + He TTE showing the relative magnitudes of excitation to different energy levels in the He^+ ion. All theoretical curves are from the 4BTTE model[3] with a plane wave for the incident projectile, Hylleraas wave function for the helium atom, and Coulomb wave for the scattered projectile. Theoretical results: — excitation to the $n = 2$ level; - - - excitation to the $n = 3$ level; ··· excitation to the $n = 4$ level.

Consequently, the theoretical cross sections must be summed over all possible excited states in order to be compared with experiment.

Figure 5 shows differential cross sections for the helium ion being left in either the $n = 2, 3$ or 4 states. From this figure, it can be seen that the contribution from the $n = 4$ state is more than one order of magnitude smaller than the contribution from the $n = 2$ state. It has also been found that contributions from angular momentum states higher than p-states are negligible. Because of this, the results shown here include only s and p excited states for $2 \leq n \leq 4$.

Figure 6 shows the FDCSs for the helium ion being left in a particular angular momentum state. The FDCSs corresponding to the individual angular momentum states for the $n=2, n=3$, and $n=4$ levels of the He^+ ion all exhibit similar behavior, so only the $n=2$ states are shown here. Clearly the $n=2$ sum of FDCSs most closely resembles the $2s$ FDCS. However, at small scattering angles for 50 keV and 75 keV, the $2p0$ contribution dominates, while at large scattering angles for all energies, the $2s$ contribution dominates. One possible explanation for this can be obtained from a classical point of view. Small scattering angles correspond to large impact parameters. At a large distance from the nucleus, the radial part of the $2p0$ He^+ wave function is larger than the $2s$ He^+ wave function, making the $2p0$ FDCS larger than the $2s$ FDCS for

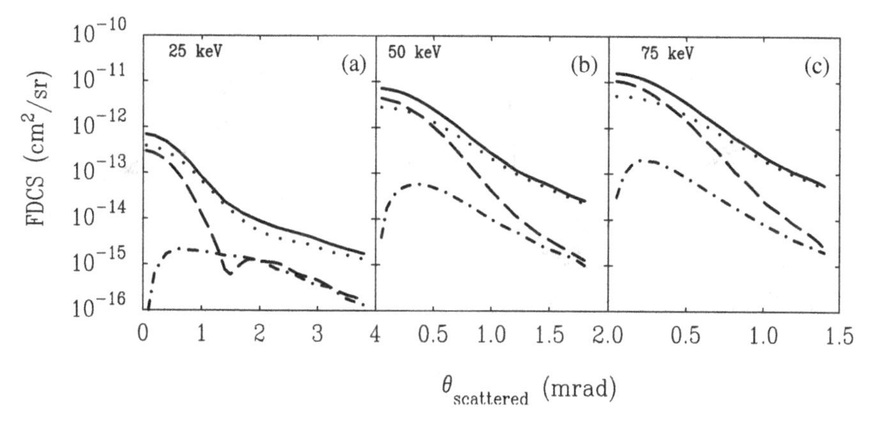

Fig. 6. FDCS as a function of the lab projectile scattering angle for p + He TTE showing the contributions of excitation to a particular angular momentum state of the He^+ ion. All theoretical curves are from the 4BTTE model with a plane wave for the incident projectile, Hylleraas wave function for the helium atom, and a Coulomb wave for the scattered projectile. Theoretical results: Solid curve shows the sum of $2s$, $2p0$, and $2p1$ contributions; $\cdots$, excitation to the $2s$ state; - - -, excitation to the $2p0$ state; dash ·-·-·-, excitation to the $2p1$ state.

angles less than a particular value. Similarly, large scattering angles correspond to small impact parameters, and at a small distance from the nucleus, the $2s$ radial wave function is larger than the $2p0$ radial wave function.

Examination of Fig. 6 also shows that the $2p1$ contribution to the FDCS sum is negligible since it is one to two orders of magnitude smaller than either the $2s$ or $2p0$ contribution. Interestingly, there is a zero in the $2p1$ FDCS at $\theta_{scattered} = 0$ mrad, but not in the $2p0$ FDCS. The $2p0$ and $2p1$ radial wave functions are identical, and the only difference between these two calculations is the angular part of the He^+ wave function. One possible reason for the $2p1$ FDCS being much smaller than either the $2s$ or $2p0$ FDCS is that for the $2p1$ case, the final state magnetic quantum number is different than the initial state magnetic quantum number. Initially, both electrons in the helium atom are in the ground state ($1s^2$) with $m = 0$. For SC, the final state is also the ground state ($1s$) and $m = 0$. For TTE to either the $2s$ or $2p0$ state, m is also equal to 0, but for TTE to the $2p1$ state, $m = 1$.

Also seen in Fig. 6 is a dip in the 25 keV $2p0$ FDCS. This minimum does not appear in either the 50 keV or 75 keV FDCSs. However, the dip does slightly resemble the dip in the SC FDCS, but appears here at a larger

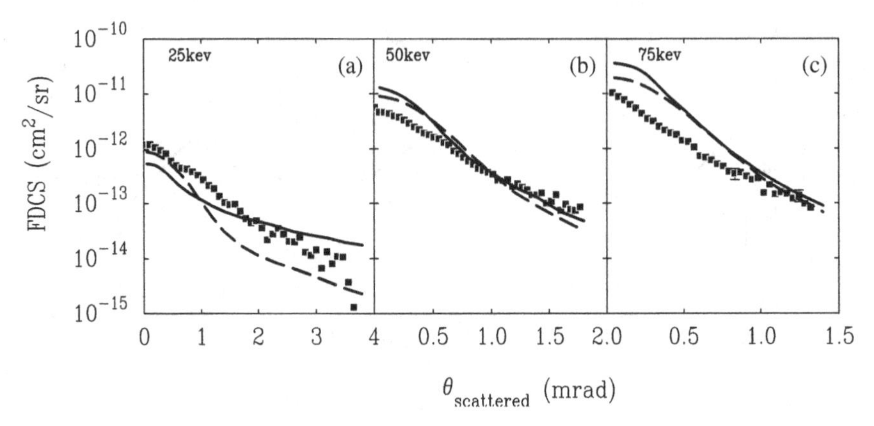

Fig. 7. FDCS as a function of the lab projectile scattering angle for p + He TTE showing the effect of the incident projectile-target atom interaction. Experiment: ■ results of Hasan *et al.*[28] for the incident projectile energies shown in the figure. Theoretical results: — 4BTTE model[3] with an eikonal wave function for the incident projectile, Hylleraas wave function for the helium atom, and Coulomb wave for the scattered projectile; - - - 4BTTE model[3] with a plane wave for the incident projectile, Hylleraas wave function for the helium atom, and Coulomb wave for the scattered projectile.

scattering angle. Unlike SC, though, the minimum in the $2p0$ FDCS does not go away when the nuclear term in the perturbation is removed.

Figures 7 through 10 show various calculations from the 4BTTE model compared with the experimental results of Hasan *et al.*[28] In each plot, a different 2-body interaction is examined. Figure 7 shows the effect of the initial state projectile-atom interaction on the FDCS. Even though the target atom is neutral, the interaction of the projectile with the constituent particles of the target atom can be included through the use of an eikonal initial state (EIS) wave function. Asymptotically, the EIS wave function is a plane wave since the projectile will 'see' a net neutral charge. However, as the projectile approaches the target, the net Coulomb force will no longer be zero due to the charge separation, meaning that the individual Coulomb interactions need to be taken into account. The eikonal logarithmic phase factor (see equation (38)) is the asymptotic form of the three Coulomb interactions between the projectile and target helium atom. The eikonal wave function is typically used for high energy projectiles, and is considered a valid approximation when the ratio Z_{nuc}/v_i is less than 1. For the three energies studied here, this ratio ranges between 1.2 (75 keV) and 2 (25 keV), pushing the limit of the eikonal's validity. Also, while the EIS wave function has been extremely successful in the treatment of 3-body problems, Belkic[23]

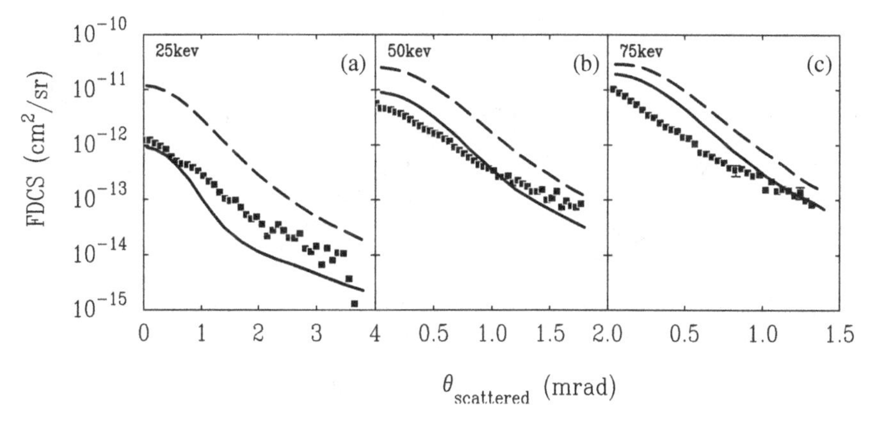

Fig. 8. FDCS as a function of the lab projectile scattering angle for p + He TTE showing the effect of the scattered projectile-residual ion interaction. Experiment: ■ results of Hasan *et al.*[28] for the incident projectile energies shown in the figure. Theoretical results: — 4BTTE model[3] with an plane wave for the incident projectile, Hylleraas wave function for the helium atom, and Coulomb wave for the scattered projectile; - - - 4BTTE model[3] with a plane wave for the incident projectile, Hylleraas wave function for the helium atom, and plane wave for the scattered projectile.

has shown that it may not be appropriate for the 4-body problem. In Fig. 7, the use of the eikonal wave function has a fairly small effect, with the largest change being observed at small scattering angles. Not surprisingly, the biggest change in shape occurs at the lowest energy, where the eikonal approximation is expected to be the least valid. However, for each energy, the eikonal wave function produced results in better agreement with the shape of the experimental data than the results using a plane wave.

Figure 8 shows results for different treatments of the outgoing hydrogen atom, which is in the field of the He^+ ion. Asymptotically, the He^+ ion has a charge of 1, while the hydrogen atom is neutral. The simplest choice for the outgoing hydrogen wave function is a plane wave, which matches asymptotic boundary conditions. However, the dynamics of the collision occur at small projectile-ion separations, and the use of a Coulomb wave for the scattered proton in the field of the He^+ ion should also be considered. Results for both of these approximations are shown in Fig. 8.

It is clear from Fig. 8 that in order to achieve the correct order of magnitude compared to experiment, a Coulomb wave is required. However, the use of a plane wave or Coulomb wave for the scattered projectile results in virtually the same shape for the FDCS. A comparison of these two calculations shows that the results become more similar as the projectile

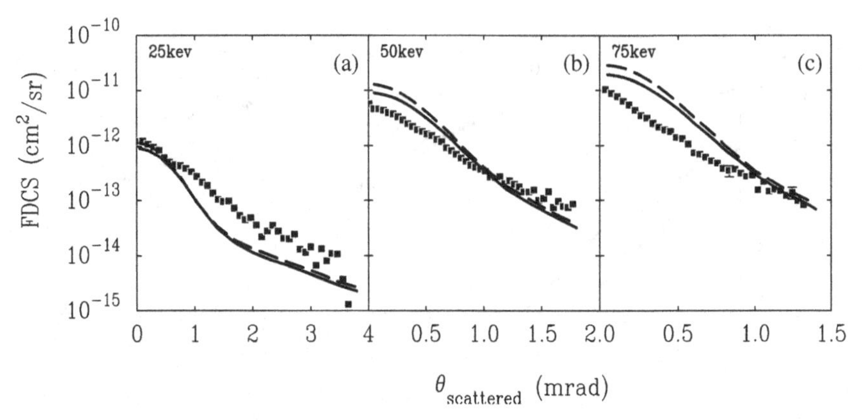

Fig. 9. FDCS as a function of the lab projectile scattering angle for p + He TTE showing the effect of electron correlation in the target atom wave function. Experiment: ■ results of Hasan *et al.*[28] for the incident projectile energies shown in the figure. Theoretical results: — 4BTTE model[3] with a plane wave for the incident projectile, Hylleraas wave function for the helium atom, and Coulomb wave for the scattered projectile; - - - 4BTTE[3] model with a plane wave for the incident projectile, Hartree-Fock wave function for the helium atom, and Coulomb wave for the scattered projectile.

energy increases. This effect is expected because a faster projectile spends less time in the field of the ion than a slower projectile.

Figure 9 shows the effect of electron-electron correlation in the target helium atom. There are two physical mechanisms that can cause both atomic electrons to change state. In the first mechanism, the projectile 'hits' one target electron and then sequentially 'hits' the second electron. In the second mechanism, the projectile 'hits' one electron, and this electron then 'hits' the second electron. The electron-electron interaction in the initial state is manifested through correlation in the helium atom wave function. This initial state electron-electron interaction is, in principle, included 'exactly' in the initial atomic wave function, and consequently, there is no electron-electron interaction in the perturbation. Because all of the initial state electron-electron interaction is contained in the atomic wave function, it would seem likely that correlation would play an important role in a 4-body process. However, the results displayed in Fig. 9 show that this expectation is incorrect.

In order to study the effect of electron-electron correlation, two different initial state atomic helium wave functions are used. The Hylleraas wave function includes both radial and angular correlation between the two atomic electrons very accurately, while the Hartree-Fock wave function is a product wave function that treats the two atomic electrons independently. The Hartree-Fock wave function has some radial correlation included indirectly because the electron-electron interaction is included in the formation of the potential from the wave functions. However, the Hartree-Fock wave function has no angular correlation. The results in Fig. 9 show that there is very little difference between the calculation using the Hylleraas wave function and the calculation using the Hartree-Fock wave function, except at small scattering angles. This indicates that electron-electron correlation is not important in the TTE process. It has been previously shown that correlation is not important for integrated cross sections,[31–35] and the results here indicate that it is also not important for differential cross sections.

Figure 10 shows the effect of the projectile-nuclear interaction on the FDCS. This term in the perturbation corresponds to nuclear scattering, and is expected to be most important at large scattering angles. Recall that at large scattering angles, the projectile penetrates the electron cloud and elastically scatters from the nucleus. Excluding the projectile-nuclear term from the calculation for SC caused a more rapid decrease of the FDCS as the scattering angle increases. However, this effect is not observed for TTE. Dropping the projectile-nuclear term caused the cross sections to decrease

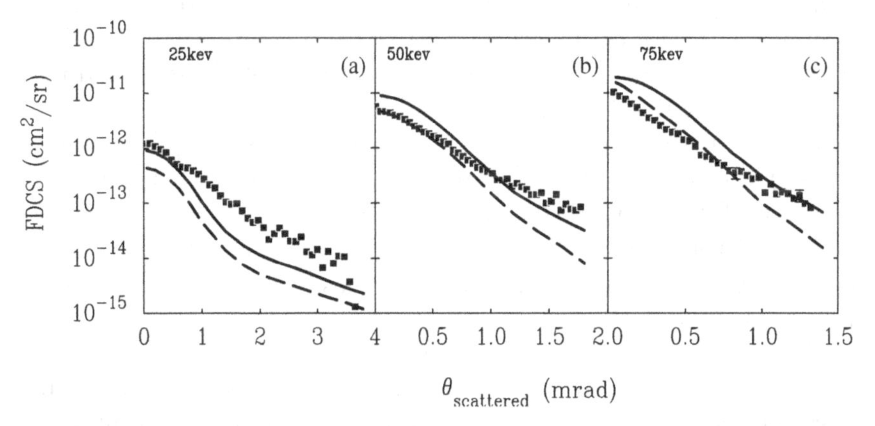

Fig. 10. FDCS as a function of the lab projectile scattering angle for p + He TTE showing the effect of the projectile-nuclear interaction. Experiment: ■ results of Hasan *et al.*[28] for the incident projectile energies shown in the figure. Both theoretical curves are from the 4BTTE model[3] with a plane wave for the incident projectile, Hylleraas wave function for the helium atom, and Coulomb wave for the scattered projectile. Theoretical results: — all three terms in the perturbation; - - - without the projectile-nuclear term in the perturbation.

at all three incident projectile energies, but the shape did not change, in sharp contrast to the SC results shown in Fig. 4

4. Four-Body Double Capture (4BDC) Model

Another 4-body charge transfer process is that of double electron capture. From a theoretical standpoint, the SC, TTE, and DC models are quite similar. For DC the FDCS is again differential only in projectile scattering angle, and is given by equation (19). The corresponding T-Matrix is given by equation (28). For the SC, TTE, and DC, the choice of how to treat the initial state wave functions is the same. For the TTE calculations, both a plane wave and an eikonal wave function were studied. However, because it has been shown by Belkić[23] that the EIS may not be appropriate for the 4-body problem, only results for an initial state plane wave are presented here.

The final state wave function for DC, however, is different from that of SC or TTE, and is given by

$$\eta_{\vec{k}_f}^{(-)} = \chi_{\vec{k}_f}^{(-)}(\vec{R}_f)\psi_{H^-}(\vec{r}_{12}, \vec{r}_{13}), \tag{64}$$

where $\psi_{H^-}(\vec{r}_{12}, \vec{r}_{13})$ is the wave function for the outgoing H^- ion, and $\chi_p^f(\vec{R}_f)$ is the scattered projectile wave function. The calculations are again performed in the center of mass frame using Jacobi coordinates. However, for DC, $\vec{R}_f$ is now the relative vector between the alpha particle and the center of mass of the H^- ion.

Because it has been shown that it is important to satisfy asymptotic boundary conditions,[23] the final projectile should not be treated as a plane wave. Therefore, the Coulomb wave of equation (40) is used, where the Sommerfeld parameter is now $\gamma = \frac{Z_{\mathrm{H}^-} Z_{\mathrm{He}^{++}}}{v_{\mathrm{H}^-}}$. The quantities Z_{H^-} and $Z_{\mathrm{He}^{++}}$ are the charges of the H^- and He^{++} ions, and v_{H^-} is the center of mass speed of the H^- ion.

Unlike SC or TTE, the effects of electron-electron correlation in both the initial and final bound-state wave functions can be studied in DC. In order to examine the effect of electron-electron correlation, two different wave functions are used. In the initial state, either an analytic Hartree-Fock wave function[6] or a 20 parameter Hylleraas[7] wave function is used for the helium atom. In the final state, either a two parameter variational wave function[36] or a 20 parameter Hylleraas[7] wave function is used for the H^- ion.

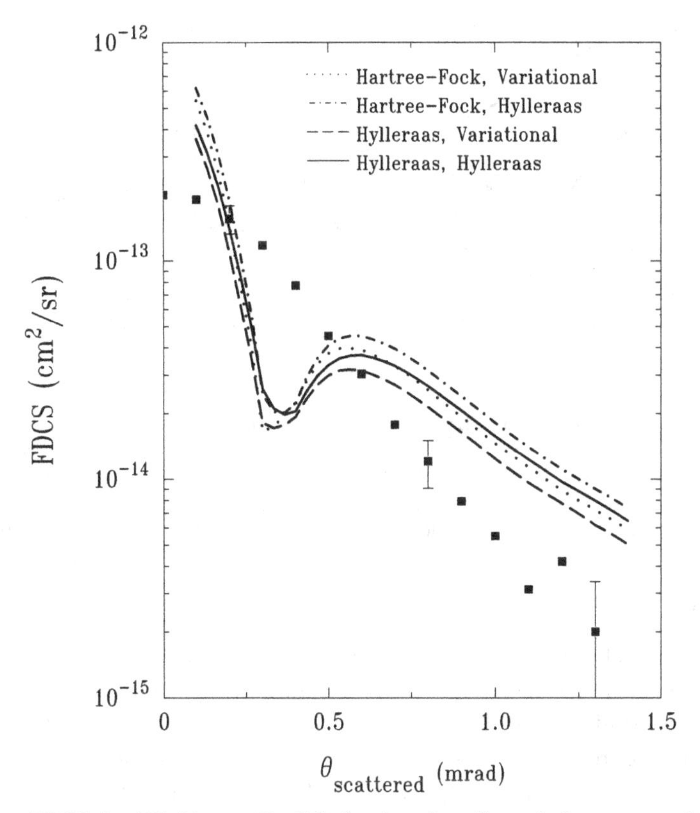

Fig. 11. FDCS for 75 keV p + He DC showing the effect of electron correlation in the target atom and the scattered ion. Experiment: ■ results of Schulz *et al.*[22] All calculations are the 4BDC model[37] with a plane wave for the incident projectile and Coulomb wave for the scattered projectile. The labels in the figure indicate the helium atom and H^- wave functions respectively. All calculations have been divided by 100.

Figure 11 shows the effect of electron-electron correlation in both the initial and final states. The four calculations that are shown present all possibilities of using either a correlated or uncorrelated bound state wave function in the initial and final state. All of the calculations shown have similar shapes and magnitudes, which indicates that similar to TTE, electron-electron correlation is not important in the DC process.

Again, the effect of the projectile-nucleus term in the perturbation of equation (30) is studied, and the results for DC are shown in Fig. 12. Recall that for SC, a minimum was observed when the projectile-nuclear term was included in the perturbation (Fig. (4)). This minimum was attributed to a cancellation of terms in the perturbation. For TTE, however, no

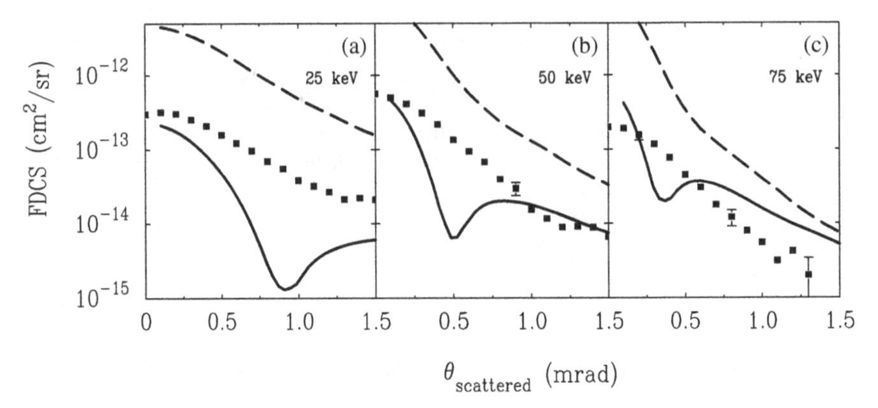

Fig. 12. FDCS as a function of projectile scattering angle for p + He DC. Experiment: ■ results of Schulz *et al.*[22] for the incident projectile energies shown in the figure. Both theoretical curves are from the 4BDC model[37] with a plane wave for the incident projectile, Hylleraas wave functions for the helium atom and H^- ion, and a Coulomb wave for the scattered projectile. Theoretical results: — all three terms in the perturbation; - - - without the projectile-nuclear term in the perturbation. Both calculations have been divided by 100.

minimum was observed when the projectile-nuclear term was included in the calculation. Now for the case of DC, a minimum is again observed in the FDCS, and similarly to SC, excluding this term from the perturbation results in the removal of this minimum. If a cancellation of terms in the perturbation was the only cause of the minimum in the SC and DC results, then a similar minimum should appear in the TTE results, as well, because the perturbation is exactly the same in all three cases. However, no minimum is observed in the TTE differential cross sections, and the only difference between the SC and TTE results is the final state of the residual ion. Because a summation over all possible excited states of the helium ion is required to compare the 4BTTE model to experiment, it is possible that the minimum is smoothed out by the summation. However, examination of the individual angular momentum states for the TTE process showed no such minimum. If the minimum was produced by a cancellation of terms in the perturbation, that cancellation should occur independent of the final state wave function. These results show that the cancellation is determined by the final state wave function, and not a cancellation between terms in the perturbation.

It is important to note that the experimental results for DC are about three orders of magnitude smaller than those for SC. Also, while the 4BTTE model does a reasonable job of predicting the magnitude of the SC and

TTE results, the 4BDC model overestimates experiment by two orders of magnitude. The reason for this magnitude discrepancy between theory and experiment is not yet understood. However, Belkić[23] has shown that for integrated cross sections, it is important to include the full Coulomb interactions between all the particles both in the initial and final states; it is not adequate to use asymptotic forms of these interactions, such as the EIS. Consequently, it is possible that the magnitude problem may be related to the fact that not all of the full Coulomb interactions have been included in the initial and final states of the models presented here.

Schulz *et al.*[22] have found some structure in the experimental results for the ratio of DC to SC, and they suggested that this structure represents important physical effects that can only be seen in the ratios. Harris *et al.*[37] have presented corresponding calculations using the 4BDC model. These results are shown in Fig. 13, which presents the ratio of DC to SC differential cross sections as a function of scattering angle.

The structure in the DC to SC ratio is partly predicted by theory, and the features can be traced to the absolute DC or SC FDCS. For the ratio, there is little similarity between theory and experiment when the projectile-nuclear term is included in the perturbation. This is a consequence of the unphysical minima in the theoretical SC and DC cross sections, which

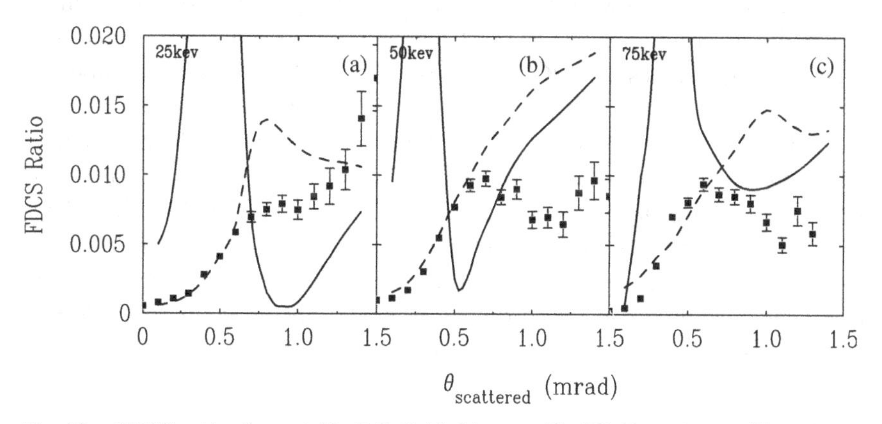

Fig. 13. FDCS ratios for p + He DC divided by p + He SC. Experiment: ■ results of Schulz *et al.*[22] for the incident projectile energies shown in the figure. The theoretical curves are from the 4BTTE[3] and 4BDC models[37] with a plane wave for the incident projectile, Hylleraas wave functions for the helium atom and H^- ion, and a Coulomb wave for the scattered projectile. Theoretical results: — all three terms in the perturbation divided by (a) 100, (b) 35, (c) 25; - - - without the projectile-nuclear term in the perturbation divided by 10 in all three panels.

are not observed in the experimental measurements. When the projectile-nuclear term is excluded from the perturbation, there is no minimum in either the SC or DC FDCS, and the shape of the theoretical results is similar to experiment. In the theoretical DC to SC ratios, the structure can be traced to a change in the slope of the absolute FDCS. Recall that these changes in slope have been attributed to a transition from electronic scattering to nuclear scattering. Consequently, it is quite likely that the structure observed in the experimental ratio data is an indication of this transition.

5. Conclusion

In conclusion, a 4-body model for charge transfer collisions has been presented here, and applied to the FDCS for SC, TTE, and DC. This model uses a fully numerical method that makes it possible to put in and take out different interactions within the same model. Results from this model correctly predict the order of magnitude of the FDCS for SC and TTE, but not for DC when all three terms are included in the perturbation. For SC, the theory predicts a minimum as a function of scattering angle that is not found in the experiment, and qualitative shape agreement (but not absolute value) when the projectile-nuclear term is omitted. For SC, the best agreement between the absolute experiment and theory was found for small scattering angle, while for TTE the best agreement was found for large scattering angles. For DC, the shapes of the FDCS were very similar to those of SC, with an unphysical minimum as a function of scattering angle when all three terms are included in the perturbation and qualitative shape agreement (but not absolute magnitude) when the projectile-nuclear term is omitted. Surprisingly, it was found that the electron-electron correlation in either the initial or final bound state had little effect on the results. There is obviously a need for an improved theoretical treatment, and the most logical candidate is a theoretical model that includes all the Coulomb interactions fully in both the initial and final channels.

Acknowledgments

This Research was supported by the National Science Foundation through Grant No. PHY-1068237 and the Extreme Science and Engineering Discovery Environment (XSEDE) resources provided by the Texas Advancement Computing Center (Grant No.TG-MCA07S029).

References

1. B. H. Bransden, *Atomic Collision Theory*, 2nd edn. The Benjamin/Cummings Publishing Co., Ing., (1983).
2. C. J. Joachain, *Quantum Collision Theory*, New York: North-Holland, (1975).
3. A. L Harris, J. L. Peacher, D. H. Madison, and J. Colgan, *Phys. Rev.* A **80**, 062707-1–062707-6 (2009).
4. M. R. C. McDowell and J. P. Coleman, *Introduction to the Theory of Ion-Atom Collisions*, North-Holland, (1970).
5. D. S. F. Crothers and J. F. McCann *J. Phys. B: At. Mol. Opt. Phys* **16**, 3229–3242 (1983).
6. F. W. Byron and C. J. Joachain *Phys. Rev.* **146**, 1–8 (1966).
7. J. F. Hart and G. Herzberg, *Phys. Rev.* **106**, 79–82 (1957).
8. B. H. Bransden and C. J. Joachain, *Physics of Atoms and Molecules*, 2nd edn. Prentice Hall, New York, (2003).
9. Dž. Belkić, R. Gayet, and A. Salin, *Phys. Rep.* **56**, 279–369 (1979).
10. B.H. Bransden and D.P. Dewangan, *Adv. Atomic Mol. Opt. Phys.* **25**, 343–374 (1988).
11. B. H. Bransden and M. R. C. McDowell, *Charge Exchange and the Theory of Ion-Atom Collisions*, The international series of monographs in physics Clarendon, Oxford, (1992).
12. D. S. F. Crothers and L. J. Dubé, *Adv. Atomic Mol. Opt. Phys.* **30**, 287–337 (1993).
13. D. P. Dewangan and J. Eichler, *Phys. Rep.* **247**, 59–219 (1994).
14. Dž. Belkić, *J. Comput. Meth. Sci. Eng.* **1**, 1 (2001).
15. Dž. Belkić, *Principles of Quantum Scattering Theory*, Institute of Physics Publishing, Bristol, (2003).
16. Dž. Belkić, *Quantum Theory of High-Energy Ion-Atom Collisions*, Taylor and Francis, London, (2008).
17. J. R. Oppenheimer, *Phys. Rev.* **31**, 349–356 (1928).
18. H. C. Brinkman and H. A. Kramers, *Proc. Acad. Sci. Amsterdam* **33**, 973–984 (1930).
19. J. D. Jackson and H. Schiff, *Phys. Rev.* **89**, 359–365 (1953).
20. T. F. Tuan and E. Gerjuoy, *Phys. Rev.* **117**, 756–763 (1960).
21. Dž. Belkić and A. Salin, *J. Phys. B: At. Mol. Opt. Phys.* **11**, 3905–3911 (1978).
22. M. Schulz, T. Vajnai, and J. A. Brand, *Phys. Rev.* A **75**, 022717-1–022717-6 (2007).
23. Dž. Belkić, *J. Math. Chem.* **47**, 1420–1467 (2010).
24. J. C. Y. Chen and P. J. Kramer, *Phys. Rev.* A **5**, 1207–1217 (1972).
25. Y. B. Band, *Phys. Rev.* A **8**, 2857–2865 (1973).
26. N. C. Sil, B. C. Saha, H. P. Saha, and P. Mandal, *Phys. Rev.* A **19**, 655–674 (1979).
27. K. Omidvar, *Phys. Rev.* A **12**, 911–926 (1975).
28. A. Hasan, B. Tooke, M. Zapukhlyak, T. Kirchner, and M. Schulz, *Phys. Rev.* A **74**, 032703-1–032703-5 (2006).

29. M. Zapukhlyak, T. Kirchner, A. Hasan, B. Tooke, and M. Schulz, *Phys. Rev.* A **77**, 012720-1–012720-9 (2008).
30. M. S. Schöffler, Ph.D. thesis, University of Frankfurt am Main, 2006 (unpublished).
31. Dž. Belkić, *Physica Scripta* **47**, 18–23 (1993).
32. D. S. F Crothers and R. McCarroll, *J. Phys. B: At. Mol. Opt. Phys.* **20**, 2835 (1987).
33. Dž. Belkić, *Phys. Rev.* A **47**, 189–200 (1993).
34. R. Schuch, E. Justiniano, H. Vogt, G. Deco, and N. Gruen, *J. Phys. B: At. Mol. Opt. Phys.* **24**, L133–L138 (1991).
35. K. M. Dunseath and D. S. F. Crothers, *J. Phys. B: At. Mol. Opt. Phys.* **24**, 5003–5022 (1991).
36. S. Chandrasekhar, *Rev. Mod. Phys.* **16**, 301–306 (1944).
37. A. L. Harris, J. L. Peacher, and D. H. Madison, *Phys. Rev.* A, **82**, 022714-1–022714-7 (2010).

Chapter 5

Distorted Wave Methodologies for Energetic Ion-Atom Collisions

Sharif D. Kunikeev
Department of Chemistry, University of Rhode Island,
Kingston, Rhode Island, 02881, USA
skunikeev@chm.uri.edu

The progress in developing distorted wave methods for heavy ion high-energy collisions (ionization, charge transfer) is reviewed. Special attention is devoted to a proper formulation of the Coulomb boundary conditions for the scattering states in the system of three charged particles. The asymptotic behavior of the existing analytical solutions for the three-body scattering states is analyzed. Our primary focus in the second part of the presentation is on a detailed account of scattering effects into the continuum revealed in differential cross-sections in both direct and resonant ionization of atoms by fast ion impact, as well as in processes of charge transfer.

1. Introduction

The fundamental elementary processes accompanying ion-atom collisions, such as direct and resonance ionization as well as charge transfer, are characterized by the presence of two or more particles interacting into the continuum in the entrance and exit channels of the corresponding reactions. Scattering considered as a classical or quantum phenomenon usually involves a comparison between the full dynamics of interacting particles and a "free" dynamics of asymptotically free particles. In a time-dependent quantum-mechanical formulation, it involves studying time evolution of certain scattering states of an interacting system, namely those states that appear to be "asymptotically free" in the distant past and/or future. While in the energy-dependent formulation, one deals with the scattering state solutions of the stationary Schrödinger equation for the system of interacting particles into the continuum that satisfy appropriate boundary conditions. The connection between these two formulations is implemented by the Möller operators[1] which are the strong limits of a product of the full and free time evolution operators.

From a theoretical point of view, to fully describe the above processes one needs first to find the scattering states by solving the Schrödinger equation for the system of several active particles interacting into the continuum, then use these states to calculate the amplitudes and the differential cross sections for the processes under study. Thus, in the case of two charged particles into continuum, the exact two-body Coulomb scattering state is well known analytically and from its asymptotic form one can easily derive the two-body Coulomb scattering amplitude and the Rutherford differential cross section. For three or more charged particles into the continuum an exact scattering state is not known and approximate, perturbative or numerical methods should be applied in order to derive an appropriate solution.

The aim of the present work is to review those analytic methods that are based on the known two-body Coulombic solutions and their possible extensions to three-particle continuum states. First, we present a detailed account of the exact plane and spherical wave solutions for two charged particles into continuum together with the corresponding asymptotic solutions obtained within the high-energy Wentzel-Kramers-Brillouin (WKB) and eikonal approximations. Then, using the WKB and eikonal representations appropriate three-particle scattering states with the local momenta are developed. Here, the concept of three-particle local momenta, which depend on coordinates and momenta of all three particles, turns out to be useful in order to incorporate effects of dynamic correlations due to the third particle into the continuum state of a given two-particle subsystem. The origin of the concept of a local momentum can be traced back to the quasi-classical description of interacting particle's dynamics where the momentum of an interacting particle differs from its asymptotic value when a particle moves along a free-motion asymptote. The continuum distorted wave (CDW) states with the local momenta are shown to obey the proper Coulomb boundary conditions and satisfy the Schrödinger equation up to the second order terms at large interparticle separations. Further, making use of the idea of analytic continuation we present a possible generalization of the CDW function with the local momenta to the case when two particles are bound and the third one is into continuum. If local momenta are replaced by asymptotic ones, then the CDW functions are represented by a product of independent two-body Coulomb distortion factors. A brief note on the CDW functions with asymptotic momenta is given in Ref. 2. Note another line of refinements of the CDW functions advocated in Refs. 3, 4, where the products of charges in the Sömmerfeld parameters have been replaced by effective position- or momentum-dependent ones.

In the second half of the paper, the CDW functions with the local momenta are applied to calculate amplitudes and differential cross sections for the direct and resonance ionization processes in Sections 4 and 5. Our primary focus here is to demonstrate a relationship between the features observed in the electron ionization spectra and the corresponding contributions coming from the plane and spherical wave terms present in the CDW functions.

The bound state CDW function with the local momenta has not been tried in applications. In Section 6, we present some results of charge exchange calculations obtained with the simplified CDW function where local, coordinate dependent momenta are replaced by the asymptotic ones. In particular, using the continuum intermediate state (CIS) method, we demonstrate for the first time that the Thomas peak in differential cross sections for electron transfer in fast ion-atom collisions is due to an interference effect between plane and spherical wave contributions to the amplitude. Throughout this chapter only hybrid-type distorted wave methods are employed with the CDW function for one scattering state alone and the first Born approximation for the other state, as in the CIS approximation.

Atomic units are used throughout the paper unless otherwise stated.

2. Two-Body Coulomb Scattering

In this section we present a detailed account of the two-body Coulomb scattering states: the known exact solutions and basic asymptotic methods, such as WKB and eikonal ones, which are useful tools in developing appropriate high-energy approximations in three- or few-body scattering problems.

2.1. *Quantum-Mechanical Plane and Spherical Waves*

Let us consider a system of two non-relativistic quantum particles with charges Z_1 and Z_2 and masses m_1 and m_2. The Hamiltonian that describes particle's relative motion in the center of mass has the form

$$H = \hat{K} + V_c = -\frac{1}{2m_{12}}\nabla_{\vec{r}}^2 + \frac{Z_1 Z_2}{r}, \tag{1}$$

where $m_{12} = m_1 m_2/(m_1 + m_2)$ is particle's reduced mass and $\vec{r} = (x, y, z)$ their relative coordinates. In the parabolic coordinates

$$\begin{cases} \kappa = r + z, \\ \eta = r - z, \\ \varphi = \arctan\left(\dfrac{y}{x}\right), \end{cases} \tag{2}$$

the Schrödinger equation takes the form

$$\left\{ \frac{4}{\kappa + \eta} \left[\frac{\partial}{\partial \kappa}\left(\kappa \frac{\partial}{\partial \kappa}\right) + \frac{\partial}{\partial \eta}\left(\eta \frac{\partial}{\partial \eta}\right) \right] \right.$$
$$\left. + \frac{1}{\kappa\eta}\frac{\partial^2}{\partial \varphi^2} + k^2 - \frac{4Z_1 Z_2 m_{12}}{\kappa + \eta} \right\} \psi^{(\pm)}(\vec{k}, \vec{r}) = 0. \tag{3}$$

Here, $\vec{k}$ is the asymptotic momentum and z-axis is taken to be parallel to $\vec{k}$. The *post* or *prior* forms of the solution, $\psi^{(-)}$ or $\psi^{(+)}$, contain, respectively, *incoming* or *outgoing* spherical waves at asymptotically large distances. There is a relationship between the *post* and *prior* forms:

$$\psi^{(-)}(\vec{k}, \vec{r}) = [\psi^{(+)}(-\vec{k}, \vec{r})]^*. \tag{4}$$

Substituting the ansatz

$$\psi^{(+)}(\vec{k}, \vec{r}) = \exp(\mathrm{i}k(\kappa - \eta)/2) F(\eta) \tag{5}$$

into equation (3), one gets the confluent hypergeometric equation for the distortion factor F:

$$z\frac{\mathrm{d}^2 F}{\mathrm{d}z^2} + (\gamma - z)\frac{\mathrm{d}F}{\mathrm{d}z} - \alpha F = 0, \tag{6}$$

where parameters take the values

$$z = \mathrm{i}\zeta, \quad \zeta = k\eta, \quad \alpha = -\mathrm{i}\nu,$$
$$\nu = \frac{Z_1 Z_2 m_{12}}{k}, \quad \gamma = 1, \tag{7}$$

The confluent hypergeometric function

$$F(\nu, \zeta) = f_c^{(+)}(\nu)\, {}_1F_1(-\mathrm{i}\nu, 1, \mathrm{i}\zeta),$$
$$f_c^{(\pm)}(\nu) = \exp(-\pi\nu/2)\, \Gamma(1 \pm \mathrm{i}\nu), \tag{8}$$

is a well-known solution to equation (6):[5]

$$
{}_1F_1(\alpha,\gamma,z) = \frac{\Gamma(\gamma)}{\Gamma(\gamma-\alpha)\Gamma(\alpha)} \int_0^1 dt\ \exp(zt) t^{\alpha-1}(1-t)^{\gamma-\alpha-1},
$$
$$
\operatorname{Re}\gamma > \operatorname{Re}\alpha > 0 \tag{9}
$$

that is regular at origin. In (7) and (8), we have introduced the Coulomb interaction (Sömmerfeld) parameter ν and the Coulomb normalization factor $f_c^{(+)}$. Using Taylor's expansion for the exponential in (9), one gets a series expansion in ascending powers of z

$$
{}_1F_1(\alpha,\gamma,z) = \sum_{n=0}^{\infty} \frac{(\alpha)_n\, z^n}{(\gamma)_n n!}.
$$
$$
(\alpha)_0 = 1, \quad (\alpha)_n = \alpha \cdot \cdots \cdot (\alpha+n-1), \quad n = 1,2,\ldots \tag{10}
$$

The regular solution can be expressed in terms of two irregular ones[6]

$$
\begin{aligned}
{}_1F_1(\alpha,\gamma,z) &= \frac{\Gamma(\gamma)}{\Gamma(\gamma-\alpha)}(-z)^{-\alpha} G(\alpha,\alpha-\gamma+1,-z) \\
&\quad + \frac{\Gamma(\gamma)}{\Gamma(\alpha)} \exp(z) z^{\alpha-\gamma} G(\gamma-\alpha,1-\alpha,z),
\end{aligned} \tag{11}
$$

where

$$
G(\alpha,\beta,z) = \frac{\Gamma(1-\beta)}{2\pi \mathrm{i}} \int_C dt \exp(t) \left(1+\frac{t}{z}\right)^{-\alpha} t^{\beta-1}. \tag{12}
$$

The contour C in the complex plane starts from $\operatorname{Re} t = -\infty$ and the lower branch of C goes below the real axis to $\operatorname{Re} t = 0$, then it passes round $t = 0$ and the upper branch goes to $\operatorname{Re} t = -\infty$ above the real axis. In raising $-z$ and z to powers in (11) we must take the arguments which have the smallest absolute value. From (12) we have for $G(\alpha,\beta,z)$ the asymptotic series

$$
G(\alpha,\beta,z) = 1 + \frac{\alpha\beta}{1!z} + \frac{\alpha(\alpha+1)\beta(\beta+1)}{2!z^2} + \cdots. \tag{13}
$$

Making use of expansion (11), the distortion function can be split into the two parts:

$$
F(\nu,\zeta) = F_p(\nu,\zeta) + F_s(\nu,\zeta), \tag{14}
$$

where the plane- (p) and spherical-wave (s) parts have the form

$$F_p(\nu,\zeta) = \exp(\mathrm{i}\nu\ln\zeta)G(-\mathrm{i}\nu,-\mathrm{i}\nu,-\mathrm{i}\zeta), \tag{15}$$

$$F_s(\nu,\zeta) = -\frac{\nu}{\zeta}\frac{\Gamma(1+\mathrm{i}\nu)}{\Gamma(1-\mathrm{i}\nu)}\exp(\mathrm{i}\zeta-\mathrm{i}\nu\ln\zeta)G(1+\mathrm{i}\nu,1+\mathrm{i}\nu,\mathrm{i}\zeta). \tag{16}$$

All three functions in (14) are exact solutions to equation (6), two of them being linearly independent. By virtue of (13), it is easy to check that F_p and F_s behave asymptotically as distorted plane and spherical waves at $\zeta\to\infty$.

For non-integral γ, equation (6) has also the particular solution $z^{1-\gamma}$ ${}_1F_1(\alpha-\gamma+1,2-\gamma,z)$, which is linearly independent of (9). Using expansion (11), we get

$$\begin{aligned} z^{1-\gamma}{}_1{}_1F_1(\alpha-\gamma+1,2-\gamma,z) = &-(-1)^{\gamma}\frac{\Gamma(2-\gamma)}{\Gamma(1-\alpha)}(-z)^{-\alpha} \\ &\times G(\alpha,\alpha-\gamma+1,-z) \\ &+\frac{\Gamma(2-\gamma)}{\Gamma(\alpha-\gamma+1)}\exp(z)z^{\alpha-\gamma} \\ &\times G(\gamma-\alpha,1-\alpha,z). \end{aligned} \tag{17}$$

From equations (11) and (17) we express $G(\alpha,\alpha-\gamma+1,-z)$ in terms of ${}_1F_1(\alpha,\gamma,z)$ and ${}_1F_1(\alpha-\gamma+1,2-\gamma,z)$; we then put $\gamma=1+\varepsilon$, and pass to the limit $\varepsilon\to 0$, resolving the indeterminacy by L'Hospital's rule. It gives the following expansion in ascending powers of z:

$$\begin{aligned} G(\alpha,\alpha,-z) = -\frac{z^{\alpha}}{\Gamma(\alpha)}\Bigg\{&\ln(z){}_1F_1(\alpha,1,z) \\ &+\sum_{n=0}^{\infty}\frac{[\psi(\alpha+n)-2\psi(n+1)](\alpha)_n\, z^n}{(n!)^2}\Bigg\}, \end{aligned} \tag{18}$$

where ψ denotes the logarithmic derivative of the gamma function: $\psi(\alpha) = \Gamma'(\alpha)/\Gamma(\alpha)$.

Using this expansion, we derive the following expressions for the plane

$$\begin{aligned} F_p(\nu,\zeta) = -f_c^{(+)}(\nu)\frac{\exp(2\pi\nu)-1}{2\pi\mathrm{i}}\Bigg\{&\left[\frac{\pi}{2}\mathrm{i}+\ln\zeta\right]{}_1F_1(-\mathrm{i}\nu,1,\mathrm{i}\zeta) \\ &+\sum_{n=0}^{\infty}[\psi(n-\mathrm{i}\nu)-2\psi(n+1)]\frac{(-\mathrm{i}\nu)_n(\mathrm{i}\zeta)^n}{(n!)^2}\Bigg\}, \end{aligned} \tag{19}$$

and spherical waves

$$F_s(\nu,\zeta) = f_c^{(+)}(\nu)\frac{\exp(2\pi\nu)-1}{2\pi \mathrm{i}}\exp(\mathrm{i}\zeta)\left\{\left[-\frac{\pi}{2}\mathrm{i}+\ln\zeta\right]{}_1F_1(1+\mathrm{i}\nu,1,-\mathrm{i}\zeta)\right.$$
$$\left.+\sum_{n=0}^{\infty}[\psi(n+1+\mathrm{i}\nu)-2\psi(n+1)]\frac{(1+\mathrm{i}\nu)_n(-\mathrm{i}\zeta)^n}{(n!)^2}\right\}. \tag{20}$$

In the limit $\zeta \to 0$, we observe weak logarithmic divergencies

$$F_p(\nu,\zeta)|_{\zeta\to 0} \sim -f_c^{(+)}(\nu)\frac{\exp(2\pi\nu)-1}{2\pi \mathrm{i}}$$
$$\times\left\{\frac{\pi}{2}\mathrm{i}+\ln\zeta+\psi(-\mathrm{i}\nu)-2\psi(1)\right\}, \tag{21}$$

$$F_s(\nu,\zeta)|_{\zeta\to 0} \sim f_c^{(+)}(\nu)\frac{\exp(2\pi\nu)-1}{2\pi \mathrm{i}}$$
$$\times\left\{-\frac{\pi}{2}\mathrm{i}+\ln\zeta+\psi(1+\mathrm{i}\nu)-2\psi(1)\right\}. \tag{22}$$

However, their sum is seen to be free of divergencies

$$F(\nu,\zeta)|_{\zeta\to 0} \sim f_c^{(+)}(\nu)\frac{\exp(2\pi\nu)-1}{2\pi \mathrm{i}}\{\psi(1+\mathrm{i}\nu)-\psi(-\mathrm{i}\nu)-\pi \mathrm{i}\} = f_c^{(+)}(\nu), \tag{23}$$

where a reflection formula[5]

$$\psi(\alpha)-\psi(1-\alpha) = -\pi\cot(\pi\alpha) \tag{24}$$

has been used.

In Figs. 1 and 2 we have plotted the probability distributions $P_{cw}^{(p+s)} = |F(\nu,\zeta)|^2$ of a particle in the Coulomb wave that is propagated, respectively, in the attractive ($\nu = -1$) and repulsive ($\nu = 1$) Coulomb fields. We also show separate contributions from the distorted plane (p), $P_{cw}^{(p)} = |F_p(\nu,\zeta)|^2$, and the distorted spherical (s) waves, $P_{cw}^{(s)} = |F_s(\nu,\zeta)|^2$. Using the asymptotic expansion (13), we obtain up to the second order terms the following asymptotic for $P_{cw}^{(p)} \sim 1+\nu^2/\zeta^2 \to 1$ and $P_{cw}^{(s)} \sim \nu^2/\zeta^2 \to 0$ and the plane wave is seen to become dominant over the spherical one as $\zeta \to \infty$. Observe that independent of the sign of ν, we have $P_{cw}^{(p)} > 1$ at $\zeta \gg 1$. Although equations (19) and (20) that define plane and spherical wave contributions look more complicated than its coherent sum (8), behaviors demonstrated by the plane and spherical wave contributions alone look simpler. Namely, they are monotonic as compared to the oscillatory structure revealed in the total Coulomb wave. Evidently, this structure

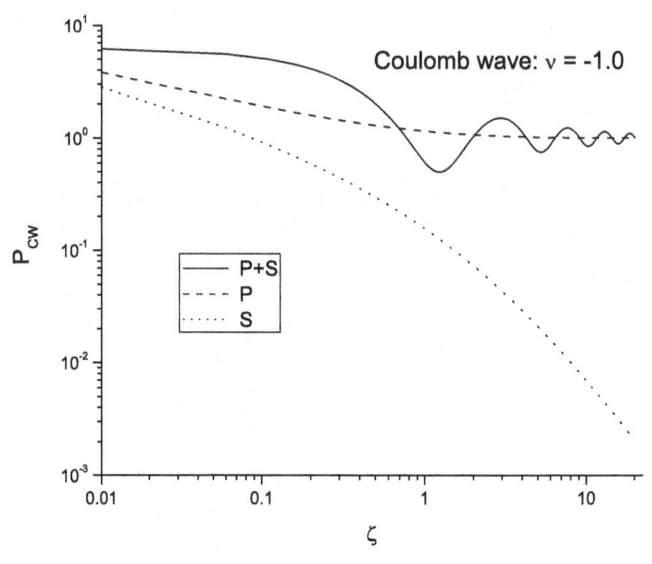

Fig. 1. The QM probability distribution P_{CW} of a particle in the Coulomb wave (solid line) is plotted as a function of $\zeta = kr - \vec{k} \cdot \vec{r}$ variable. The Sömmerfeld's Coulomb interaction parameter $\nu = -1$. Also, shown are separate contributions from the plane (dashed) and spherical (dotted line) wave parts.

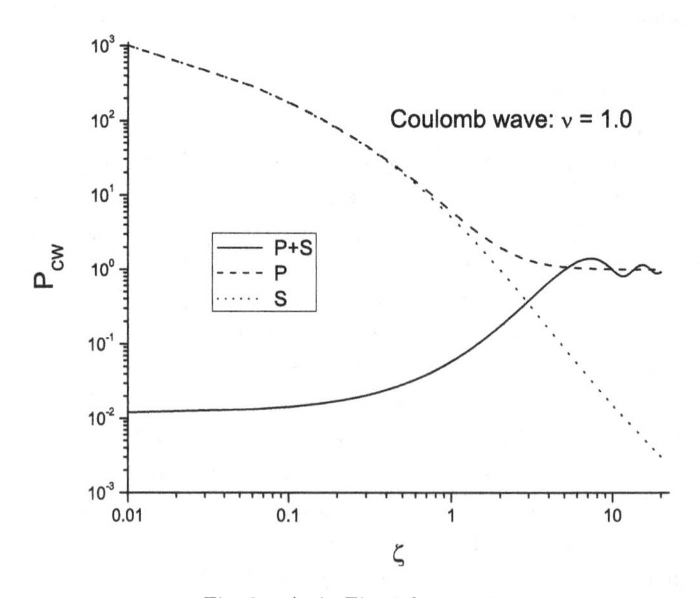

Fig. 2. As in Fig. 1 for $\nu = 1$.

is due to an interference effect between the plane and spherical waves. At intermediate and small values of $\zeta \lesssim 1$, the plane and spherical wave contributions are seen to be comparable in magnitude. Notice a counterintuitive behavior of $P_{cw}^{(p)}$ and $P_{cw}^{(s)}$ in the repulsive Coulomb field (Fig. 2) at $\zeta \lesssim 1$. As $\zeta \to 0$, both $P_{cw}^{(p)}$ and $P_{cw}^{(s)}$ show an exponential growth, while their coherent sum $P_{cw}^{(p+s)}$ as a result of destructive interference decays to an exponentially small value:

$$P_{cw}^{(p+s)}|_{\zeta\to 0} \sim |f_c^{(+)}(\nu)|^2 = \frac{2\pi\nu}{\exp(2\pi\nu) - 1}. \tag{25}$$

Notice that it is also possible to derive the corresponding analytic expressions in terms of the hypergeometric functions ${}_2F_1$[5] for the plane- and spherical waves in the momentum representation: $\psi_{p,s}^{(\pm)}(\vec{p}, \vec{k}\,) = \langle \exp(i\vec{p}\cdot \vec{r})|\exp(i\vec{k}\cdot\vec{r})F_{p,s}^{(\pm)}(\vec{k}, \vec{r})\rangle$. These expressions can be readily obtained from matrix elements of the type equation (204), which will be evaluated in Section 6. We do not show these expressions here since in the following we will be working in the coordinate representation.

2.2. *The WKB and Eikonal Representations*

In deriving the corresponding WKB representations, it is convenient to rewrite the Schrödinger equation keeping explicitly Planck's constant $\hbar$. We are looking for an asymptotic solution of the form

$$\psi = \exp\left(\frac{S}{\varepsilon}\right) \sum_{n=0}^{\infty} a_n \varepsilon^n \tag{26}$$

where we have introduced the formal asymptotic parameter $\varepsilon = -i\hbar$. In the end of calculation, we have to put $\hbar = 1$, i.e., $\varepsilon = -i$.

After substituting (26) into the Schrödinger equation, we get the system of equations

$$\frac{(\nabla S)^2}{2m_{12}} + V_c = E, \tag{27}$$

$$2(\nabla S)\cdot\nabla a_0 + (\nabla^2 S)a_0 = 0, \tag{28}$$

$$2(\nabla S)\cdot\nabla a_k + (\nabla^2 S)a_k = -\nabla^2 a_{k-1}, \quad k = 1, 2, \ldots \tag{29}$$

where the total energy $E = k^2/(2m_{12})$, with $\vec{k}$ being the asymptotic momentum. The first equation for the phase or action function S is called the Hamilton-Jacobi equation, whereas the second, homogeneous one is for the zeroth order amplitude a_0. It is called the continuity equation. The higher order amplitudes a_k, $k = 1, 2, \ldots$ are derived from the inhomogeneous recurrence equations in the third line.

In parabolic coordinates, equation (27) has the form

$$\frac{4\kappa}{\kappa+\eta}\left(\frac{\partial S}{\partial \kappa}\right)^2 + \frac{4\eta}{\kappa+\eta}\left(\frac{\partial S}{\partial \eta}\right)^2 + \frac{1}{\kappa\eta}\left(\frac{\partial S}{\partial \varphi}\right)^2 = k^2 - \frac{2Z_1Z_2m_{12}}{\kappa+\eta}. \quad (30)$$

As an ansatz, we take the function

$$S = \frac{k(\kappa-\eta)}{2} + \Phi(\eta), \quad (31)$$

where the first term is the plane-wave phase and Φ is the phase due to interaction. Substituting this ansatz, we get a quadratic equation with respect to the derivative $\mathrm{d}\Phi/\mathrm{d}\zeta$:

$$\left(\frac{\mathrm{d}\Phi}{\mathrm{d}\zeta}\right)^2 - \frac{\mathrm{d}\Phi}{\mathrm{d}\zeta} = -\frac{\nu}{\zeta}. \quad (32)$$

The quadratic equation has two solutions

$$\frac{\mathrm{d}\Phi_\pm}{\mathrm{d}\zeta} = \frac{1 \pm w}{2},$$
$$w = \sqrt{1 - \frac{4\nu}{\zeta}}. \quad (33)$$

Integrating this equation, we get

$$\Phi_\pm = \frac{\zeta}{2}(1 \pm w) \pm \nu \ln\frac{w-1}{w+1} + \varphi_{0\pm}, \quad (34)$$

where $\varphi_{0\pm}$ is the integration constant.

Assuming a_k's are functions of η, equations (28) and (29) can be rewritten as

$$a'_{0\pm} + p_\pm a_{0\pm} = 0, \quad (35)$$

$$a'_{k\pm} + p_\pm a_{k\pm} = -\frac{1}{2\varkappa_\pm}(a'_{(k-1)\pm} + \eta a''_{(k-1)\pm}), \quad k = 1, 2, \ldots \quad (36)$$

where

$$\varkappa_\pm = \eta S'_\pm = \pm\frac{k\eta}{2}w,$$
$$p_\pm = \frac{k}{4\varkappa_\pm} + \frac{1}{2}(\ln\varkappa_\pm)', \quad (37)$$

and primes denote differentiation with respect to η: $f' = df/d\eta$, $f'' = d^2f/d\eta^2$. Integrating the homogeneous equation, we get

$$a_{0\pm} = C_{0\pm} \exp\left(-\int \mathrm{d}\eta\, p\right) = C_{0\pm}\left[\frac{2}{\pm\zeta w}\left(\frac{w-1}{w+1}\right)^{\pm 1}\right]^{1/2}, \tag{38}$$

where $C_{0\pm}$ is the integration constant. Taking into account the asymptotic expansion

$$w|_{\zeta\to\infty} \sim 1 - \frac{2\nu}{\zeta}, \tag{39}$$

we find that the $(+)$ and $(-)$ solutions correspond, respectively, to the spherical and plane-wave QM solutions (16) and (15). By comparing their asymptotic, one can easily find integration constants that give the WKB solutions with correct asymptotic behavior:

$$C_{0\pm}\exp(\mathrm{i}\varphi_{0\pm}) = \left(\mp\frac{\nu}{2}\right)^{1/2} \exp\left(\pm\mathrm{i}\nu(1-\ln(-\nu))\right) \begin{cases} \exp(2\mathrm{i}\arg\Gamma(1+\mathrm{i}\nu)) \\ 1 \end{cases}. \tag{40}$$

Similarly, we can solve equations for the higher order amplitudes by using the method of constant variation. For example, for the first order amplitude we get solutions of the form

$$a_{1\pm} = -\exp\left(-\int \mathrm{d}\eta\, p_\pm\right)\int \mathrm{d}\eta \exp\left(\int \mathrm{d}\eta\, p_\pm\right)\frac{a'_{0\pm} + \eta a''_{0\pm}}{2\varkappa_\pm} + C_{1\pm}a_{0\pm}, \tag{41}$$

where the first term is a particular solution, whereas the second one is an arbitrary solution of the homogeneous equation, with $C_{1\pm}$ being an arbitrary constant. Making use of equation (35), the first term can be reduced to an integral such that $a_{1\pm}$ reads as

$$a_{1\pm} = a_{0\pm}I_{1\pm}, \tag{42}$$

where

$$I_{1\pm} = J_{1\pm} + C_{1\pm}, \tag{43}$$

$$J_{1\pm} = \frac{1}{2}\int \frac{\mathrm{d}\eta}{\varkappa_\pm}(p_\pm + \eta p'_\pm - \eta p_\pm^2). \tag{44}$$

The zeroth-order WKB approximation is legitimate only if the first order term is small compared with the zeroth-order one: $|a_{1\pm}| \ll |a_{0\pm}|$. From (42) we get a quantitative *quasi-classicality* condition:

$$|I_{1\pm}| \ll 1. \tag{45}$$

By changing integration variable η to w:

$$\begin{aligned} \eta &= -\frac{4\nu}{k}\frac{1}{w^2-1}, \\ \mathrm{d}\eta &= \frac{4\nu}{k}\frac{2w\mathrm{d}w}{(w^2-1)^2}, \end{aligned} \tag{46}$$

the integral is evaluated to

$$J_{1\pm}(w) = -\frac{1}{8\nu}\begin{cases} \ln w + w + \dfrac{3}{4w} - \dfrac{3}{4w^2} - \dfrac{5}{12w^3}, \\ 4\ln(w+1) - 3\ln w - w \\ \quad -\dfrac{3}{4w} - \dfrac{4}{w+1} - \dfrac{3}{4w^2} + \dfrac{5}{12w^3}. \end{cases} \tag{47}$$

Let us choose the constant $C_{1\pm} = -J_{1\pm}(1)$ and expand equation (47) in Taylor's series around $w = 1$:

$$J_{1\pm}(w) = J_{1\pm}(1) + \frac{\mathrm{d}J_{1\pm}(1)}{\mathrm{d}w}(w-1) + \frac{1}{2}\frac{\mathrm{d}^2 J_{1\pm}(1)}{\mathrm{d}w^2}(w-1)^2 + \cdots. \tag{48}$$

Further, calculating derivatives of (47), we get

$$\begin{aligned} \frac{\mathrm{d}J_{1+}(1)}{\mathrm{d}w} &= -\frac{1}{2\nu}, \\ \frac{\mathrm{d}J_{1-}(1)}{\mathrm{d}w} &= \frac{\mathrm{d}^2 J_{1-}(1)}{\mathrm{d}w^2} = 0, \\ \frac{\mathrm{d}^3 J_{1-}(1)}{\mathrm{d}w^3} &= \frac{3}{4\nu}. \end{aligned} \tag{49}$$

Finally, from (39), (43), (48) and (49) we obtain asymptotic estimates for

$$I_{1+} \sim \frac{1}{\zeta}, \quad I_{1-} \sim -\frac{4\nu^2}{\zeta^3}, \tag{50}$$

and, hence, from (45) we can determine regions where the WKB approximation is expected to be valid.

Observe that the WKB solutions are singular at caustics points. In the *attractive* field $\nu < 0$, the inequality $w > 1$ is satisfied; $w \sim \zeta^{-1/2}$ and $a_{0\pm} \sim \zeta^{-1/4}$ are singular at caustics point $\zeta = 0$.

In the *repulsive* field $\nu > 0$, we get two caustics points: $\zeta_1 = 4\nu$ and $\zeta_2 = 0$. If $\zeta > \zeta_1$, we have inequalities $0 < w < 1$ valid and as $\zeta \to \zeta_1$, $w \to 0$ and $a_{0\pm} \sim (\zeta - \zeta_1)^{-1/4}$ is singular. In the interval $0 < \zeta < \zeta_1$, w takes imaginary values $w \to \pm \mathrm{i}|w|$ and $a_{0\pm} \sim \zeta^{-1/4}$ is singular at $\zeta \to 0$. We can analytically continue the WKB solutions found in the outer region $\zeta > \zeta_1$ to the inner, classically inaccessible region between the two caustics points $0 < \zeta < \zeta_1$ by making a substitution $w \to \pm \mathrm{i}|w|$. As a result of continuation, we get two types of the *complex* WKB solutions in the inner region

$$F_{\pm}^{\mathrm{WKB}} = a_{0\pm} \exp(\mathrm{i}\Phi_{\pm}) \to f_{\pm}^{WKB} = \left[\frac{1}{\zeta|w|}\right]^{1/2} \exp(\Omega_{\pm}), \tag{51}$$

where the complex phase is

$$\Omega_{\pm} = \mathrm{i}\frac{\zeta}{2} \pm \mathrm{i} \arctan|w| \pm \frac{\zeta}{2} \mp 2\nu \arctan|w|. \tag{52}$$

One solution, f_{+}^{WKB}, is seen to be exponentially small, whereas the other, f_{-}^{WKB}, is exponentially large. In the previous section, we have seen that the QM plane and spherical waves are exponentially large in the inner region but their coherent sum is exponentially small. We can get the WKB analogues of the plane (p) and spherical (s) waves in this region by taking appropriate linear combinations of (51):

$$F_p^{WKB} = c_+ f_+^{WKB} + c_- f_-^{WKB}, \tag{53}$$

$$F_s^{WKB} = c_+ f_+^{WKB} - c_- f_-^{WKB}, \tag{54}$$

$$F^{WKB} = F_p^{WKB} + F_s^{WKB} = 2c_+ f_+^{WKB}. \tag{55}$$

In Figs. 3 and 4 we compare the corresponding quasi-classical (WKB) and QM solutions, respectively, in the attractive ($\nu = -1$) and repulsive ($\nu = 1$) fields. The fitting coefficients in (53)–(55) are chosen to be $c_+ = 0.60$, $c_- = 0.84$. The asymptotic behavior of the WKB amplitudes at $\zeta \to \infty$

$$|a_{0p}^{WKB}|^2 \sim 1 + \frac{\nu^2}{\zeta^2}, \tag{56}$$

$$|a_{0s}^{WKB}|^2 \sim \frac{\nu^2}{\zeta^2}, \tag{57}$$

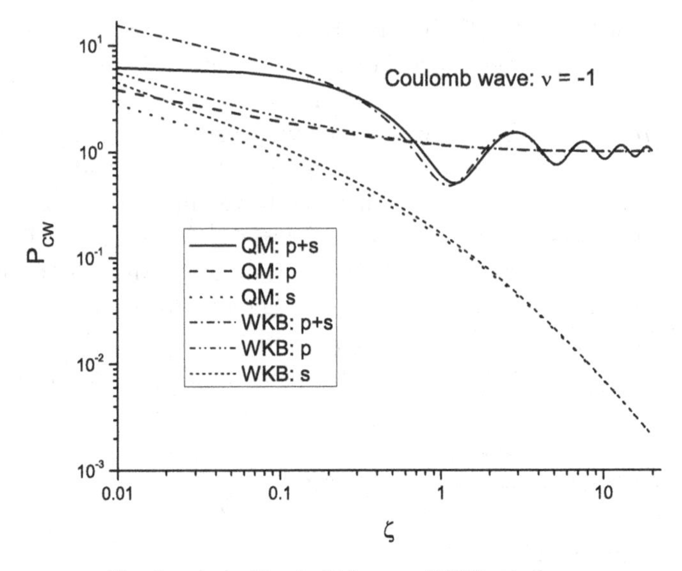

Fig. 3. As in Fig. 1: QM versus WKB solutions.

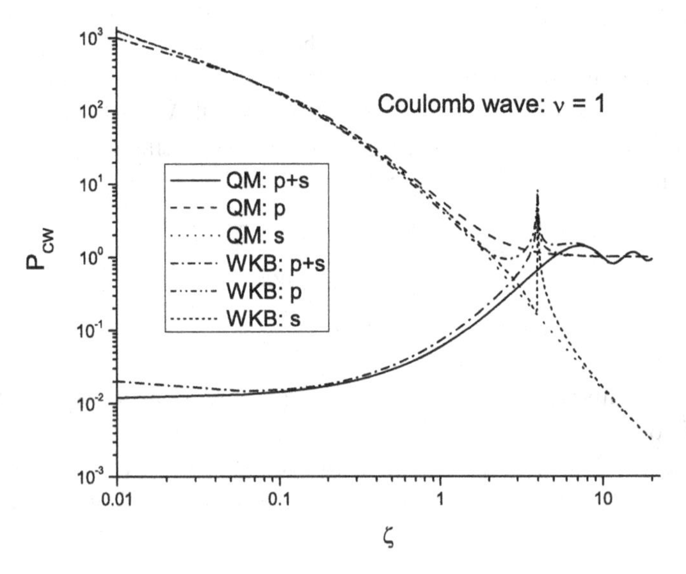

Fig. 4. As in Fig. 3 for $\nu = 1$.

is seen to be the same as the QM ones. The discrepancies between the WKB and exact QM results are most pronounced near the caustics points where the WKB approximation fails. Observe the qualitative difference between the WKB and QM results near the caustics point $\zeta = 4\nu$ in the repulsive case. The WKB solutions show a singular behavior, whereas the QM solutions behave smoothly in the region.

The existence of caustics points is intimately related to the (symplectic) geometry of classical integral trajectories in the phase space. The complete solutions[7,8]

$$\tilde{S}(\vec{r},\vec{k},t) = S(\vec{r},\vec{k}\,) - \frac{k^2}{2m_{12}}t \tag{58}$$

of the time-dependent Hamilton-Jacobi equation generate corresponding families of integral trajectories. By solving the ordinary differential equations

$$\dot{\vec{r}}(t) = \frac{1}{m_{12}}\nabla_{\vec{r}} S(\vec{r}(t),\vec{k}\,) \tag{59}$$

or inverting the algebraic equation with respect to $\vec{r}(t)$

$$\vec{q} + \frac{\vec{k}}{m_{12}}t = \nabla_{\vec{k}} S(\vec{r}(t),\vec{k}\,), \tag{60}$$

where $\vec{q}$ is a constant position vector, we get an integral trajectory $\vec{r}(\vec{q},\vec{k},t)$ as a function of time t and $(\vec{q},\vec{k}\,)$ parameters. For example, if we neglect interaction, $V \equiv 0$, the plane wave $S_0(\vec{r},\vec{k}) = \vec{k}\cdot\vec{r}$ is a complete solution. Then, from equation (60) we get rectilinear lines $\vec{r}(t) = \vec{q} + \frac{\vec{k}}{m_{12}}t$ as integral trajectories.

In general, according to implicit function theorem, equation (60) is solvable for $\vec{r}(t)$ at least locally under the condition $\det \frac{\partial^2 S}{\partial\vec{k}\partial\vec{r}} \neq 0, \infty$. Upon substitution $\vec{r}(\vec{q},\vec{k},t)$ into equation (60), it reduces to an identity. Let us differentiate both sides of this identity with respect to $\vec{q}$:

$$I = \frac{\partial^2 S(\vec{r}(\vec{q},\vec{k},t),\vec{k}\,)}{\partial\vec{k}\partial\vec{r}}\frac{\partial\vec{r}(\vec{q},\vec{k},t)}{\partial\vec{q}}, \tag{61}$$

where I is an identity matrix. To eliminate any ambiguities in above partial differential notations, it should be noted that in calculating the partial derivatives the arguments of S function are assumed to be $\vec{r}$ and $\vec{k}$. Only after the derivatives are taken should one put $\vec{r} = \vec{r}(\vec{q},\vec{k},t)$. In a more concise notation, it could be rewritten as

$$\frac{\partial^2 S(\vec{r}(\vec{q},\vec{k},t),\vec{k}\,)}{\partial\vec{k}\partial\vec{r}} = \left.\frac{\partial^2 S(\vec{r},\vec{k}\,)}{\partial\vec{k}\partial\vec{r}}\right|_{\vec{r}=\vec{r}(\vec{q},\vec{k},t)}. \tag{62}$$

From (61) we get

$$\frac{\partial\vec{r}(\vec{q},\vec{k},t)}{\partial\vec{q}} = \left(\frac{\partial^2 S(\vec{r}(\vec{q},\vec{k},t),\vec{k}\,)}{\partial\vec{k}\partial\vec{r}}\right)^{-1}. \tag{63}$$

Let us consider the Jacobian

$$J(\vec{q},\vec{k},t) = \det\frac{\partial\vec{r}(\vec{q},\vec{k},t)}{\partial\vec{q}}. \tag{64}$$

Evidently, the time derivative of J can be written as

$$\frac{\partial J}{\partial t} = \sum_{i=1}^{n} J_i, \tag{65}$$

where $n=3$ is the system dimensionality. The determinant J_i is obtained from J by replacing the entries of the i-th row with $\partial^2 r_i/\partial t\partial q_j$ $(j=1,\ldots,n)$. By virtue of (59)

$$\frac{\partial r_i(\vec{q},\vec{k},t)}{\partial t} = \frac{1}{m_{12}}\frac{\partial S(\vec{r}(\vec{q},\vec{k},t),\vec{k}\,)}{\partial r_i} \tag{66}$$

and, hence,

$$\frac{\partial^2 r_i}{\partial t\partial q_j} = \frac{1}{m_{12}}\sum_{k=1}^{n}\frac{\partial^2 S}{\partial r_i\partial r_k}\frac{\partial r_k}{\partial q_j}. \tag{67}$$

As a consequence, we have that only the i-th term in the sum of (67) is linearly independent of other rows in the matrix of determinant J_i. Therefore,

$$J_i = \frac{1}{m_{12}}\frac{\partial^2 S}{\partial r_i^2}J \tag{68}$$

and

$$\frac{\partial J}{\partial t} = \frac{1}{m_{12}}J\Delta S, \tag{69}$$

where

$$\Delta = \nabla^2 = \sum_{i=1}^{n} \frac{\partial^2}{r_i^2}. \tag{70}$$

For the inverse value J^{-1} we get equation

$$\frac{\partial J^{-1}}{\partial t} + \frac{1}{m_{12}} J^{-1} \Delta S = 0. \tag{71}$$

On the other hand, let us multiply the continuity equation (28) by the amplitude a_0 and consider it on the integral trajectories

$$\begin{aligned}(\nabla_{\vec{r}} S(\vec{r}(\vec{q},\vec{k},t),\vec{k}\,)) \cdot \nabla_{\vec{r}} a_0^2(\vec{r}(\vec{q},\vec{k},t),\vec{k})& \\ + (\nabla_{\vec{r}}^2 S(\vec{r}(\vec{q},\vec{k},t),\vec{k}))a_0^2(\vec{r}(\vec{q},\vec{k},t),\vec{k}\,) = 0.& \end{aligned} \tag{72}$$

Differentiating the square of amplitude with respect to time, we get

$$\frac{\partial a_0^2(\vec{r}(\vec{q},\vec{k},t),\vec{k}\,)}{\partial t} = \nabla_{\vec{r}} a_0^2(\vec{r}(\vec{q},\vec{k},t),\vec{k}\,) \cdot \frac{\partial \vec{r}(\vec{q},\vec{k},t)}{\partial t}. \tag{73}$$

Using (59), it can be rewritten as

$$m_{12} \frac{\partial a_0^2(\vec{r}(\vec{q},\vec{k},t),\vec{k})}{\partial t} = (\nabla_{\vec{r}} S(\vec{r}(\vec{q},\vec{k},t),\vec{k}\,)) \cdot \nabla_{\vec{r}} a_0^2(\vec{r}(\vec{q},\vec{k},t),\vec{k}\,). \tag{74}$$

Therefore, on integral trajectories the continuity equation can be recast as

$$\frac{\partial a_0^2}{\partial t} + \frac{1}{m_{12}} a_0^2 \Delta S = 0. \tag{75}$$

Comparing (71) and (75), we find that J^{-1} and a_0^2 are solutions of the same equation and up to a constant they should be equal

$$a_0^2 = C J^{-1}. \tag{76}$$

Using (63), it can be rewritten as

$$a_0^2 = C \det \frac{\partial^2 S(\vec{r},\vec{k}\,)}{\partial \vec{k} \partial \vec{r}}. \tag{77}$$

This equation provides us with a general recipe to calculate the amplitude if a complete solution of the Hamilton-Jacobi equation is known. Let us apply it to the above solutions (31) and (34). We have

$$\frac{\partial^2 S_\pm(\vec{r},\vec{k}\,)}{\partial k_i \partial r_j} = \delta_{ij} + \frac{\partial^2 \Phi_\pm(\vec{r},\vec{k})}{\partial k_i \partial r_j}, \tag{78}$$

where δ_{ij} is the Kronecker symbol and

$$\frac{\partial^2 \Phi_\pm(\vec{r},\vec{k})}{\partial k_i \partial r_j} = \frac{1 \pm w}{2}(\hat{k}_i \hat{r}_j - \delta_{ij}) \pm \frac{1-w^2}{4w}\left(\hat{k}_i + \frac{\hat{k}_i - \hat{r}_i}{1-\cos\theta}\right)(\hat{r}_j - \hat{k}_j), \quad (79)$$

where $\cos\theta = \vec{k}\cdot\vec{r}/(kr)$ and $\hat{k}_i = k_i/k$, $\hat{r}_j = r_j/r$.

The determinant (77) is a function invariant with respect to rotations of the system of coordinates. To simplify matrix entries we can choose the coordinate system such that $k_i = (0,0,k)$ $(i = 1,2,3)$ and $r_j = (r\sin\theta, 0, r\cos\theta)$ $(j = 1,2,3)$ so that $\hat{k}_i = (0,0,1)$ and $\hat{r}_j = (\sin\theta, 0, \cos\theta)$. With this choice, the matrix takes the form

$$\frac{\partial^2 S_\pm(\vec{r},\vec{k})}{\partial k_i \partial r_j} = \begin{pmatrix} a_{11} & 0 & a_{13} \\ 0 & a_{22} & 0 \\ a_{31} & 0 & a_{33} \end{pmatrix}, \quad (80)$$

where

$$\begin{aligned} a_{11} &= \frac{\pm w - 1}{4(\pm w)}\{(\pm w + 1)\cos\theta - (\pm w - 1)\}, \\ a_{33} &= \frac{1}{2(\pm w)}\{(\pm w + 1)\cos\theta + (\pm w - 1)\}, \\ a_{22} &= \frac{1-(\pm w)}{2}, \quad a_{13} = \frac{1-(\pm w)^2}{4(\pm w)}\sin\theta, \\ a_{31} &= \frac{\pm w + 1}{2(\pm w)}\sin\theta. \end{aligned} \quad (81)$$

The result of determinant calculation

$$\begin{aligned} \det \frac{\partial^2 S_\pm(\vec{r},\vec{k})}{\partial k_i \partial r_j} &= a_{22}(a_{11}a_{33} - a_{13}a_{31}) \\ &= -\frac{(\pm w - 1)^2}{4(\pm w)}. \end{aligned} \quad (82)$$

Taking the square of (38), we get

$$\begin{aligned} a_{0\pm}^2 &= \frac{2C_{0\pm}^2}{\zeta(\pm w)}\left(\frac{w-1}{w+1}\right)^{\pm 1} \\ &= -\frac{C_{0\pm}^2}{2\nu}\frac{(w^2-1)}{(\pm w)}\left(\frac{w-1}{w+1}\right)^{\pm 1} \\ &= -\frac{2C_{0\pm}^2}{\nu}\frac{(\pm w - 1)^2}{4(\pm w)}, \end{aligned} \quad (83)$$

which up to a constant coincides with (82).

If we differentiate the Hamilton-Jacobi equation (27) with respect to the asymptotic momentum $\vec{k}$, we get an equation

$$\frac{\partial^2 S}{\partial \vec{k} \partial \vec{r}} \frac{\partial S}{\partial \vec{r}} = \vec{k}, \tag{84}$$

that determines a linear relationship between the local momentum $\vec{k}(\vec{r}) = \nabla_{\vec{r}} S(\vec{r})$ and its asymptotic value $\vec{k}$.

It allows a simple physical interpretation. Let us define a vector field of local momenta $\vec{k}(\vec{r})$ attached to each point $\vec{r}$ in the coordinate space. In case of the two-body Coulomb scattering we have a two-fold vector field $\vec{k}_{\pm}(\vec{r})$ corresponding to the plane and spherical wave boundary conditions. Then, the integral trajectories are curves tangent to this vector field. At a fixed asymptotic momentum $\vec{k}$, equation (84) defines a fixed vector field of local momenta

$$\vec{k}(\vec{r}, \vec{k}) = \left(\frac{\partial^2 S(\vec{r}, \vec{k})}{\partial \vec{k} \partial \vec{r}} \right)^{-1} \vec{k} \tag{85}$$

if the matrix is non-singular. Let us consider a set of classical particles in an elementary volume $\mathrm{d}^3 \vec{r}_t$ moving along an integral trajectory $\vec{r}(\vec{q}, \vec{k}, t)$. Without loss of generality we can assume that the distribution of particles in a small volume is uniform. Then, the number of particles is proportional to the volume. The continuity equation guarantees the conservation of the number of particles in $\mathrm{d}^3 \vec{r}_t$ while moving along the integral trajectories. From (63) we get

$$\begin{aligned} \mathrm{d}^3 \vec{r}_t &= \left| \det \frac{\partial \vec{r}(\vec{q}, \vec{k}, t)}{\partial \vec{q}} \right| d^3 \vec{q} \\ &= \frac{\mathrm{d}^3 \vec{q}}{\left| \det \frac{\partial^2 S(\vec{r}(\vec{q}, \vec{k}, t), \vec{k})}{\partial \vec{k} \partial \vec{r}} \right|}, \end{aligned} \tag{86}$$

that is, along the trajectory the quantity

$$a_0^2(\vec{r}_t, \vec{k}) d^3 \vec{r}_t \tag{87}$$

is invariant and, therefore, for two different moments t and t' we have

$$a_0^2(\vec{r}_t, \vec{k}) d^3 \vec{r}_t = a_0^2(\vec{r}_{t'}, \vec{k}) d^3 \vec{r}_{t'}. \tag{88}$$

As a consequence of this relationship, which includes the normalization factors, the corresponding quasi-classical wavefunction with the amplitude

a_0 will be properly normalized, except, maybe, in the regions near the caustics points.

At caustics, as we have seen the amplitudes become singular. It means that the elementary volume gets squeezed at caustics region: $d^3\vec{r}_t \to 0$. In the attractive Coulombic field, the set of caustics points is a one-dimensional half-axis of focusing or attraction points $\vec{r}_{foc}$, where integral trajectories coming from an asymptotic region converge due to attraction to the Coulomb center. They are defined by

$$\begin{aligned} \vec{r}_{foc} &= \{\vec{r} : r = z\} \\ &= \{\vec{r} : x = y = 0,\ z > 0\}. \end{aligned} \tag{89}$$

In the repulsive Coulombic field, the caustics points $\vec{r}_{env}$ are localized on a two-dimensional envelope surface that separates the classically accessible and inaccessible regions. The envelope surface is defined by

$$\begin{aligned} \vec{r}_{env} &= \left\{\vec{r} : r = z + \frac{4\nu}{k}\right\} \\ &= \left\{\vec{r} : z = \frac{x^2 + y^2 - \left(\frac{4\nu}{k}\right)^2}{2\left(\frac{4\nu}{k}\right)}\right\}. \end{aligned} \tag{90}$$

A more detailed description of the geometry of integral trajectories corresponding to the *post* form of the plane and spherical quasi-classical wavefunctions is given in Refs. 9–11.

Since we know the exact solution (8) and have an explicit integral representation (9), it is also instructive to derive the asymptotic WKB solutions applying the method of stationary phase to the integral. To this end, we rewrite the integral representation of the solution as

$$\begin{aligned} F(\nu, \zeta) &= \frac{\exp(-\pi\nu/2)}{\Gamma(-\mathrm{i}\nu)} I(\nu, \zeta), \\ I(\nu, \zeta) &= \int_0^1 \mathrm{d}t\, f(t) \exp(\mathrm{i}\zeta\varphi(t)), \\ \varphi(t) &= t + \frac{\nu}{\zeta}[\ln(1-t) - \ln t], \\ f(t) &= \frac{1}{t}\left[\frac{t}{1-t}\right]^{\varepsilon}, \end{aligned} \tag{91}$$

where $\varepsilon > 0$ is a regularization parameter, which should be taken in the limit $\varepsilon \to 0$ in the end of calculation.

In the stationary phase approximation, we have

$$I(\nu, \zeta \to \infty) \sim \sum_{t_c} \sqrt{\frac{2\pi}{\zeta|\varphi''(t_c)|}} f(t_c) \exp\left[\mathrm{i}\zeta\varphi(t_c) + \mathrm{i}\frac{\pi}{4}\mathrm{sign}\,\varphi''(t_c)\right], \quad (92)$$

where critical points t_c are solutions to the stationary phase equation:

$$\varphi'(t_c) = 0. \quad (93)$$

From (93) we get two solutions

$$t_{c\pm} = \frac{1 \pm w}{2}. \quad (94)$$

It is easy to check that the ($\pm$)-solutions (92) coincide with the above found WKB ($\pm$)-solutions (34) and (38).

In the regular stationary phase formula (92), it is assumed that $\varphi''(t_c) \neq 0$. At critical points, we have

$$\varphi''(t_{c\pm}) = \mp\frac{\zeta}{\nu}w. \quad (95)$$

This explains appearance of the singularity $\sim 1/\sqrt{w}$ in the WKB formulas near the caustic point $\zeta_1 = 4\nu$, since at this point in the repulsive case we get $w = 0$. To avoid a breakdown of the WKB formulas in this region, one could include higher order terms in the expansion for the phase function. For example, up to the third order terms, we get an expansion for

$$\varphi(t) \approx \varphi(t_{c\pm}) + \frac{1}{2}\varphi''(t_{c\pm})(t - t_{c\pm})^2 + \frac{1}{6}\varphi'''(t_{c\pm})(t - t_{c\pm})^3, \quad (96)$$

where

$$\begin{aligned} \varphi'''(t_{c\pm}) &= 2\frac{\zeta}{\nu}\left[3 - \frac{\zeta}{\nu}\right] \\ &= -8 \text{ at } \zeta = 4\nu \end{aligned} \quad (97)$$

is not equal to zero at caustics point. We do not consider further this third order correction in the present work.

Physically, the *eikonal* approximation is based on the assumption that the energy term E is dominant over the potential one V. It is expected to be applicable in the case of high-energy scattering. Let us first consider the Hamilton-Jacobi equation in the zeroth approximation with the potential energy being totally neglected. The energy eikonal equation

$$(\nabla_{\vec{r}} S_0)^2 = k^2 \tag{98}$$

is quadratic in S_0 and has two well-known solutions: one is a plane-wave eikonal $S_{0p} = \vec{k} \cdot \vec{r}$ and the other a spherical-wave one $S_{0s} = kr$. We are looking for an approximate solution of the Hamilton-Jacobi equation in the form $S^{eik}_{p,s} = S_{0p,s} + R_{p,s}$, where $R_{p,s}$ is assumed to be a small correction to the first term due to potential interaction. Substituting this form into (27), we get

$$\frac{1}{m_{12}}(\nabla_{\vec{r}} S_{0p,s}) \cdot (\nabla_{\vec{r}} R_{p,s}) + \frac{(\nabla_{\vec{r}} R_{p,s})^2}{2m_{12}} = -V_c(\vec{r}). \tag{99}$$

Mathematically, eikonal equation is a linearized version of the WKB one. Within the eikonal approximation we neglect the second, quadratic term in the left-hand-side of (99). It assumes that

$$|\nabla_{\vec{r}} S_{0p,s}| \gg |\nabla_{\vec{r}} R_{p,s}|. \tag{100}$$

The linearized eikonal equation with the Coulomb potential in the right-hand-side can be easily solved in parabolic coordinates. However, there is no need to repeat these calculations here since we already did this while solving the more complicated quadratic equation (32). In the case of the plane-wave eikonal, we should neglect in (32) the quadratic term, thus, obtaining the Coulombic logarithm as a solution

$$\frac{\mathrm{d}R_p^{(\log)}}{\mathrm{d}\zeta} = \frac{\nu}{\zeta},$$
$$R_p^{(\log)} = \nu \ln \zeta. \tag{101}$$

Similarly, for the spherical-wave eikonal we get

$$\frac{\mathrm{d}R_s^{(\log)}}{\mathrm{d}\zeta} = -\frac{\nu}{\zeta},$$
$$R_s^{(\log)} = -\nu \ln \zeta. \tag{102}$$

Further, we can calculate the eikonal amplitudes making use of equation (77). The eikonal results for the amplitudes read

$$|a_{0p}^{eik}|^2 = 1 - \left(\frac{\nu}{\zeta}\right)^2 \left[1 + \left(1 - \frac{\nu}{\zeta}\right)(1 - \cos\theta)\right]$$
$$= 1 + O\left[\left(\frac{\nu}{\zeta}\right)^2\right] \tag{103}$$

and

$$|a_{0s}^{eik}|^2 = \left(\frac{\nu}{\zeta}\right)^2 \left[1 + \frac{\nu}{\zeta}(1 - \cos\theta)\right]$$
$$= \left(\frac{\nu}{\zeta}\right)^2 \left[1 + O\left(\frac{\nu}{\zeta}\right)\right]. \tag{104}$$

Comparing with the asymptotic behavior of the exact solutions (15) and (16) and the WKB solutions (56) and (57) at $\zeta \to \infty$, we find that the eikonal solutions do satisfy the Coulomb boundary conditions. However, the plane-wave amplitude (103) is defined correctly only up to the first order terms. The second order term enters with the wrong, "minus" sign. This is because the eikonal solution (101) we have obtained is of a logarithmic order. In order to get the amplitude up to the second order terms, the eikonal phase function should be determined up to the first order terms. To find this first-order correction $R_p^{(1)}$, equation (99) can be solved by applying perturbative methods.[12,13] Thus, substituting the perturbation expansion $R_p = R_p^{(\log)} + R_p^{(1)}$ into (99) and neglecting the higher-order terms, we obtain a linear equation for

$$(\nabla_{\vec{r}} S_{0p}) \cdot (\nabla_{\vec{r}} R_p^{(1)}) = -\frac{1}{2}(\nabla_{\vec{r}} R_p^{(log)})^2. \tag{105}$$

Solving this equation for $R_p^{(1)}$, we get $R_p^{(1)} = -\nu^2/\zeta$. It is easy to check that with this term included, the eikonal amplitude (103) will be defined correctly up to the second-order terms.

Observe that eikonal phases (101) and (102) show logarithmic divergencies at caustics point $\zeta = 0$, while the corresponding WKB phases remain finite at $\zeta = 0$. However, one can see that eikonal amplitudes in the repulsive case behave smoothly near the caustics point $\zeta = 4\nu$, as compared to the WKB counterparts. Also, the eikonal approximation does not reproduce an exponential suppression of particles in the inner region $\zeta < 4\nu$. Such discrepancies between eikonal and WKB approximations are a direct consequence of the key assumption made in the eikonal approximation, that is, the energy E must be much bigger than the potential energy V. At caustics and inner regions, this assumption is not valid; here, we have, respectively, $E = V$ and $V > E$.

2.3. *Coulomb Scattering Amplitude*

The standard way of getting the Coulomb or Rutherford scattering amplitude from the asymptotic of the known exact, spherical-wave solution (16) gives the following expression for the amplitude[6,14]

$$t_c = -\nu \frac{\exp[-2\mathrm{i}\nu \ln \sin(\theta/2) + 2\mathrm{i} \arg \Gamma(1 + \mathrm{i}\nu)]}{2k \sin^2(\theta/2)}, \tag{106}$$

where θ is the scattering angle.

However, the explicit computation of the matrix element[15]

$$\tilde{t}_c = \langle \exp(\mathrm{i}\vec{p} \cdot \vec{r}) | V_c(r) \exp(-\varepsilon r) | \psi^{(+)}(\vec{k}, \vec{r}) \rangle, \tag{107}$$

where ε is a small regularization parameter, gives an expression

$$\tilde{t}_c = f_c^{(+)}(\nu) \frac{4\pi Z_1 Z_2}{[|\vec{k} - \vec{p}|^2 + \varepsilon^2]^{1+\mathrm{i}\nu}} [p^2 - (k + \mathrm{i}\varepsilon)^2]^{\mathrm{i}\nu} \tag{108}$$

which does not tend to the physical Coulomb amplitude t_c in the on-energy-shell limit $p \to k$ and as $\varepsilon \to 0$. Actually, on the energy shell $p = k$ and $\hat{\vec{p}} \neq \hat{\vec{k}}$, the phase of $\tilde{t}_c \sim \exp(\mathrm{i}\nu \ln \varepsilon)$ is logarithmically divergent as $\varepsilon \to 0$.

The reason for such a poor on-energy-shell behavior lies in the fact that the plane wave $\exp(\mathrm{i}\vec{p} \cdot \vec{r})$ used as an asymptotic state in (107) does not satisfy the proper Coulomb boundary condition including the Coulombic logarithmic phase (101) as it has been derived in the eikonal approximation. However, it should not be expected that an exact on-shell amplitude could be obtained by using the asymptotic logarithmic eikonal solution instead of a plane-wave one, with an appropriate modification of the interaction potential V_c in (107) being made. This is because the integration in the matrix element is carried out over the entire coordinate space, including the region near the caustics points where the eikonal solutions are not properly defined, in particular, they are not properly normalized. On the other hand, with the explicit exact formulas for the QM states $\psi^{(\pm)}(\vec{k}, \vec{r})$ (taking into account (4)) the scattering amplitude for the process $\vec{k} \to \vec{p}$ is given by $\langle \psi^{(-)}(\vec{p}, \vec{r}) | \psi^{(+)}(\vec{k}, \vec{r}) \rangle = \langle \vec{p} | S_c | \vec{k} \rangle$, where S_c is the scattering operator. The explicit computation of this scalar product[16] leads exactly to the amplitude t_c given in (106).[14]

The logarithmic divergence in (108) is intimately related to the logarithmic divergence found in a time-parametrization of the classical Coulomb trajectory asymptotes. For example, using equations (60), (31) and (34), we get for the plane-wave integral trajectory asymptote

$$\vec{r}_-(t)|_{t\to-\infty} \sim \vec{r}_0(t) + \frac{Z_1 Z_2 m_{12}}{k^3}\vec{k}\ln k^2|t|, \tag{109}$$

where $\vec{r}_0(t) = \vec{q} + \frac{\vec{k}}{m_{12}}t$ is a free-motion asymptote.

The logarithmic distortion of the free-motion asymptote due to the long-range nature of Coulomb interaction in classical mechanics should have a similar effect on the "free evolution" operator in quantum mechanics. Following Dollard,[17,18] we define a modified "free evolution" operator $\exp[-\mathrm{i}H_{0,c}(t)]$:

$$H_{0,c}(t) = H_0 t + \operatorname{sign}(t)\frac{Z_1 Z_2 m_{12}}{|-\mathrm{i}\nabla_{\vec{r}}|}\ln(4H_0|t|), \tag{110}$$

with $H_0 = \hat{K}$. For the modified operator, in contrast with the free evolution operator $\exp[-\mathrm{i}H_0 t]$, it can be shown that the Möller operators

$$\begin{aligned} \Omega_{\pm,c} &= \lim_{t\to\mp\infty} \exp(\mathrm{i}Ht)\exp(-\mathrm{i}H_{0,c}(t)), \\ H &= H_0 + V \end{aligned} \tag{111}$$

exist in the strong sense; they are isometric and asymptotically complete, and the scattering matrix $S_c = \Omega^\dagger_{-,c}\Omega_{+,c}$ is unitary. Within the non-stationary approach developed by Dollard to the definition of the Coulomb modified S_c-matrix, the correct physical amplitude t_c has been derived, for example, in Ref. 19.

It should be emphasized that the physical two-body Coulombic scattering amplitude (106) has been derived in the energy-dependent formulation from the exact two-body solution that satisfies the spherical wave boundary condition. There is no contribution to the amplitude from the exact plane-wave type solution. In contrast, in the three-body Coulomb scattering we will find physical effects where both two-body plane- and spherical wave contributions are important. Moreover, we will observe interference effects between the plane- and spherical waves in the processes of resonance ionization and charge transfer.

3. Three-Body Coulomb Scattering

The aim of this section is to generalize analytic results and methods of the previous section to the system of three charged particles with charges Z_i and masses m_i $(i = 1, 2, 3)$. We restrict ourselves to the case of three active particles; for four-body methods in high-energy ion-atom collisions see, e.g., recent review[20] and book[21] and references therein.

3.1. *Coulomb Boundary Conditions for Three Particles into Continuum*

The Hamiltonian H, total energy E and potential energy V of the three-body system in the center of mass are assumed to be

$$H = \hat{K} + \sum_{i<j=1}^{3} V_{ij} = -\frac{1}{2m_{ij}}\nabla^2_{\vec{r}_{ij}} - \frac{1}{2\mu_k}\nabla^2_{\vec{R}_k} + \sum_{i<j=1}^{3}\frac{Z_iZ_j}{r_{ij}}, \quad (112)$$

$$E = \frac{\vec{k}^2_{ij}}{2m_{ij}} + \frac{\vec{K}^2_k}{2\mu_k}, \quad V = \sum_{i<j=1}^{3} V_{ij}, \quad (113)$$

where $m_{ij} = m_i m_j(m_i + m_j)$, $\mu_k = m_k(m_i + m_j)/M$, $M = m_1 + m_2 + m_3$ are reduced masses $(i, j, k = 1, 2, 3,\ i \neq j \neq k \neq i)$; $X = (\vec{r}_{ij}, \vec{R}_k)$ and $K = (\vec{k}_{ij}, \vec{K}_k)$ are sets of independent canonically conjugate Jacobi coordinates and momenta. Without loss of generality, we can assume that $m_2 = 1$.

Since the exact three-body scattering solutions to the Schrödinger equation $(H - E)\Psi^{(-)} = 0$ are not known (to be definite a *post* form of the state will be derived further), we first within the WKB approach seek to find approximate solutions of the corresponding Hamilton-Jacobi equation

$$\frac{1}{2m_{ij}}(\nabla_{\vec{r}_{ij}} S)^2 + \frac{1}{2\mu_k}(\nabla_{\vec{R}_k} S)^2 = E - V. \quad (114)$$

Separating out the three-body plane-wave eikonal S_{0p},

$$\begin{aligned} S &= S_{0p} + \Phi, \\ S_{0p} &= \vec{k}_{ij}\cdot\vec{r}_{ij} + \vec{K}_k\cdot\vec{R}_k, \end{aligned} \quad (115)$$

after substitution into (114) we get

$$\frac{\vec{k}_{ij}}{m_{ij}}\cdot\nabla_{\vec{r}_{ij}}\Phi + \frac{\vec{K}_k}{\mu_k}\cdot\nabla_{\vec{R}_k}\Phi + \frac{1}{2m_{ij}}(\nabla_{\vec{r}_{ij}}\Phi)^2 + \frac{1}{2\mu_k}(\nabla_{\vec{R}_k}\Phi)^2 = -V. \quad (116)$$

Within the eikonal approximation neglecting the quadratic terms, we obtain as a solution the sum over pairwise Coulomb logarithms:

$$\Phi^{eik} = \sum_{i<j=1}^{3} \Phi_{ij}^{eik},$$

$$\Phi_{ij}^{eik} = \Phi^{eik}(\nu_{ij}, \xi_{ij}) = -\nu_{ij} \ln(\xi_{ij}),$$

$$\nu_{ij} = \frac{Z_i Z_j m_{ij}}{k_{ij}}, \quad \xi_{ij} = k_{ij} r_{ij} + \vec{k}_{ij} \cdot \vec{r}_{ij}. \tag{117}$$

This was first stated as an ansaz in Ref. 22 and attributed to Redmond. Exploiting the additivity principle and the correspondence between the two-body Coulomb eikonal, WKB and QM solutions

$$F_{p,s}^{eik}(\nu_{ij}, \xi_{ij}) \leftrightarrow F_{p,s}^{WKB}(\nu_{ij}, \xi_{ij}) \leftrightarrow F_{p,s}^{QM}(\nu_{ij}, \xi_{ij}), \tag{118}$$

where subindices p and s denote plane and spherical parts of the solution, one can suggest as a generalization of this eikonal solution the factorized QM *post* form solution[22–24]:

$$\Psi^{(-)} = \exp(\mathrm{i}S_{0p}) \prod_{i<j=1}^{3} F_{ij}^{(-)}, \tag{119}$$

$$F_{ij}^{(-)} = f_c^{(-)}(\nu_{ij})\, {}_1F_1(\mathrm{i}\nu_{ij}, 1, -\mathrm{i}\xi_{ij}).$$

This form of the CDW function, which was derived directly as a QM solution in Ref. 23, has been extensively employed in computations of various atomic collision processes, such as single ionization of atoms by ion[23,25–28] or electron impact, for progress in $(e, 2e)$-processes, see, e.g. Refs. 29, 30, as well as in double-photoionization, $(\gamma, 2e)$-processes.[31] It is known in the literature under various acronyms, such as the CDW (in ion-atom), 3C or BBK (in electron-atom collisions) wavefunction.

Let us consider asymptotic and normalization properties of the state (119) in more detail. Using a separation of the continuum distortion factors into the plane and spherical parts, $F_{ij}^{(-)} = F_{pij}^{(-)} + F_{sij}^{(-)}$, the state can be split into the four parts:

$$\Psi^{(-)} = \Psi_{3p}^{(-)} + \Psi_{2ps}^{(-)} + \Psi_{p2s}^{(-)} + \Psi_{3s}^{(-)}, \tag{120}$$

where subindices $3p$, $2ps$ and so on indicate how many plane or spherical-wave parts are grouped into the corresponding term. Correspondingly, the terms in (120) can be called triple-plane, single-spherical, double-spherical and triple-spherical terms. If the particles are in the region of configuration space $\Omega_0 = \{X : r_{ij} \sim R \gg 1\}$, where they are all well separated, then (120) is an expansion into a small parameter $1/R$, where $\Psi_{3p}^{(-)} = O(1)$ is a leading term, while $\Psi_{2ps}^{(-)} = O(1/R)$, $\Psi_{p2s}^{(-)} = O(1/R^2)$, and $\Psi_{3s}^{(-)} = O(1/R^3)$ are terms of the first, second, and third orders, respectively. We confine ourselves to the study of the first two leading terms.

The first term can be rewritten as

$$\Psi_{3p}^{(-)} = A_{3p} \exp(\mathrm{i}S_{0p} + \mathrm{i}\Phi_{3p}), \tag{121}$$

where the amplitude and phase

$$\begin{aligned} A_{3p} &= a_{p23}a_{p12}a_{p13}, \\ \Phi_{3p} &= \Phi_{p23} + \Phi_{p12} + \Phi_{p13} \end{aligned} \tag{122}$$

are defined, respectively, as a product of amplitudes and as a sum of phases of the corresponding pairwise distortion factors.

Let us consider a WKB analogue of (121). Then, the phase $\Phi_{3p} \equiv \Phi_{3p}^{(0)}$ should be an approximate solution of the Hamilton-Jacobi equation (116). Substituting $\Phi_{3p}^{(0)}$ into (116), we get for the residual term $\delta[\Phi_{3p}^{(0)}]$ the following expression

$$\begin{aligned} \delta[\Phi_{3p}^{(0)}] = &-(\nabla_{\vec{r}_{23}}\Phi_{p23}) \cdot (\nabla_{\vec{r}_{12}}\Phi_{p12}) - \frac{1}{m_3}(\nabla_{\vec{r}_{23}}\Phi_{p23}) \cdot (\nabla_{\vec{r}_{13}}\Phi_{p13}) \\ &+ \frac{1}{m_1}(\nabla_{\vec{r}_{12}}\Phi_{p12}) \cdot (\nabla_{\vec{r}_{13}}\Phi_{p13}). \end{aligned} \tag{123}$$

The smallness of the residual term $\delta[\Phi_{3p}^{(0)}]$ can be a measure of how close a particular asymptotic solution is to an exact one. If $\delta[\Phi_{3p}^{(0)}] \equiv 0$, then $\Phi_{3p}^{(0)}$ is an exact solution. Using the explicit WKB expressions for the pairwise plane and spherical phases Φ_p and Φ_s such that $\Phi_p + \Phi_s = -\xi$, we calculate their gradients

$$\begin{aligned} \nabla_{\vec{r}}\Phi_{p,s} &= k(\hat{\vec{k}} + \hat{\vec{r}})\frac{\pm u - 1}{2}, \\ u &= \sqrt{1 - \frac{4\nu}{\xi}}, \end{aligned} \tag{124}$$

where the (+)- and (−)-signs correspond to the plane and spherical waves, respectively. In the asymptotic region $\xi \to \infty$, we obtain

$$\nabla_{\vec{r}}\Phi_p \sim -\frac{\nu k}{\xi}(\hat{\vec{k}} + \hat{\vec{r}}), \tag{125}$$

$$\nabla_{\vec{r}}\Phi_s \sim -k(\hat{\vec{k}} + \hat{\vec{r}}) \tag{126}$$

and

$$\begin{aligned}\delta[\Phi_{3p}^{(0)}] \sim &-\frac{Z_1 Z_3 Z_2^2 m_{23} m_{12}}{\xi_{23}\xi_{12}}(\hat{\vec{k}}_{23} + \hat{\vec{r}}_{23}) \cdot (\hat{\vec{k}}_{12} + \hat{\vec{r}}_{12}) \\ &- \frac{Z_1 Z_3^2 Z_2 m_{23} m_{13}}{m_3 \xi_{23}\xi_{13}}(\hat{\vec{k}}_{23} + \hat{\vec{r}}_{23}) \cdot (\hat{\vec{k}}_{13} + \hat{\vec{r}}_{13}) \\ &+ \frac{Z_1^2 Z_3 Z_2 m_{12} m_{13}}{m_1 \xi_{12}\xi_{13}}(\hat{\vec{k}}_{12} + \hat{\vec{r}}_{12}) \cdot (\hat{\vec{k}}_{13} + \hat{\vec{r}}_{13}) \\ = &\, O\left(\frac{1}{\xi^2}\right)\end{aligned} \tag{127}$$

if all $\xi_{ij} \sim \xi \gg 1$.

One can see that $\Phi_{3p}^{(0)}$ in equation (122) is an asymptotic solution in the region Ω_0 that satisfies the Hamilton-Jacobi equation (116) up to a second order residual term. However, in the region $\Omega_{ij} = \{X : r_{ij} \sim 1,\, r_{ki},\, r_{kj} \sim R \gg 1\}$, where a pair of particles (i, j) is not well separated but it is far away from the third particle k, the residual term $\delta[\Phi_{3p}^{(0)}]$ turns out to be of the first order. Further, to be specific we consider the region Ω_{23}; a similar analysis can be readily done in the regions Ω_{12}, Ω_{13}.

We seek an asymptotic solution in the region Ω_{23} that satisfies the Hamilton-Jacobi equation up to a second order residual term. To this end, we substitute the ansatz

$$\Phi_{3p}^{(23)} = \tilde{\Phi}_{p23} + \Phi_{p12} + \Phi_{p13} \tag{128}$$

where $\tilde{\Phi}_{p23}(X)$ is an unknown function and Φ_{p12}, Φ_{p13} are the corresponding two-body WKB solutions, into (116). As a result, we get an (exact) equation for $\tilde{\Phi}_{p23}$:

$$\begin{aligned}&\frac{1}{2m_{23}}(\nabla_{\vec{r}_{23}}\tilde{\Phi}_{p23})^2 + \frac{\vec{k}_{23}(X)}{m_{23}} \cdot (\nabla_{\vec{r}_{23}}\tilde{\Phi}_{p23}) + V_{23} \\ &+ \frac{1}{2\mu_1}(\nabla_{\vec{R}_1}\tilde{\Phi}_{p23})^2 + \frac{\vec{K}_1(X)}{\mu_1} \cdot (\nabla_{\vec{R}_1}\tilde{\Phi}_{p23}) \\ = &\, \frac{1}{m_1}(\nabla_{\vec{r}_{12}}\Phi_{p12}) \cdot (\nabla_{\vec{r}_{13}}\Phi_{p13}),\end{aligned} \tag{129}$$

where the three-body local momenta were introduced

$$\vec{k}_{23}(X) = \vec{k}_{23} + m_{23}\nabla_{\vec{r}_{12}}\Phi_{p12} + (1 - m_{23})\nabla_{\vec{r}_{13}}\Phi_{p13},$$
$$\vec{K}_1(X) = \vec{K}_1 - \nabla_{\vec{r}_{12}}\Phi_{p12} + \nabla_{\vec{r}_{13}}\Phi_{p13}. \tag{130}$$

In Refs. 32, 33, the local momenta have been defined in the eikonal approximation.

In Ω_{23}, the right-hand side of equation (129) is of the second order and can be neglected. Then, the function

$$\tilde{\Phi}_{p23} = \Phi_{p23}(\vec{k}_{23}(X), \vec{r}_{23}), \tag{131}$$

where Φ_{p23} is the Coloumb WKB plane-wave solution for the pair $(2,3)$, with the asymptotic momentum $\vec{k}_{23}$ being replaced by $\vec{k}_{23}(X)$, will satisfy equation (129) up to the second order terms. It can be shown that the residual term will be proportional to the derivatives

$$\frac{\partial\vec{k}_{23}(X)}{\partial X} = \left(\frac{\partial\vec{k}_{23}(X)}{\partial\vec{r}_{23}}, \frac{\partial\vec{k}_{23}(X)}{\partial\vec{R}_1}\right) = O\left(\frac{1}{R^2}\right). \tag{132}$$

Observe that the correction introduced in the asymptotic momentum $|\vec{k}_{23}(X) - \vec{k}_{23}| \sim 1/R$ is of the first order in Ω_{23}. If we neglect this correction, $\Phi_{3p}^{(23)}$ will coincide with $\Phi_{3p}^{(0)}$. Similarly, we can define asymptotic solutions $\Phi_{3p}^{(12)}$ and $\Phi_{3p}^{(13)}$, respectively, with the three-body momenta $\vec{k}_{12}(X)$ and $\vec{k}_{13}(X)$ in the regions Ω_{12} and Ω_{13}. Moreover, we can define a unified asymptotic solution

$$\Phi_{3p} = \Phi_{p23}(\vec{k}_{23}(X), \vec{r}_{23}) + \Phi_{p12}(\vec{k}_{12}(X), \vec{r}_{12}) + \Phi_{p13}(\vec{k}_{13}(X), \vec{r}_{13}), \tag{133}$$

which satisfies equation (116) up to the second order terms in the unified region $\Omega_0 \cup \Omega_{23} \cup \Omega_{12} \cup \Omega_{13}$.

The single-spherical phases are obtained from the triple-plane phase (122) by replacing one of the pair's plane-wave phase by a spherical one. Say, for the pair $(2,3)$ we obtain

$$\Phi_{2ps}^{(23)} = \Phi_{s23}(\vec{k}_{23}, \vec{r}_{23}) + \Phi_{p12}(\vec{k}_{12}, \vec{r}_{12}) + \Phi_{p13}(\vec{k}_{13}, \vec{r}_{13}). \tag{134}$$

The residual term $\delta[\Phi_{2ps}^{(23)}]$ is obtained by replacing $\Phi_{p23} \to \Phi_{s23}$ in (123). From (125) and (126) we get that $\delta[\Phi_{2ps}^{(23)}] = O(1/\xi)$ is of the first order. Further, in order to find the second-order solution we can employ the same anzatz (128), where the unknown function is replaced by $\tilde{\Phi}_{s23}$, and derive exactly the same equation (129) where $\tilde{\Phi}_{p23} \to \tilde{\Phi}_{s23}$. Then, the asymptotic solution

$$\tilde{\Phi}_{s23} = \Phi_{s23}(\vec{k}_{23}(X), \vec{r}_{23}) \tag{135}$$

will satisfy equation up to the second order terms. Finally, as a natural extension we define a unified solution

$$\Phi_{2ps}^{(23)} = \Phi_{s23}(\vec{k}_{23}(X), \vec{r}_{23}) + \Phi_{p12}(\vec{k}_{12}(X), \vec{r}_{12}) + \Phi_{p13}(\vec{k}_{13}(X), \vec{r}_{13}), \tag{136}$$

which satisfies equation (116) up to the second order terms in the unified region $\Omega_0 \cup \Omega_{23} \cup \Omega_{12} \cup \Omega_{13}$. Obviously, similar representations can be obtained for other single-spherical phases $\Phi_{2ps}^{(12)}$ and $\Phi_{2ps}^{(13)}$.

To calculate the WKB amplitudes we can employ the explicit formula (77). A sketch of the proof we gave for this formula in the two-body section is quite general. It can be easily extended to the three-body case, since it does not depend on a particular dimensionality or potential of the system. The formula is exact as far as we know a complete exact solution S of (114). Using this formula with an asymptotic complete solution, S will give us an asymptotically correct amplitude.

As an example, let us apply this formula to computation of the leading amplitude A_{3p}. We consider a system of three particles which is relevant to ion-atom collisions, namely, we denote by particles 1 and 3, respectively, the incident and residual target ions with charges $Z_1 = Z_p$, $Z_3 = Z_t$ and masses $m_1 = m_p$, $m_3 = m_t \gg 1$, and by particle 2 the ejected electron with $Z_2 = -1$ and $m_2 = 1$. To simplify further calculations we replace in (133) the three-body local momenta $\vec{k}_{ij}(X)$ by their asymptotic values $\vec{k}_{ij}$. Then, we have

$$\begin{aligned} A_{3p}^2 &= \left| \det \frac{\partial^2 S_{3p}(X,P)}{\partial X \partial P} \right| \\ &= \left| \det \begin{pmatrix} I + \dfrac{\partial^2 \Phi_{3p}}{\partial \vec{r}_{23} \partial \vec{k}_{23}} & \dfrac{\partial^2 \Phi_{3p}}{\partial \vec{r}_{23} \partial \vec{K}_1} \\ \dfrac{\partial^2 \Phi_{3p}}{\partial \vec{R}_1 \partial \vec{k}_{23}} & I + \dfrac{\partial^2 \Phi_{3p}}{\partial \vec{R}_1 \partial \vec{K}_1} \end{pmatrix} \right|, \end{aligned} \tag{137}$$

where I is the 3-by-3 identity matrix. Using linear relations between Jacobi coordinates

$$\begin{aligned} \vec{r}_{12} &= m_{23}\vec{r}_{23} - \vec{R}_1, \\ \vec{r}_{13} &= (1 - m_{23})\vec{r}_{23} + \vec{R}_1, \end{aligned} \tag{138}$$

we obtain the following expressions for the block matrices

$$\frac{\partial^2\Phi_{3p}}{\partial\vec{r}_{23}\partial\vec{k}_{23}} = \sigma_{23} + m_{12}m_{23}\sigma_{12} + \frac{m_{13}}{m_3(m_3+1)}\sigma_{13}, \tag{139}$$

$$\frac{\partial^2\Phi_{3p}}{\partial\vec{r}_{23}\partial\vec{K}_1} = -\frac{m_{12}m_{23}}{\mu_1}\sigma_{12} + \frac{m_{13}}{\mu_1(m_3+1)}\sigma_{13}, \tag{140}$$

$$\frac{\partial^2\Phi_{3p}}{\partial\vec{R}_1\partial\vec{k}_{23}} = -m_{12}\sigma_{12} + \frac{m_1}{m_1+m_3}\sigma_{13}, \tag{141}$$

$$\frac{\partial^2\Phi_{3p}}{\partial\vec{R}_1\partial\vec{K}_1} = \frac{m_{12}}{\mu_1}\sigma_{12} + \frac{m_{13}}{\mu_1}\sigma_{13}, \tag{142}$$

where we have denoted by σ_{ij} the matrices

$$\sigma_{ij} = \frac{\partial^2\Phi_{pij}(\vec{k}_{ij},\vec{r}_{ij})}{\partial\vec{r}_{ij}\partial\vec{k}_{ij}} \tag{143}$$

due to interaction between particles i and j. If interaction between particles i and j is switched off, then $a_{ij}\equiv 0$. Differentiating (124) with respect to $\vec{k}$, we get

$$\frac{\partial^2\Phi_p(\vec{k},\vec{r})}{\partial\vec{r}\partial\vec{k}} = \frac{u-1}{2}\Bigg\{I + \hat{\vec{r}}\otimes\hat{\vec{k}} - \frac{u+1}{2u(1+\cos\theta)}(\hat{\vec{r}}+\hat{\vec{k}})\otimes(\hat{\vec{r}}+(2+\cos\theta)\hat{\vec{k}}\,)\Bigg\}, \tag{144}$$

where $\cos\theta = \hat{\vec{r}}\cdot\hat{\vec{k}}$ and $\otimes$ denotes a tensor or Kronecker product. Note that in the asymptotic region $\xi\to\infty$, the matrix entries

$$\frac{\partial^2\Phi_p(\vec{k},\vec{r})}{\partial\vec{r}\partial\vec{k}} \sim u-1 = -\frac{4\nu}{\xi(u+1)} = O\left(\frac{\nu}{\xi}\right) \tag{145}$$

are of the first order. Moreover, for heavy ion particles 1 and 3, the matrix $\sigma_{13}\sim 1/m_{13}\ll 1$ is small. If we neglect small contributions of the order $O(1/m_{1,3})$, equation (137) can be reduced to

$$A_{3p}^2 = |\det(I+\sigma_{23}+\sigma_{12})|. \tag{146}$$

Using explicit expressions equations (143) and (144) for matrices σ_{23}, σ_{12} and the formula for determinants in terms of traces

$$\det(I+\sigma) = 1 + \operatorname{tr}\sigma - \frac{1}{2}[\operatorname{tr}\sigma^2 - (\operatorname{tr}\sigma)^2] + \frac{1}{6}(\operatorname{tr}\sigma)^3 - \frac{1}{2}\operatorname{tr}\sigma\operatorname{tr}\sigma^2 + \frac{1}{3}\operatorname{tr}\sigma^3, \tag{147}$$

after some lengthy algebra one can get an explicit expression for (146). In general, this expression cannot be factorized into a product of the squares of amplitudes

$$A_{3p,f}^2 = a_{p23}^2 a_{p12}^2 = |\det(I+\sigma_{23})\det(I+\sigma_{12})| \\ = |\det(I+\sigma_{23}+\sigma_{12}+\sigma_{23}\sigma_{12})| \tag{148}$$

as defined in equation (122) ($a_{p13} = 1$ up to terms of order $O(1/m_{1,3})$). Taking into account the asymptotic $\sigma \sim \nu/\xi$ and that

$$\operatorname{tr}\sigma_{ij} = \frac{(u_{ij}-1)^2}{4u_{ij}}(3\cos\theta_{ij}+1) = O\left(\frac{\nu_{ij}^2}{\xi_{ij}^2}\right), \tag{149}$$

$$u_{ij} = \sqrt{1-\frac{4\nu_{ij}}{\xi_{ij}}}, \quad \cos\theta_{ij} = \hat{\vec{r}}_{ij}\cdot\hat{\vec{k}}_{ij},$$

we obtain for their difference

$$A_{3p}^2 - A_{3p,f}^2 = -\operatorname{tr}(\sigma_{23}\sigma_{12}) + O\left(\frac{\nu^3}{\xi^3}\right). \tag{150}$$

After some lengthy algebra, the leading second-order term can be rewritten as

$$-\operatorname{tr}(\sigma_{23}\sigma_{12}) = -\frac{(u_{23}-1)(u_{12}-1)}{4}g, \tag{151}$$

where

$$\begin{aligned}
g &= g_1 - g_2 + g_3,\\
g_1 &= 3+\cos\theta_{23} - \frac{u_{23}+1}{4u_{23}}(3\cos\theta_{23}+1)\\
&\quad + \cos\theta_{12} - \frac{u_{12}+1}{4u_{12}}(3\cos\theta_{12}+1) + (\hat{\vec{r}}_{12}\cdot\hat{\vec{k}}_{23})(\hat{\vec{k}}_{12}\cdot\hat{\vec{r}}_{23}),\\
g_2 &= \frac{u_{23}+1}{2u_{23}(1+\cos\theta_{23})}(\hat{\vec{r}}_{12}\cdot(\hat{\vec{r}}_{23}+(2+\cos\theta_{23})\hat{\vec{k}}_{23}))(\hat{\vec{k}}_{12}\cdot(\hat{\vec{r}}_{23}+\hat{\vec{k}}_{23}))\\
&\quad + \frac{u_{12}+1}{2u_{12}(1+\cos\theta_{12})}(\hat{\vec{r}}_{23}\cdot(\hat{\vec{r}}_{12}+(2+\cos\theta_{12})\hat{\vec{k}}_{12}))(\hat{\vec{k}}_{23}\cdot(\hat{\vec{r}}_{12}+\hat{\vec{k}}_{12})),\\
g_3 &= \frac{(u_{23}+1)(u_{12}+1)}{4u_{23}u_{12}(1+\cos\theta_{23})(1+\cos\theta_{12})}\\
&\quad \times [(\hat{\vec{r}}_{12}+\hat{\vec{k}}_{12})\cdot(\hat{\vec{r}}_{23}+(2+\cos\theta_{23})\hat{\vec{k}}_{23})]\\
&\quad \times [(\hat{\vec{r}}_{23}+\hat{\vec{k}}_{23})\cdot(\hat{\vec{r}}_{12}+(2+\cos\theta_{12})\hat{\vec{k}}_{12})]
\end{aligned} \tag{152}$$

is seen to be an elementary but rather involved function of coordinates and momenta of all three particles. In the limit $u_{23}, u_{12} \to 1$, g is reduced to a kinematic or geometric factor that depends only on the angles between momentum and position vectors of all three particles.

By virtue of (149), amplitude distortions are seen to be effects of the second order and the difference between the squared amplitudes A^2_{3p}, and $A^2_{3p,f}$ is of the second order too. The general reason why the A^2_{3p} amplitude is not factorizable is that the coordinate and momentum variables in the subsystems $(1,2)$ and $(2,3)$ are not independent.

As a natural refinement of the WKB representations, one can replace in the above formulae, using the correspondence (118), the WKB phases $\Phi_{p,s,ij}$ by QM ones. However, with such replacements the formulae will not be elementary anymore.

3.2. *Coulomb Boundary Conditions for Two Bound Particles and the Third Particle Into Continuum*

In considering the bound state problem, we assume the same three-body system, an incident ion — particle 1, a residual-target ion — particle 3, and an electron — particle 2; but now two particles, say, 2 and 3 are in a bound state $\psi_i(\vec{r}_{23})$. The CDW approximation, introduced by Cheshire[34] for electron capture in proton-hydrogen collisions, represents the state of a bound target electron as a product of the unperturbed electronic wavefunction $\psi_i(\vec{r}_{23})$ and the continuum distortion function $F^{(+)}(\nu_{12}, \zeta_{12})$ in the *prior* form. A similar Vainstein-Presnyakov-Sobelman (VPS) wavefunction has been suggested independently by Vainshtein *et al.*[35] for electron-impact excitation. Formally, the bound CDW or VPS wavefunction can be obtained from the corresponding three-body continuum state (119), written in the *prior* form, by an analytic continuation when the two body continuum state $\psi^{(+)}(\vec{k}_{23}, \vec{r}_{23})$ is replaced by a bound state $\psi_i(\vec{r}_{23})$ and the momenta in the distortion factors for pairs $(1,2)$ and $(1,3)$ are defined using linear relations (138) between coordinates. Differentiating them with respect to time (assuming that coordinates are on the classical trajectory asymptotes), we obtain linear relations between the asymptotic velocities: $\vec{v}_{12} = m_{23}\vec{v}_{23} - \vec{v}_1$, $\vec{v}_{13} = (1 - m_{23})\vec{v}_{23} + \vec{v}_1$, where $\vec{v}_1 = \vec{v}$ is the velocity of projectile with respect to the electron-residual target ion subsystem and we should put $\vec{v}_{23} = 0$ in a bound state. Then, we have $\vec{k}_{12} = -\vec{v}$, $\vec{k}_{13} = m_{13}\vec{v}$ and the bound CDW wavefunction takes the form

$$\Psi_i^{(+)} = \exp(\mathrm{i}\vec{K}_1\vec{R}_1)F^{(+)}(\vec{k}_{12}, \vec{r}_{12})F^{(+)}_{eik}(\vec{k}_{13}, \vec{r}_{13})\psi_i(\vec{r}_{23}), \tag{153}$$

where distortion factor for the internuclear interaction of heavy ions is included in the eikonal approximation.

In a bound state, the electron is exponentially confined to a region close to the target ion and, therefore, we have to define a bound wavefunction satisfying proper boundary conditions in the asymptotic region Ω_{23}. However, as we know, the state (119) satisfies the Schrödinger equation up to the first order terms in Ω_{23}. A similar asymptotic behavior for the residual term in Ω_{23} we find for the bound CDW state (153). To include these first order terms in the continuous case, we have modified the wavefunction for the subsystem $(2, 3)$ by introducing a local three-body momentum $\vec{k}_{23}(X)$ into the subsystem. In the bound case, the problem is qualitatively different — the spectrum in the subsystem is discrete near the energy of the bound state and, therefore, to include these first order effects we have to consider introducing the local momentum into a set of discrete states. Following the work,[36] we suggest the ansatz:

$$\Psi_i^{(+)} = \Psi_{ip}^{(+)} + \Psi_{is}^{(+)}, \tag{154}$$

$$\Psi_{ip,s}^{(+)} = \exp(\mathrm{i}\vec{K}_1\vec{R}_1)F_{p,s}^{(+)}(\vec{k}_{12}, \vec{r}_{12})F_{eik}^{(+)}(\vec{k}_{13}, \vec{r}_{13})\psi_{p,s}^{(+)}(\vec{r}_{23}, \vec{R}_1), \tag{155}$$

where the unknown subsystem wavefunctions $\psi_{p,s}^{(+)}$ satisfy the exact [up to $O(1/m_{1,3})$ terms] equations:

$$(\hat{h}_{23} - \epsilon_i - \mathrm{i}\vec{k}_{12}^{(p,s)}(\vec{r}_{12}) \cdot \nabla_{\vec{r}_{23}} - \mathrm{i}\vec{v} \cdot \nabla_{\vec{R}_1})\psi_{p,s}^{(+)}(\vec{r}_{23}, \vec{R}_1) = 0, \tag{156}$$

where

$$\hat{h}_{23} = -\frac{1}{2}\nabla_{\vec{r}_{23}}^2 + V_{23} \tag{157}$$

is the atomic Hamiltonian and

$$\vec{k}_{12}^{(p,s)}(\vec{r}_{12}) = -\mathrm{i}\nabla_{\vec{r}_{12}} \ln F_{p,s}^{(+)}(\vec{k}_{12}, \vec{r}_{12}) \tag{158}$$

denotes an effective *complex* momentum introduced into the electron-target subsystem by the electron-projectile continuum distortion factor. It is this three-body operator containing the effective momentum that establishes a coupling between the electron-target and electron-projectile subsystems. If we neglect this operator, then equation (156) has a solution $\psi_{p,s}^{(+)} = \psi_i$ (ϵ_i denotes the binding energy of the initial state ψ_i: $\hat{h}_{23}\psi_i = \epsilon_i\psi_i$) and equation (154) retrieves the CDW function (153) when the relationship $F = F_p + F_s$ is applied.

In the WKB approximation, we can write the real and imaginary parts of the momenta $\vec{k}_{12}^{(p,s)}(\vec{r}_{12}) = \mathrm{Re}\,\vec{k}_{12}^{(p,s)}(\vec{r}_{12}) + \mathrm{i}\,\mathrm{Im}\,\vec{k}_{12}^{(p,s)}(\vec{r}_{12})$ as

$$\begin{aligned} \mathrm{Re}\,\vec{k}_{12}^{(p,s)}(\vec{r}_{12}) &= -\frac{v}{2}(\hat{\vec{r}}_{12} + \hat{\vec{v}})(\pm w_{12} - 1), \\ \mathrm{Im}\,\vec{k}_{12}^{(p,s)}(\vec{r}_{12}) &= \frac{v\nu_{12}(\hat{\vec{r}}_{12} + \hat{\vec{v}})}{(\zeta_{12} w_{12})^2} \frac{1 \mp w_{12}}{1 \pm w_{12}}, \\ w_{12} &= \sqrt{1 - \frac{4\nu_{12}}{\zeta_{12}}}, \quad \nu_{12} = -\frac{Z_1}{v}, \quad \zeta_{12} = v r_{12} + \vec{v} \cdot \vec{r}_{12}. \end{aligned} \tag{159}$$

From (159) we get in the asymptotic region $\Omega_{23}(r_{23} \sim 1 \ll R_1)$:

$$\begin{aligned} \mathrm{Re}\,\vec{k}_{12}^{(p)}(-\vec{R}_1(t))|_{t\to-\infty} &\sim -\frac{Z_1}{v^2|t|}\hat{\vec{v}} = O(R_1^{-1}), \\ \mathrm{Im}\,\vec{k}_{12}^{(p)}(-\vec{R}_1(t))|_{t\to-\infty} &\sim \frac{Z_1^2}{4v^5|t|^3}\hat{\vec{v}} = O(R_1^{-3}), \\ \mathrm{Re}\,\vec{k}_{12}^{(s)}(-\vec{R}_1(t))|_{t\to-\infty} &\sim 2\vec{v} = O(1), \\ \mathrm{Im}\,\vec{k}_{12}^{(s)}(-\vec{R}_1(t))|_{t\to-\infty} &\sim \frac{1}{|t|}\hat{\vec{v}} = O(R_1^{-1}), \end{aligned} \tag{160}$$

where $\vec{R}_1(t) = \vec{b} + \vec{v}t$, $\vec{b} \cdot \vec{v} = 0$, with b, t being an impact parameter and time of motion. Observe that asymptotically the real parts of the plane and spherical momenta are the leading terms in comparison with the corresponding imaginary parts. The imaginary plane and spherical momenta are, respectively, of the third and first orders. The coupling operators with the real components of momenta are Hermitian ones, while those with the imaginary parts are non-Hermitian. The non-Hermitian operators introduce a non-unitarity in the time dynamics described by equation (156).

Let us expand the state $\psi_{p,s}^{(+)}$ into a basis of the target-atom wavefunctions ψ_α ($\hat{h}_{23}\psi_\alpha = \epsilon_\alpha \psi_\alpha$):

$$\psi_{p,s}^{(+)} = \sum_\alpha a_\alpha^{(p,s)}(\vec{R}_1(t))\psi_\alpha(\vec{r}_{23}), \tag{161}$$

where expansion coefficients satisfy the system of differential equations

$$\begin{aligned} \mathrm{i}\dot{a}_\alpha^{(p,s)}(t) - (\epsilon_\alpha - \epsilon_i + d_{\alpha\alpha}^{(p,s)}(t))a_\alpha^{(p,s)}(t) &= \sum_{\beta\neq\alpha} d_{\alpha\beta}^{(p,s)}(t) a_\beta^{(p,s)}(t), \\ d_{\alpha\beta}^{(p,s)}(t) &= \langle\psi_\alpha|\vec{k}_{12}^{(p,s)}(\vec{r}_{12}) \cdot (-\mathrm{i}\nabla_{\vec{r}_{23}})|\psi_\beta\rangle. \end{aligned} \tag{162}$$

Applying the phase transformation

$$a_\alpha^{(p,s)}(t) = \exp(-\mathrm{i}\varphi_\alpha(t))c_\alpha^{(p,s)}(t),$$
$$\varphi_\alpha(t) = (\epsilon_\alpha - \epsilon_i)t + \int_{-\infty}^{t} \mathrm{d}\tau d_{\alpha\alpha}^{(p,s)}(\tau). \tag{163}$$

Equation (162) can be reduced to

$$\mathrm{i}\dot{c}_\alpha^{(p,s)}(t) = \sum_{\beta\neq\alpha} V_{\alpha\beta}^{(p,s)}(t)c_\beta^{(p,s)}(t),$$
$$V_{\alpha\beta}^{(p,s)}(t) = d_{\alpha\beta}^{(p,s)}(t)\exp(\mathrm{i}(\varphi_\alpha(t) - \varphi_\beta(t))). \tag{164}$$

The initial condition for the coefficients we put $c_\alpha^{(p,s)}(-\infty) = \delta_{\alpha i}$. The formal solution of equation (164) can be written in the matrix form using the T exp chronological operator

$$c^{(p,s)}(t) = \mathrm{T}\exp\left(-\mathrm{i}\int_{-\infty}^{t} \mathrm{d}\tau V^{(p,s)}(\tau)\right) c^{(p,s)}(-\infty), \tag{165}$$

where $c^{(p,s)}(t)$ is the column of coefficients $c_\alpha^{(p,s)}(t)$ and $V^{(p,s)}(t)$ the matrix $V_{\alpha\beta}^{(p,s)}(t)$ with zero diagonal elements. Within the first order in the interaction $V^{(p,s)}(t)$, we obtain an explicit solution

$$c^{(p,s)}(t) = \exp\left(-\mathrm{i}\int_{-\infty}^{t} \mathrm{d}\tau V^{(p,s)}(\tau)\right) c^{(p,s)}(-\infty) \tag{166}$$

which describes in a general form how the local momentum ($V^{(p,s)}$ is proportional to the local momentum) affects the dynamics of population of a set of discrete states. As mentioned above, with inclusion of the imaginary momenta into $V^{(p,s)}$ the dynamics is non-unitary, if we neglect them, equation (166) describes a unitary evolution.

In the asymptotic region Ω_{23}, we can approximate $\vec{k}_{12}^{(p,s)}(\vec{r}_{12}) \approx \vec{k}_{12}^{(p,s)}(-\vec{R}_1)$. Then, the matrix elements $d_{\alpha\beta}^{(p,s)}(t)$ are reduced to the dipole matix elements in the momentum space

$$d_{\alpha\beta}^{(p,s)}(t) = \vec{k}_{12}^{(p,s)}(-\vec{R}_1(t)) \cdot \langle\tilde{\psi}_\alpha(\vec{p})|\vec{p}|\tilde{\psi}_\beta(\vec{p})\rangle = \vec{k}_{12}^{(p,s)}(t)\cdot\vec{d}_{\alpha\beta}, \tag{167}$$
$$\tilde{\psi}_\alpha(\vec{p}) = \frac{1}{(2\pi)^{3/2}}\int \mathrm{d}^3\vec{r}\exp(-\mathrm{i}\vec{p}\cdot\vec{r})\psi_\alpha(\vec{r}).$$

Using selection rules for the dipole matrix elements, we obtain that $d_{\alpha\beta}^{(p,s)}(t) \neq 0$ only for the states $|\alpha\rangle \equiv |n_1 l_1 m_1\rangle$ and $|\beta\rangle \equiv |n_2 l_2 m_2\rangle$ such that $m_1 = m_2$ and $l_1 = l_2 \pm 1$, or $l_1 = l_2$ at $l_1 \neq 0$.

Another simplification is possible in calculating time integrals

$$\Theta_{\alpha\beta}^{(p,s)}(t) = \int_{-\infty}^{t} \mathrm{d}\tau V_{\alpha\beta}^{(p,s)}(\tau) = \vec{d}_{\alpha\beta} \cdot \int_{-\infty}^{t} \mathrm{d}\tau \exp(\mathrm{i}(\epsilon_\alpha - \epsilon_\beta)\tau) \vec{k}_{12}^{(p,s)}(\tau). \tag{168}$$

If $\epsilon_\beta \neq \epsilon_\alpha$, then, integrating by parts, we obtain

$$\Theta_{\alpha\beta}^{(p,s)}(t) = \frac{\exp(\mathrm{i}(\epsilon_\beta - \epsilon_\alpha)\tau)}{\mathrm{i}(\epsilon_\beta - \epsilon_\alpha)} \vec{d}_{\alpha\beta} \cdot \vec{k}_{12}^{(p,s)}(\tau)|_{\tau=-\infty}^{t} \tag{169}$$

$$- \frac{\vec{d}_{\alpha\beta}}{\mathrm{i}(\epsilon_\alpha - \epsilon_\beta)} \cdot \int_{-\infty}^{t} \mathrm{d}\tau \exp(\mathrm{i}(\epsilon_\alpha - \epsilon_\beta)\tau) \dot{\vec{k}}_{12}^{(p,s)}(\tau).$$

In Ω_{23}, the remaining integral can be neglected since $|\dot{\vec{k}}_{12}^{(p,s)}(t)| \ll |\vec{k}_{12}^{(p,s)}(t)|$. If $\epsilon_\alpha = \epsilon_\beta$, the time integral does not depend on the state indices α and β

$$\Theta_{\alpha\beta}^{(p,s)}(t) = \vec{d}_{\alpha\beta} \cdot \int_{-\infty}^{t} \mathrm{d}\tau \, \vec{k}_{12}^{(p,s)}(\tau) \tag{170}$$

and can be calculated using the explicit elementary expressions (159). The degenerate case is realized, for example, for the hydrogen-like states $|nlm\rangle$ with the same principle quantum number $n > 1$ and different orbital and magnetic quantum numbers l and m.

4. Coulomb Scattering Effects in Ionization Electron Spectra

As an application of the CDW theory with the local momenta reviewed in the previous section, we consider the single ionization process

$$\mathrm{I}^{Z_1} + \mathrm{A}(i) \to \mathrm{I}^{Z_1} + \mathrm{A}^+(f) + e, \tag{171}$$

where a heavy ion I^{Z_1} of charge Z_1 hits an atom $\mathrm{A}(i)$ in an intial state ψ_i, as a result of which one electron e (called active) is ionized while the other electrons (called passive) remain in the same state during the collision.

4.1. *Plane and Spherical Wave Contributions*

The single ionization process (171) provides a good benchmark for testing the quality of the CDW function. In the final state of the ionization reaction, we have three charged particles interacting into continuum: particle 1 — the scattered ion I^{Z_1}; particle 2 — the electron $e(E_e, \theta_e)$ ejected with the energy E_e and the angle of ejection θ_e; particle 3 — the residual target-ion $A^+(f)$. The three-body Coulomb interaction affects the angular and energy distributions of the reaction products, in particular, the angular and energy distributions of the ejected electron, which can be detected by measuring the corresponding double-differential cross sections (DDCS). Several ionization models[23,25–28,38] have been suggested and tested with partial success in ion-atom collisions at intermediate and high energies. In these models the final state has been the same CDW function, whereas the initial state described by various functions starting from the simplest Born approximation,[26,27] then, with increasing level of sophistication, going from eikonal[25,28] to CDW[23,34,37] and impulse (IA)[38] approximations. Numerically these models are quite advantageous since the ionization amplitudes can be calculated either analytically or reduced to a three-dimensional integration over momentum, as in the IA model.

Evaluation of the ionization amplitude with the CDW function with local momenta is, in general, a more challenging task requiring a six-dimensional integration in the coordinate space. However, using the final state CDW function with the local momenta, represented in a simplified form

$$\begin{aligned}\Psi_f^{(-)} &= \Psi_p^{(-)} + \Psi_s^{(-)}, \qquad (172)\\ \Psi_{p,s}^{(-)} &= \psi_{23}^{(-)}(\vec{k}_{23}^{(p,s)}(\vec{R}_1), \vec{r}_{23}) \exp(\mathrm{i}\vec{K}_1\vec{R}_1)\\ &\quad\times F_{p,s}^{(-)}(\nu_{12}, \xi_{12}(\vec{R}_1)) F_{eik}^{(-)}(\nu_{13}, \xi_{13}(\vec{R}_1)),\end{aligned}$$

can greatly reduce the number of integrations. With this choice for the final state wavefunction and Born function for the initial state (for details on the Born choice for the initial state and arguments on why this choice can be advantageous compared to the eikonal initial state (EIS) approximation, see discussion in Refs. 39, 40), the problem of calculating the DDCS can be simplified to a three-dimensional integration (one integration over time and two integrations over the impact parameter and the azimuthal angle)[39]:

$$\frac{\mathrm{d}^2\sigma}{\mathrm{d}E_e \mathrm{d}\Omega_e} = \sqrt{2E_e} \int \mathrm{d}^2\vec{b}\, P(\vec{b}), \qquad (173)$$

where the ionization probability $P(\vec{b})$, as a function of impact parameter vector is defined as

$$P(\vec{b}) = \left| \int_{-\infty}^{\infty} \mathrm{d}t\, A_{fi}(t, \vec{b}) \right|^2,$$

$$A_{fi}(t, \vec{b}) = A_p(t, \vec{b}) + A_s(t, \vec{b}),$$

$$A_{p,s}(t, \vec{b}) = a_{fi}(\vec{k}_{23}^{(p,s)}(\vec{R}_1), \vec{R}_1) \exp(\mathrm{i}(E_e - \epsilon_i)t)$$

$$\times F_{p,s}^{(-)*}(\nu_{12}, \xi_{12}(\vec{R}_1)) F_{eik}^{(-)*}(\nu_{13}, \xi_{13}(\vec{R}_1)),$$

$$a_{fi}(\vec{k}, \vec{R}) = \left\langle \psi_{23}^{(-)}(\vec{k}, \vec{r}) \left| \frac{-Z_1}{|\vec{r} - \vec{R}|} \right| \psi_i(\vec{r}) \right\rangle. \tag{174}$$

Here, A_p and A_s determine separate contributions to the total ionization amplitude from the plane and spherical rescattering waves generated in the electron-projectile ion subsystem (1, 2). In Eq. (174), a_{fi} is the matrix element of ionization transition from the bound initial state ψ_i into the Coulomb scattering state with the local momentum $\vec{k}_{23}^{(p,s)}(\vec{R}_1)$. Further, ϵ_i is the binding energy. Note that a_{fi} can be computed analytically.[39]

4.2. *Cusp Peak*

The electron capture to the continuum (ECC) cusp peak is a prominent feature observed in the electron spectra when the electron velocity $\vec{v}_e$ is matched to the velocity $\vec{v}$ of the projectile. Since the intensity and the shape of the ECC peak is very sensitive to the scattering details in the electron-projectile subsystem, this feature should be a stringent test for the above formulas.

In Figs. 5 and 6 we show the results of total calculation (line 1), according to equations (173) and (174), of the ECC peak at 0^0 electron ejection angle for the 100 keV/u H^+ + He and He^{2+} + He atomic collision systems, respectively, together with the available experimental data.[41] Also, the separate contributions from the plane (line 2) and spherical (line 3) waves are shown. One can see that the spherical wave contribution is dominant on the high-energy wing ($v_e > v$) of the cusp, while on the low-energy wing ($v_e < v$), the plane wave contribution is dominant. To demonstrate the effects associated with the modification of the asymptotic electron momentum by the projectile ion field, the results of a model calculation (line 4), where the local momenta $\vec{k}_{23}^{(p,s)}(\vec{R}_1)$ were set to be equal to $\vec{v}_e$ in the total calculation, are shown. One can see that both the field-modified

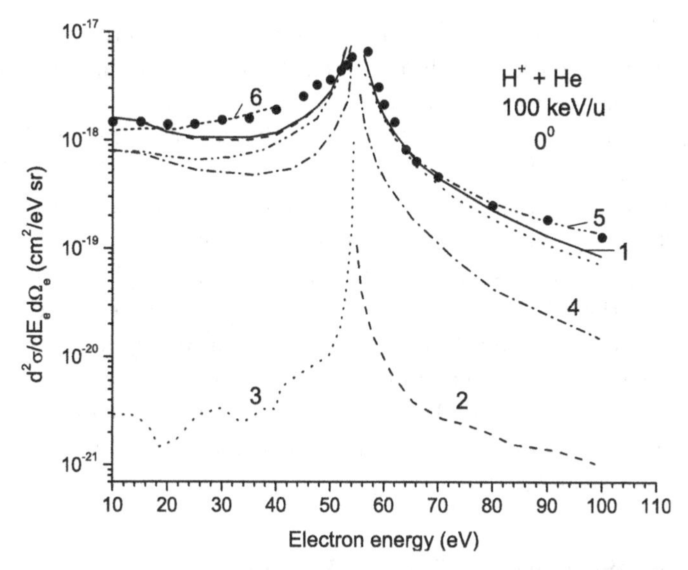

Fig. 5. Electron emission DDCS in the forward direction for 100 keV u^{-1} H^+ projectiles on He. Experimental points are from Ref. 41. Curve 1 is the total calculation according to equations (173) and (174); curves 2 and 3 represent separate contributions from the plane and spherical waves, respectively; curve 4 is the total calculation with the local momenta taken to be an asymptotic one, $\vec{k}_{23}^{(p,s)}(\vec{R}_1) \equiv \vec{v}_e$; curve 5: CDW-EIS[41]; curve 6: IA.[38]

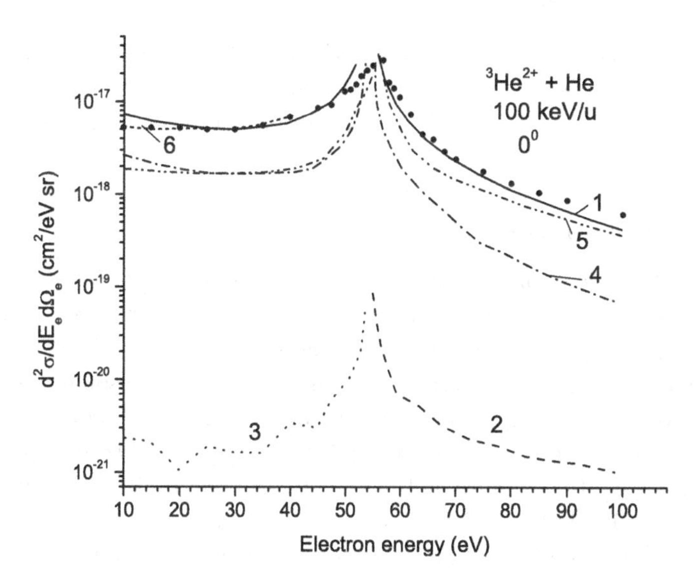

Fig. 6. As in Fig. 5 for $^3He^{2+}$ projectiles.

momenta $\vec{k}_{23}^{(p,s)}(\vec{R}_1)$ have a considerable effect on the intensity and the shape of the ECC peak, namely the intensity of the peak increases which improves essentially the agreement between the theory and experiment. For a comparison, the results of the CDW-EIS (line 5)[41] and the impulse approximation (IA) (line 6)[38] models are also presented. The CDW-EIS calculations underestimate the experimental data on the low-energy wing of the ECC peak, and the agreement is poorer as the projectile charge increases. This may be associated with an incorrect description of the region Ω_{23} given by the CDW function (119). More conclusive evidence of the importance of contributions from Ω_{23} can be obtained only if contributions from the inner region or interaction zone Ω_{int}, where all particles are close to each other, are correctly identified. However, none of the perturbative approaches realized so far can reproduce reasonably well these complicated particle dynamics in Ω_{int}.

5. Coulomb Scattering Effects in Autoionization Electron Spectra

In this section we consider an application of the CDW functions to a unified description of the two-step autoionization process

$$\mathrm{I}^{Z_1} + \mathrm{A}(i) \to \mathrm{I}^{Z_1} + \mathrm{A}^{**}(\alpha) \to \mathrm{I}^{Z_1} + \mathrm{A}^{+}(f) + e, \tag{175}$$

where an incident ion I^{Z_1} first excites the target atom $\mathrm{A}(i)$ into a quasi-stationary doubly excited autoionization state (AIS) $|\alpha\rangle = |n_1 l_1, n_2 l_2; LM\rangle$, which then decays, with one of the excited electrons e being ejected into the continuum.

5.1. *Coulomb Focusing Effect*

In a time-dependent picture of the resonance ionization (175), there is a small dimensionless parameter $\lambda_\alpha = \tau_{exc,\alpha}/\tau_{dec,\alpha} \ll 1$, where $\tau_{exc,\alpha}$ and $\tau_{dec,\alpha} = \Gamma_{\alpha 0}^{-1}$ are the effective average times for the excitation and decay of the resonance state α. The term $\Gamma_{\alpha 0}$ is the width of the resonance α. The smallness of λ_α allows us to introduce a two-step mechanism for resonance ionization: in the first step, in the limit $\tau_{exc,\alpha} \to 0$ a δ-like excitation of the resonance state α occurs at time $t = 0$ with the excitation amplitude $A_{exc,\alpha}$. Then, in the second step, the target atom in the quasistationary state α decays in the field of the scattered projectile ion into the final state of (175). Evaluation of the excitation amplitudes that involve two-electron transitions is out of the scope of the present work. For calculation of two-electron excitation amplitudes we refer, e.g., to Refs. 20, 42–44.

At the moment of resonance decay, the scattered ion and the target atom will be separated by a large distance $R \sim v\tau_{dec,\alpha} = v/\Gamma_{\alpha 0} \gg 1$. In the literature, the interaction in the final state between the ion I^{Z_1} and the products of decay is called the post-collision interaction (PCI). If an autoionization electron is ejected in the direction not too close to the direction of velocity of the scattered ion, the interaction between them will be much smaller than the kinetic energy of their relative motion. In this region of kinematic parameters, where the PCI is small compared to the kinetic energy, the eikonal approximation[45–47] should be adequate in describing the PCI effects.

There is a kinematic region, where the PCI is strong and the eikonal approximation fails. The PCI is strong if electron is ejected close to the direction of the receding ion with the velocity $v_e > v$. The CDW function (119) provides an adequate description for the strong PCI dynamics. In Refs. 48, 49, an analytic formula for the amplitude of resonance ionization (175)

$$A_{res,\alpha}(\varepsilon_\alpha) = \frac{2K_{PCI}(\varepsilon_\alpha)}{\Gamma_{\alpha 0}(\varepsilon_\alpha + \mathrm{i})} A_{dec,\alpha} A_{exc,\alpha}, \tag{176}$$

where

$$\begin{aligned} K_{PCI}(\varepsilon) &= f_c^{(+)}(\nu_{12}) f_c^{(+)}(\nu_{13}) \left(1 + \frac{a_{12}}{\varepsilon + \mathrm{i}}\right)^{\mathrm{i}\nu_{12}} \left(1 + \frac{a_{13}}{\varepsilon + \mathrm{i}}\right)^{\mathrm{i}\nu_{13}} \\ &\quad \times {}_2F_1\left(-\mathrm{i}\nu_{12}, -\mathrm{i}\nu_{13}, 1, \frac{a_{12}a_{13}}{(\varepsilon + a_{12} + \mathrm{i})(\varepsilon + a_{13} + \mathrm{i})}\right), \qquad (177) \\ a_{12} &= \frac{2}{\Gamma_{\alpha 0}} \left(k_{12}v - \vec{k}_{12} \cdot \vec{v}\right), \quad a_{13} = \frac{2}{\Gamma_{\alpha 0}} \left(k_{13}v + \vec{k}_{13} \cdot \vec{v}\right), \\ \varepsilon_\alpha &= \frac{2(E_e - E_{\alpha 0})}{\Gamma_{\alpha 0}}, \end{aligned}$$

has been obtained, with the CDW function being used for the final state. Here, $A_{exc,\alpha}$ and $A_{dec,\alpha}$ are the amplitudes for excitation and decay of an isolated AIS α; $E_{\alpha 0}$ and $\Gamma_{\alpha 0}$ are the energy and width of the α-th AIS; ${}_2F_1$ is the hypergeometric function.[5] Similar expressions for the resonance amplitude have been derived with an analogous CDW wavefunction in Refs. 50, 51.

In Eq. (176), $K_{PCI}(\varepsilon)$ is the coefficient that takes into account effects of the strong PCI. If we put $K_{PCI} = 1$, then equation (176) gives a symmetric Lorentzian profile. Observe that the kinematic parameters a_{12}, a_{13} are proportional to $R \gg 1$ and $a_{13} = 4m_{13}vR \gg 1$. If a_{12}, $a_{13} \to \infty$, then the argument of ${}_2F_1$ goes to 1, and since

$$ {}_2F_1(-\mathrm{i}\nu_{12}, -\mathrm{i}\nu_{13}, 1, 1) = \frac{\Gamma(1+\mathrm{i}(\nu_{12}+\nu_{13}))}{\Gamma(1+\mathrm{i}\nu_{12})\Gamma(1+\mathrm{i}\nu_{13})}, \tag{178} $$

the amplitude (176) goes over into the eikonal one

$$ A_{res,\alpha} \to A_{res,\alpha}^{(eik)} = f_c^{(+)}(\nu)\frac{2\exp(\mathrm{i}\nu_{12}\ln a_{12} + \mathrm{i}\nu_{13}\ln a_{13})}{\Gamma_{\alpha 0}(\varepsilon_\alpha + \mathrm{i})^{1+\mathrm{i}\nu}} A_{dec,\alpha}A_{exc,\alpha}, \tag{179} $$

where $\nu = \nu_{12} + \nu_{13}$.

In the kinematic region of the strong PCI, $v_e > v$ and $\theta_e \to 0$, the parameter $a_{12} = 2R(|\vec{v}_e - \vec{v}| - v_e \cos\theta_e + v) \to 0$ can be small. In this region two qualitatively new effects, absent in the eikonal approximation, have been theoretically predicted[49]: (i) the intensity of the resonance increases sharply at small ejection angles as a result of the capture of an autoionization electron into the continuum of the scattered ion (an analogue of the ECC peak in the resonance ionization) and (ii) an additional peak appears in the low-energy wing of the resonance profile at small ejection angles $\theta_e \approx 1-5^0$ as a result of the rescattering of some autoionization electrons by a scattered ion. Later, the first effect has been observed in[52] and interpreted as being the result of Coulomb focusing of electrons in the field of the receding ion and the second effect has been detected in[53] and interpreted as resulting from the interference of two coherent amplitudes corresponding to two different classical trajectories of the autoionization electron in the field of the receding ion. However, it is not quite clear how to explain the appearance of this inteference structure directly from the explicit quantum-mechanical amplitude (176). In some sense equation (176), producing an inteference structure, works as a "black box."

5.2. *Interference Effects between Plane and Spherical Distorted Waves*

A more general quantum-mechanical model for the strong PCI kinematic region has been developed in Refs. 54–59. Based on the CDW function with the local momenta (172), this model takes into account the PCI not only in the final state but also in the intermediate quasistationary state. The amplitude of the resonance ionization in this model can be represented as

$$
\begin{aligned}
A_{res,\alpha} &= A^{(p)}_{res,\alpha} + A^{(s)}_{res,\alpha} \\
A^{(p,s)}_{res,\alpha} &= A_{exc,\alpha}(-\mathrm{i}) \int_0^\infty \mathrm{d}t\, A_{dec,\alpha}(\vec{k}^{(p,s)}_{23}(\vec{R}(t))) \\
&\quad \times \exp(\mathrm{i}E_e t - \mathrm{i}\int_0^{t} \mathrm{d}\tau\, E_{c\alpha}(\tau)) \\
&\quad \times F^{(-)*}_{p,s}(\nu_{12}, \xi_{12}(\vec{R}(t))) F^{(-)*}_{eik}(\nu_{13}, \xi_{13}(\vec{R}(t))). \qquad (180)
\end{aligned}
$$

Here,

$$
A_{dec,\alpha}(\vec{k}) = \langle \hat{A}[\psi^{(-)}_{\vec{k}} \psi_{tf}] | V_{ee} | \Phi_\alpha \rangle \qquad (181)
$$

is the amplitude for the autoionization state Φ_α to decay into the final state $\hat{A}[\psi^{(-)}_{\vec{k}} \psi_{tf}]$, as a result of which an electron with momentum $\vec{k}$ is in the continuum and the residual target ion is in the state ψ_{tf}; $\hat{A}$ is the (anti)symmetrization operator with respect to the electron coordinates; V_{ee} is the interelectron interaction operator; E_e is the energy of the autoionization electron; $E_{c\alpha}(t) = E_\alpha(t) - (\mathrm{i}/2)\Gamma_\alpha(t)$ is the time-dependent complex energy of the α-autoionization resonance, taking into account the PCI in the intermediate state; and $\vec{R}(t) = \vec{v}t$.

Similar to the direct ionization amplitude (174), the resonance amplitude (180) is split into the plane and spherical wave contributions. The distortion factors $F^{(-)*}_{p,s}$ account for the scattering of the particle pair (1,2) in the final state with the plane- and spherical wave boundary conditions. The autoionization decay probability determined by the matrix element (181) depends on time and position of the scattered ion: at a fixed time t the autoionization electron undergoes a transition from a bound to a continuum state with momentum $\vec{k}^{(p)}_{23}(t)$ or $\vec{k}^{(s)}_{23}(t)$, depending on the plane- or spherical wave boundary conditions are imposed.

The integrands in equation (180) are proportional to the phase factors $\exp(\mathrm{i}S_{p,s}(t))$, where

$$
S_{p,s}(t) = E_e t - \int_0^t \mathrm{d}\tau\, E_{c\alpha}(\tau) - \Phi_{p,s}(\nu_{12}, \xi_{12}(\vec{v}t)) - \Phi_{eik}(\nu_{13}, \xi_{13}(\vec{v}t)). \qquad (182)
$$

The points t_c, where the phase is stationary, $\dot{S}_{p,s}(t_c) = 0$, determine the region making the main contribution to the time integral, i.e. the time interval when the transition probabilities are maximum. The stationary phase condition can be expressed as

$$E_{fp,s}(t_c) = E_{c\alpha}(t_c), \tag{183}$$

where

$$E_{fp,s}(t) = E_e - \dot{\Phi}_{p,s}(\nu_{12}, \xi_{12}(\vec{v}t)) - \dot{\Phi}_{eik}(\nu_{13}, \xi_{13}(\vec{v}t)) \tag{184}$$

is the time-dependent energy of the autoionization electron in the final state, i.e. transitions occur at the moment the electronic terms in the intermediate (resonance) and final states cross. The transition points are displaced from the real axis into the complex plane, since the intermediate state is quasistationary.

If the effect of the PCI in the intermediate state is neglected, then the energy and width of the resonance state do not depend on time and are atomic parameters, defined for an isolated atom such that $E_{c\alpha 0} = E_{\alpha 0} - (\mathrm{i}/2)\Gamma_{\alpha 0}$ is a complex energy parameter. Using the explicit WKB expressions for the $\Phi_{p,s}$ phase functions and $E_{c\alpha 0}$ as an atomic parameter, the stationary phase equation simplifies to

$$\pm\sqrt{1 - 4\nu_{12}/(a_{12}((\Gamma_{\alpha 0}/2)t)} = 1 + 2(\varepsilon_\alpha + \mathrm{i})/a_{12} + 2\nu_{13}/(a_{12}((\Gamma_{\alpha 0}/2)t). \tag{185}$$

This quadratic equation with respect to t can be solved analytically giving the stationary phase points $t_{p,s}$.[55]

Taking the decay matrix elements in (180) out of the sign of integral at the stationary phase points t_p and t_s, we obtain

$$A^{(p,s)}_{res,\alpha} = -\frac{2\mathrm{i}}{\Gamma_{\alpha 0}} A_{exc,\alpha}\, A_{dec,\alpha}(\vec{k}^{(p,s)}_{23}(t_{p,s}))I_{p,s}(\varepsilon_\alpha), \tag{186}$$

where the remaining integrals

$$I_{p,s}(\varepsilon) = \int_0^\infty \mathrm{d}\tau\, \exp[(\mathrm{i}\varepsilon - 1)\tau]\tau^{\mathrm{i}\nu_{13}} F^{(-)*}_{p,s}(\nu_{12}, a_{12}\tau) \tag{187}$$

can be calculated analytically[60]

$$I_p(\varepsilon) = \frac{\exp(\pi\nu_{12}/2)}{(1-\mathrm{i}\varepsilon)^{1+\mathrm{i}\nu_{13}}} \frac{\Gamma^2(1+\mathrm{i}\nu_{13})}{\Gamma(1+\mathrm{i}\nu_{13}-\mathrm{i}\nu_{12})} \times {}_2F_1\left(-\mathrm{i}\nu_{12}, 1+\mathrm{i}\nu_{13}, 1+\mathrm{i}\nu_{13}-\mathrm{i}\nu_{12}, \frac{1-\mathrm{i}(a_{12}+\varepsilon)}{1-\mathrm{i}\varepsilon}\right), \tag{188}$$

$$
\begin{aligned}
I_s(\varepsilon) &= \mathrm{i}\nu_{12}\exp(2\mathrm{i}\arg\Gamma(1+\mathrm{i}\nu_{12})) \\
&\times \frac{\exp(\pi\nu_{12}/2)}{(1-\mathrm{i}(a_{12}+\varepsilon))^{1+\mathrm{i}\nu_{13}}}\frac{\Gamma^2(1+\mathrm{i}\nu_{13})}{\Gamma(2+\mathrm{i}\nu_{13}+\mathrm{i}\nu_{12})} \\
&\times {}_2F_1\left(1+\mathrm{i}\nu_{12}, 1+\mathrm{i}\nu_{13}, 2+\mathrm{i}\nu_{13}+\mathrm{i}\nu_{12}, \frac{1-\mathrm{i}\varepsilon}{1-\mathrm{i}(a_{12}+\varepsilon)}\right). \qquad (189)
\end{aligned}
$$

If we put the local momenta in the decay amplitude $\vec{k}_{23}^{(p,s)} \equiv \vec{v}_e$, then the resonance amplitude (180) will coincide exactly with the amplitude (176) (if the internuclear interaction is taken in the eikonal approximation $a_{13} \to \infty$). The separation (180) of the resonance amplitude makes it possible to investigate in detail the relative contribution of the plane and spherical waves to the electron intensity as well as the possible interference between them.

In Fig. 7 we show the electron energy spectra near the $(2s^2)^1S$ resonance of the helium atom, excited in a collision with ions with velocity $v < v_e$, the velocity of autoionization electrons, at ejection angle $\theta_e = 5^0$, calculated using equations (180), (186) for the amplitude: curve 1 — the total calculation, curves 2 and 3 — separate contributions from the plane and spherical waves, respectively. The results are presented for 10-keV $^3He^+$ (bottom) and $^3He^{2+}$ (top panel) projectile ions. To fully incorporate the influence of the electronic structure of projectile on the electron spectra, a generalization of the CDW function[48–51] has been proposed in Ref. 61 that gives a more precise quantum-mechanical description of the emitted autoionization electron-partially-stripped-projectile ion interaction. One can see that a distructive interference between plane and spherical waves results in the appearence of a dip in the resonance profile. With increasing the charge of the projectile ion, the interference dip becomes even more pronounced. The experimental data for 10-keV $^3He^+$ – He atomic system of Ref. 53 are also presented in Fig. 7. For comparison purposes, the theoretical profile (curve 1) has been convolved with a Gaussian energy-resolution function with FWHM = 0.3 eV. The result of convolving is presented by curve 4. One can see that the less pronounced dip present in the theoretical profile disappears in the convolved one (bottom panel), but the more pronounced interference dip in the top panel survives the convolution procedure.

However, it should be noticed that the above numerical example which conceives the observed interference structure in the spectra as a result of interference between the plane and spherical waves is not fully consistent with the interpretation given by Swenson *et al.* in the quasi-classical model.[53] This is the case for the part where there is a cluster (or continuum)

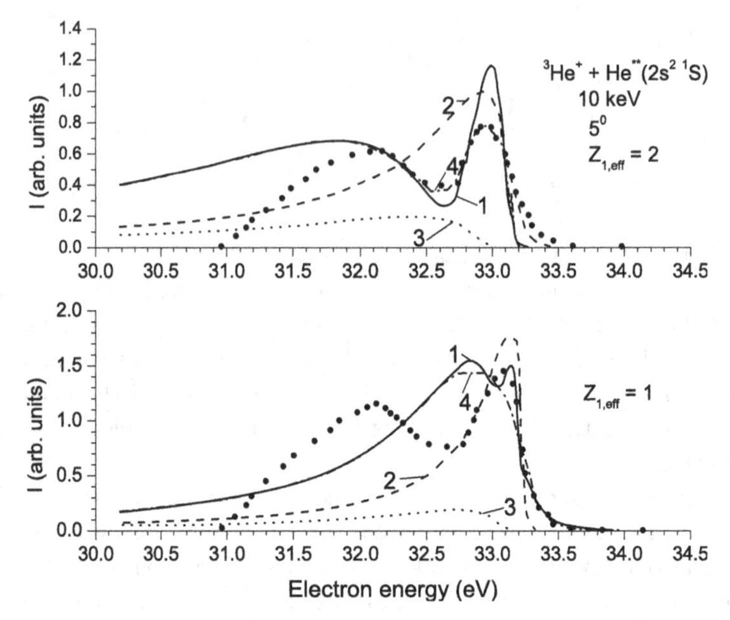

Fig. 7. Energy dependence of the relative intensity $I(E_e)$ of electrons ejected at angle $\theta_e = 5^0$ during the decay of the autoionization $(2s^2)^1S$ state excited in a helium atom in a collision with 10-keV $^3He^+$ ions. The results of calculation according to equation (180) with effective $^3He^+$ ion charges $Z_{1,eff} = 1$ (bottom) and $Z_{1,eff} = 2$ (top panel) are shown: curve 1 — total calculation; curves 2 and 3 — separate plane and spherical wave contributions, respectively; curve 4 was obtained by convolving the theoretical profile (curve 1) and a Gaussian energy-resolution function with FWHM = 0.3 eV. The experimental points are taken from Ref. 53.

of classical trajectories corresponding to either the plane or spherical waves. Therefore, it is not quite clear how to define the two interfering Coulomb paths, as in Ref. 53, and to make a viable analogy with the electron scattering interference experiment on the double-slit screen.

5.3. *Unitarized Post-Collision Interaction Models*

Let us consider the differential in angle yield of the autoionization electrons, the area under the resonance profile. If we integrate the resonance profile over the energy, we get for the Coulomb focusing (CF) coefficient the following expression[62,63]

$$K_{CF} = \frac{2\pi\nu_{12}}{\exp(2\pi\nu_{12}) - 1} \exp\{2\nu_{12} \arctan(a_{12}/2)\} \times {}_2F_1(-i\nu_{12}, i\nu_{12}, 1, a_{12}^2/(a_{12}^2 + 4)). \tag{190}$$

K_{CF} describes the change in the differential resonance yield due to the PCI. As $a_{12} \to \infty$, that corresponds to the eikonal description of the PCI, $K_{CF} \to 1$. The PCI in the eikonal approximation does not change the differential resonance yield, $K_{CF} \equiv 1$, but does redistribute the autoionization electrons in energy. In the kinematic region of the strong PCI, at $v_e > v$ and small ejection angles θ_e, the parameter $a_{12} \sim 1$ and the differential yield enhances, $K_{CF} > 1$ at $\nu_{12} < 0$. Moreover, in the forward direction $\theta_e = 0$ the yield enhances as $K_{CF} = |f_c^{(-)}(\nu_{12})|^2$.

However, it should be noticed that equation (190) does not reproduce the CF effect rigorously. Rather, it predicts the differential yield enhancement at all ejection angles, thus violating the unitarity condition. This is because in deriving this formula we have assumed that the scattered ion has no influence on the decay probability of the AIS α. In other words, the decay takes place with the same width $\Gamma_{\alpha 0}$, and the autoionization electrons have the same energy distribution as in the absence of the scattered ion. The distribution of the ejected electrons changes as a result of interaction of the charged particles during their subsequent motion after the decay of the AIS (interaction in the final state). Taking into account the interaction in the intermediate state results in a modification of the time-dependent complex energy of the αth AIS, $E_{c\alpha}(t) = E_\alpha(t) - (\mathrm{i}/2)\Gamma_\alpha(t)$, in equation (180). As a result, we obtain a unitarized version for the CF coefficient:

$$K_{CF} = \Gamma_{\alpha 0} \int_0^\infty dt \exp\left(- \int_0^t \mathrm{d}\tau\, \Gamma_\alpha(\tau) \right) |F^{(-)}(\nu_{12}, a_{12}(\Gamma_{\alpha 0}/2)t)|^2 \quad (191)$$

If we put $\Gamma_\alpha(\tau) \to \Gamma_{\alpha 0}$ in (191), then evaluation of the integral results in equation (190). In (191), the instantaneous width $\Gamma_\alpha(\tau)$ is integrated with respect to time over the interval $(0, t)$. Therefore, it is more convenient to work with the width averaged over the interval $(0, t)$, which is more directly related to the physical characteristic of the decay process:

$$\begin{aligned} \bar{\Gamma}_{\alpha=LM}(R = vt) &= \frac{1}{t} \int_0^t \mathrm{d}\tau\, \Gamma_\alpha(\tau) \\ &= \Gamma_{\alpha 0} \int \mathrm{d}\Omega_e\, |Y_{LM}(\Omega_e)|^2 \Phi(t, \Omega_e), \end{aligned} \quad (192)$$

where we have introduced the width distortion function due to interaction in the intermediate state[54]

$$\Phi(t, \Omega_e) = \frac{1}{t} \int_0^t \mathrm{d}\tau |F^{(-)}(\nu_{12}, a_{12}(\Gamma_{\alpha 0}/2)\tau)|^2. \quad (193)$$

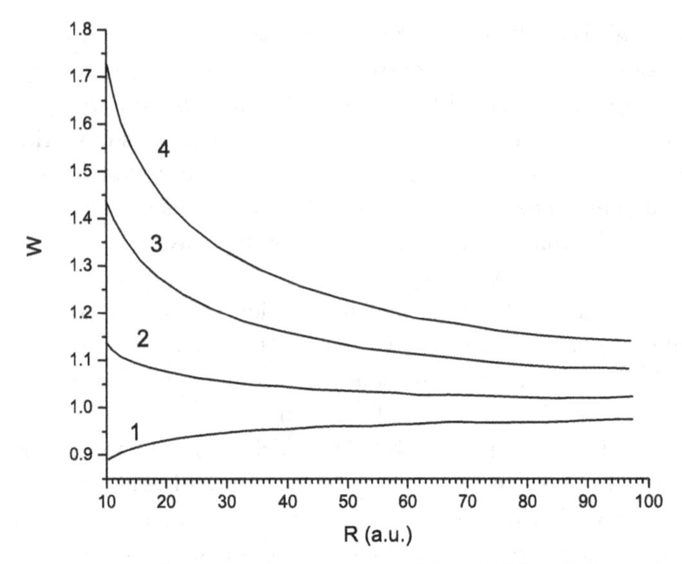

Fig. 8. Dependence on the internuclear separation R (a.u.) of the relative width $W(R)$ of the $(2s^2)^1$S resonance of the helium atom decaying in the Coulomb field of scattered ions with different charges Z_1 and velocity $v = 1.0$ a.u. Curves 1–4 correspond to calculations with the charges $Z_1 = -1$, 1, 5 and 10, respectively.

In Fig. 8, we present the results of calculations in accordance with equation (192) of the relative width $W(R) = \bar{\Gamma}_{LM}(R)/\Gamma_0$ of the $(2s^2)^1S$ resonance of the helium atom (the unperturbed resonance energy and width are $E_0 = 1.222$ a.u. and $\Gamma_0 = 0.005$ a.u., respectively) excited in a collision with ions having velocity $v = 1.0$ a.u. and possessing charges $Z_1 = \pm 1, 5, 10$ of both signs and different magnitudes. One can see that with increasing charge Z_1 the influence of distortion of the resonance width by the Coulomb field of the scattered ion increases. Depending on the sign of the charge Z_1, the width of the resonance increases in the field of a positive charge, but decreases in the field of a negative charge. The increase (decrease) of the resonance width in the case of positive (negative) charge Z_1 leads by virtue of (191) to a decrease (increase) in the differential and the total integrated resonance yields. This compensates for the increase (decrease) resulting from allowance for the final-state interaction in a such way that the total integrated yield of the resonance is unchanged when simultaneous allowance is made for interaction in the final state and in the resonance state.

Figure 9 illustrates this property of unitarity, namely, it shows the angular dependence of K_{CF} equation (191) calculated for the $(2s^2)^1S$ resonance of the helium atom decaying in the Coulomb field of the scattered

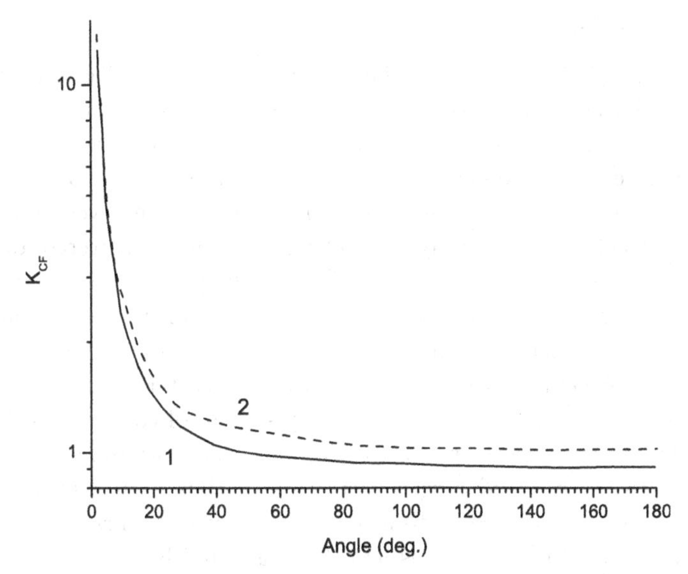

Fig. 9. Angular dependence of the Coulomb focusing coefficient K_{CF} for the $(2\mathrm{s}^2)^1\mathrm{S}$ resonance of the helium atom decaying in the Coulomb field of receding ion with charge $Z_1 = 5$ and velocity $v = 0.36$ a.u. Curves 2 and 1 are calculations within the non-unitarized (equation [190]) and unitarized (equation [191]) models, respectively.

ion with charge $Z_1 = 5$ and velocity $v = 0.36$ a.u., which corresponds to 10 keV energy of He^+ ion. One can see that additional allowance for interaction in the intermediate state results in a decrease of the angle-dependent yield of autoionization electrons such that the integrated (over angle) yield remains the same.

6. Effects of the Continuum Distortion in Charge Transfer

In this section we consider an application of the bound state CDW function to the process of electron capture

$$\mathrm{I}^{Z_1} + (\mathrm{A}^+, e)_i \to (\mathrm{I}^{Z_1}, e)_f + \mathrm{A}^+, \tag{194}$$

where the incident ion I^{Z_1} impinges the atom $(\mathrm{A}^+, e)_i$ in a bound state ψ_{ti} and as a result of collision captures an electron e forming the scattered ion (atom) $(\mathrm{I}^{Z_1}, e)_f$ in a bound state ψ_{pf}. Here, one can distinguish three active particles involved in the reaction: 1 — the incident ion I^{Z_1}, 2 — the captured electron e, and 3 — the residual target ion A^+.

6.1. *Thomas Peak*

It is now widely believed that the Thomas mechanism for electron capture plays a vital role at high velocity collisions. In 1927 Thomas[64] gave a classical treatment of the capture by a fast point projectile of a bound electron whose orbital velocity v_e is much less than the velocity of the projectile v. The two-step mechanism he suggested involved a double scattering, whereby the nearly free electron is first scattered off the projectile at a laboratory angle of 60^0, for which it attains a velocity equal to that of the projectile, and then elastically off the target nucleus to redirect this velocity vector in the direction of the projectile. In the classical two-collision treatment, the projectile is scattered to the Thomas angle $(\theta_T)_{lab} = \sqrt{3}/(2m_1)$ in the laboratory system of coordinates. In the quantum treatment, it corresponds to a second-order Born process in which the projectile-electron and target-electron potentials each act once, and indeed the second order Born term dominates over the first order Born term in the limit of high v as was first demonstrated by Drisco in 1955.[65] In the differential cross section, the Thomas mechanism is revealed by a peak (or shoulder) at the Thomas angle. Experimentally, the Thomas peak has been first observed in the angular distributions reported for electron capture by protons of 5.42 and 7.40 MeV from He.[66]

Several more elaborate quantum theories, such as the second Born-Faddev approximation (B2F),[67] the strong-potential Born (SPB) approximation[68] or the impulse approximation (IA),[69–71] that include second- and higher-order terms have been quite successful in reproducing the Thomas peak. Interestingly, while the CDW function for charge transfer[20,37,72] is not usually described as a two-step process, it nevertheless produces a Thomas peak at high velocities and has some similarities to the second-Born approximation in the high-velocity limit. Therefore, it is important to enlighten this feature of the CDW function in detail by tracing down how the Thomas structure appears in the differential cross section.

The *prior*-form amplitude for charge exchange can be written as

$$t_{fi}^{(-)} = \langle \psi_f^{(-)} | V_i | \phi_i \rangle, \tag{195}$$

where ϕ_i is the asymptotic initial state wavefunction represented by a product of the ground state ψ_{ti} for the target atom and the plane wave describing the relative motion of the incident ion and the target atom. Here, V_i is the interaction potential between the target atom and projectile in the entrance channel. With the bound state CDW function taken as an exact final state $\psi_f^{(-)}$, the amplitude (195) takes the form[73]

$$t_{fi}^{(-)} = -(2\pi)^{-3} Z_1 N_c \lim_{\varepsilon \to +0} \int \mathrm{d}^3\vec{s}\, \bar{\psi}_{pf}^*(\vec{s}) \\ \times I_{13}(\vec{v} + \vec{Q} - \vec{s}, \varepsilon) I_{23}(\vec{s} - \vec{v}), \tag{196}$$

which is Geltman's Coulomb-Born (CB) approximation,[37] where

$$\bar{\psi}_{pf}(\vec{s}) = (2\pi)^{-3} \int \mathrm{d}^3\vec{r}\, \psi_{pf}(\vec{r}) \frac{1}{r} \exp(-\mathrm{i}\vec{s}\cdot\vec{r}), \tag{197}$$

$$I_{13}(\vec{u}, \varepsilon) = \int \mathrm{d}^3\vec{r} \exp(\mathrm{i}\vec{r}\cdot\vec{u} - \varepsilon r)\, {}_1F_1(-\mathrm{i}\nu_{13}, 1, \mathrm{i}(k_{13}r + \vec{k}_{13}\cdot\vec{r})), \tag{198}$$

$$I_{23}(\vec{u}) = \int \mathrm{d}^3\vec{r} \exp(\mathrm{i}\vec{r}\cdot\vec{u})\, \psi_{ti}(\vec{r})\, {}_1F_1(-\mathrm{i}\nu_{23}, 1, \mathrm{i}(k_{23}r + \vec{k}_{23}\cdot\vec{r})), \tag{199}$$

and

$$N_c = f_c^{(+)}(\nu_{23}) f_c^{(+)}(\nu_{13}),$$
$$\nu_{23} = -Z_3/v, \quad \nu_{13} = Z_1 Z_3/v,$$
$$\vec{k}_{ij} = m_{ij}\vec{v}, \quad \vec{Q} = \left(-\frac{v}{2} + \frac{\Delta\epsilon_{fi}}{v}\right)\vec{n}_i - (m_{13}v\theta)\vec{n}_\varphi.$$

Here, $\vec{v}$ is projectile's velocity before collision; $\vec{n}_i = \vec{v}/v$; $\vec{Q}$ is the momentum lost by projectile ion as a result of collision; θ is the c.m. scattering angle; $\vec{n}_\varphi$ is a unit vector in the azimuthal direction such that $\vec{n}_\varphi \cdot \vec{n}_i = 0$; the defect of energy $\Delta\epsilon_{fi} = \epsilon_f - \epsilon_i$, where $\epsilon_{i(f)}$ is the bound energy of the initial (final) state.

The integral I_{13} in (196) is a δ-like function (if ${}_1F_1 \equiv 1$, then $I_{13}(\vec{u}, \varepsilon \to 0+) = (2\pi)^3\delta(\vec{u})$) with a singularity located at $\vec{s} = \vec{v} + \vec{Q}$. Using Taylor's series expansion, we obtain

$$\bar{\psi}_{pf}(\vec{s}) = \bar{\psi}_{pf}(\vec{v} + \vec{Q}\,) + \frac{\partial\bar{\psi}_{pf}(\vec{v} + \vec{Q}\,)}{\partial\vec{s}} \cdot (\vec{s} - \vec{v} - \vec{Q}\,) + \cdots \tag{200}$$

Substituting expansion (200) into (196), we get

$$t_{fi}^{(-)} = t_{peak} + \Delta t_{peak}^{(1)} + \cdots. \tag{201}$$

Here, t_{peak} and $\Delta t_{peak}^{(1)}$ are contributions from the first and second terms in (200). The first amplitude is obtained in the peaking approximation, while the next one is a correction to it. Analytic expressions for these contributions with the simplest hydrogen-like ground state ψ_{ti} have been derived in Ref. 73.

We display in Figs. 10 and 11 the results of calculations[73] obtained for the differential cross sections for electron capture by protons on helium atoms, respectively, at high (5.42 and 7.40 MeV) and intermediate (100 keV) collision energies. These correspond to the velocities $v = 14.72$, 17.20 and 2.0 a.u., together with the available experimental data.[66,74] For quantitative comparison to the data, the theoretical results have been folded with the corresponding experimental angular resolution functions. As one can see, the CB calculation, curve 1, is in a good agreement with the experimental data and reproduces well the Thomas peak seen at high energies. At intermediate energies, the Thomas peak does not appear in angular distributions and this corresponds to the data. In Fig. 10, also shown are the results of the calculation in the impulse (IA, curve 2) and eikonal (curve 3) approximations. As opposed to the eikonal approximation, the IA is seen to reproduce the Thomas peak. In Fig. 11, curves 1 and 2

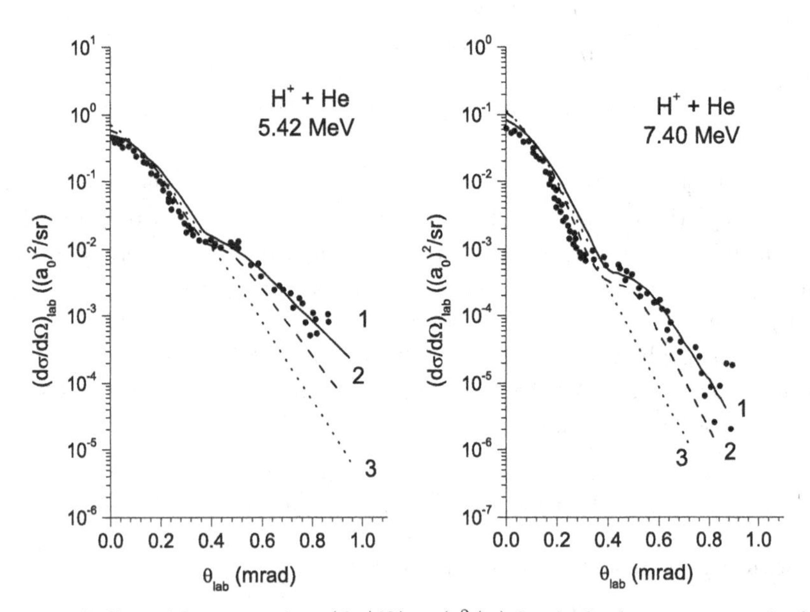

Fig. 10. Differential cross sections $(d\sigma/d\Omega)_{lab}$ (a_0^2/sr) for single-electron capture in the H^+–He collisions as a function of the scattering angle θ_{lab} (mrad) at 5.42 (left) and 7.40 MeV (right panel). a_0 is the Bohr radius. Cross sections and scattering angles are in the laboratory system. Curve 1: calculation in the peaking Coulomb-Born approximation from Eq. (201). Curves 2 and 3: calculations in the impulse and eikonal approximations, respectively. Experimental data are from Ref. 66.

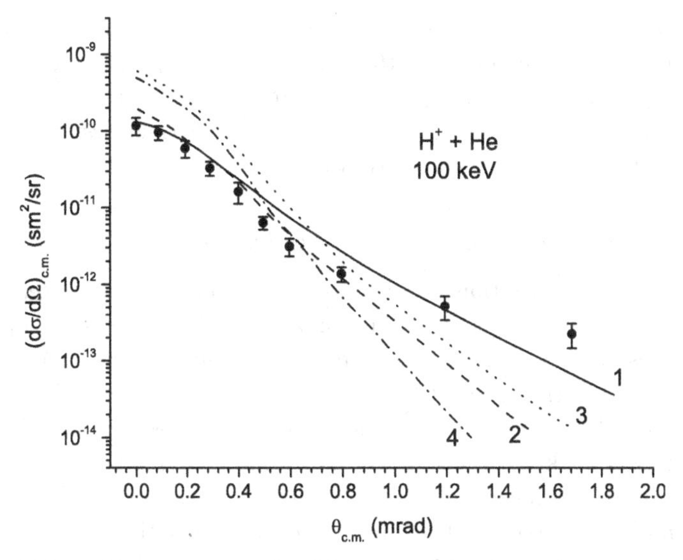

Fig. 11. Differential cross sections $(d\sigma/d\Omega)_{c.m.}$ (sm^2/sr) for single-electron capture in the H^+–He collisions as a function of the scattering angle $\theta_{c.m.}$ (mrad) at 100 KeV. Cross sections and scattering angles are in the center-of-mass system. Curves 1 and 2 are calculations in the same approximations as in Fig. 10. Curve 3: calculation in the peaking Coulomb-Born approximation from Eq. (201), neglecting the correction term. Curve 4: the OBK approximation. Experimental data are from Ref. 74.

are the same as in Fig. 10, whereas curve 3 is the CB calculation with the correction term $\Delta t^{(1)}_{peak}$ being neglected. From these results one can conclude that the peaking approximation, the first term in (201), gives a very crude estimate, at least at the intermediate energies. Curve 4 is a calculation in the first-order Oppenheimer-Brinkman-Kramers (OBK) approximation, which can be derived from the CB one if the continuum distortion factors are neglected.

Since the internuclear interaction does not affect qualitatively the Thomas peak, we can put $\nu_{13} \equiv 0$ in (198) and the amplitude is evaluated exactly to coincide with the result from the CIS method[37]

$$t^{(-)}_{fi} = t_{OBK} f^{(+)}_c(\nu_{23})(1+u)^{\mathrm{i}\nu_{23}} \left(1 + \frac{(\mathrm{i}\nu_{23}/\alpha)}{1+u}(\mathrm{i}v + \alpha u)\right) \tag{202}$$

$$u = 2\frac{\vec{v}\cdot\vec{Q} - \mathrm{i}\alpha v}{Q^2+\alpha^2},$$

where t_{OBK} is the OBK amplitude and $\alpha = Z_3$ the effective charge of the target ion. Then, at high velocities we have in the leading order $\vec{v}\cdot\vec{Q} = -v^2/2$

and $Q^2 = v^2(0.25 + m_{13}^2\theta^2)$, where θ is the scattering angle in the center-of-mass (c.m.) system. Therefore, at the Thomas angle $(\theta_T)_{c.m.} = \sqrt{3}/(2m_{13})$ we obtain $Q^2 + 2\vec{v}\cdot\vec{Q} = 0$, so that the last term in (202), containing $1+u$ in the denominator, will increase sharply at $\theta = \theta_T$, thus, resulting in the Thomas peak in the differential cross section.

6.2. *Plane and Spherical Wave Contributions*

In the resonance ionization section we have seen that an interference structure formed in the spectra of autoionization electrons at small ejection angles is a result of interference between the plane and spherical parts of the corresponding (electron-projectile) continuum distortion factor. It is highly plausible that the same interference mechanism also lies behind formation of the Thomas peak when the CDW function is used. In order to check this assumption, we have to derive separate contributions to the electron capture amplitude from the plane and spherical parts of the electron-target-ion continuum distortion factor. To further simplify calculations, we neglect the internuclear interaction, $\nu_{13} \equiv 0$. Then, the CIS amplitude (202) can be separated as

$$t_{fi}^{(-)} = t_{fi}^{(p)} + t_{fi}^{(s)} = -Z_1\tilde{\psi}_{pf}^*(\vec{v}+\vec{Q})(I_{23}^{(p)}(\vec{Q}) + I_{23}^{(s)}(\vec{Q})), \qquad (203)$$

$$I_{23}^{(p,s)}(\vec{Q}) = \int \mathrm{d}^3\vec{r}\,\exp(\mathrm{i}\vec{r}\cdot\vec{Q})\,\psi_{ti}(\vec{r})\,F_{p,s}(\nu_{23}, vr + \vec{v}\cdot\vec{r}), \qquad (204)$$

where the plane and spherical continuum distortion functions $F_{p,s}$ are defined by equations (15) and (16). Here, as opposed to (199), we have included the Coulomb normalization factor $f_c^{(+)}(\nu_{23})$ into the definition of $I_{23}(\vec{Q})$.

It is convenient to express these distortion functions via Kummer's confluent hypergeometric function U[5]:

$$F_p(\nu,\zeta) = \exp(\pi\nu/2)U(-\mathrm{i}\nu, 1, \mathrm{i}\zeta),$$
$$F_p(\nu,\zeta) = \mathrm{i}\nu\exp(\pi\nu/2)\frac{\Gamma(1+\mathrm{i}\nu)}{\Gamma(1-\mathrm{i}\nu)}\exp(\mathrm{i}\zeta)U(1+\mathrm{i}\nu, 1, -\mathrm{i}\zeta). \qquad (205)$$

For U function there exists an integral representation[5]:

$$U(a,b,z) = \frac{1}{\Gamma(a)}\int_0^\infty \mathrm{d}t\,\exp(-zt)t^{a-1}(1+t)^{b-a-1}, \qquad (206)$$
$$(\mathrm{Re}\,a > 0,\ \mathrm{Re}\,z > 0).$$

Making use of this integral representation, we can easily evaluate integrals over coordinates. Thus, with the $1s$ hydrogen-like bound state $\psi_{ti}(r) = N_\alpha \exp(-\alpha r)$, where the normalization constant $N_\alpha = \sqrt{(\alpha^3/\pi)}$, we obtain

$$I_{23}^{(p)}(\vec{Q}) = N_\alpha \exp(\pi\nu_{23}/2)\frac{1}{\Gamma(-\mathrm{i}\nu_{23})}\left(-\frac{\partial}{\partial\alpha}\right)\int_0^\infty \mathrm{d}t\,(t+1)^{\mathrm{i}\nu_{23}}t^{-\mathrm{i}\nu_{23}-1} \times \frac{4\pi}{Q^2+\alpha^2-2(\vec{Q}\cdot\vec{v}-\mathrm{i}\alpha v)t}, \tag{207}$$

$$I_{23}^{(s)}(\vec{Q}) = N_\alpha \frac{\mathrm{i}\nu_{23}}{f_c^{(-)}(\nu_{23})}\left(-\frac{\partial}{\partial\alpha}\right)\int_0^\infty \mathrm{d}t\,t^{\mathrm{i}\nu_{23}}(t+1)^{-\mathrm{i}\nu_{23}-1} \times \frac{4\pi}{Q^2+\alpha^2+2(\vec{Q}\cdot\vec{v}-\mathrm{i}\alpha v)(t+1)}. \tag{208}$$

Using an integral representation for the hypergeometric function ${}_2F_1$[5]:

$${}_2F_1(a,b,c,z) = \frac{\Gamma(c)}{\Gamma(b)\Gamma(c-b)}\int_0^\infty \mathrm{d}t\,t^{-b+c-1}(t+1)^{a-c} \times (t-z+1)^{-a}, \tag{209}$$

equations (207) and (208) can be rewritten as

$$I_{23}^{(p)}(\vec{Q}) = N_\alpha \frac{1}{f_c^{(-)}(\nu_{23})}\left(-\frac{\partial}{\partial\alpha}\right)\frac{(-4\pi)}{2(\vec{Q}\cdot\vec{v}-\mathrm{i}\alpha v)} \times {}_2F_1\left(1,1,1-\mathrm{i}\nu_{23},1+\frac{Q^2+\alpha^2}{2(\vec{Q}\cdot\vec{v}-\mathrm{i}\alpha v)}\right), \tag{210}$$

$$I_{23}^{(s)}(\vec{Q}) = N_\alpha \frac{\mathrm{i}\nu_{23}/(1+\mathrm{i}\nu_{23})}{f_c^{(-)}(\nu_{23})}\left(-\frac{\partial}{\partial\alpha}\right)\frac{4\pi}{2(\vec{Q}\cdot\vec{v}-\mathrm{i}\alpha v)} \times {}_2F_1\left(1,1,2+\mathrm{i}\nu_{23},-\frac{Q^2+\alpha^2}{2(\vec{Q}\cdot\vec{v}-\mathrm{i}\alpha v)}\right). \tag{211}$$

With the help of a linear transformation formula[5]

$${}_2F_1(a,b,c,z) = \frac{\Gamma(c)\Gamma(c-a-b)}{\Gamma(c-a)\Gamma(c-b)}{}_2F_1(a,b,a+b-c+1,1-z) + (1-z)^{c-a-b}\frac{\Gamma(c)\Gamma(a+b-c)}{\Gamma(a)\Gamma(b)} \times {}_2F_1(c-a,c-b,c-a-b+1,1-z) \tag{212}$$

and an elementary expression for

$$ {}_2F_1(a,b,b,z) = (1-z)^{-a} \tag{213} $$

it is easy to check that their coherent sum can be reduced to

$$ I_{23}^{(p)}(\vec{Q}\,) + I_{23}^{(s)}(\vec{Q}\,) = N_\alpha f_c^{(+)}(\nu_{23}) \left(-\frac{\partial}{\partial\alpha}\right) \frac{4\pi}{Q^2+\alpha^2} \times \left[1 + \frac{2(\vec{Q}\cdot\vec{v} - \mathrm{i}\alpha v)}{Q^2+\alpha^2}\right]^{\mathrm{i}\nu_{23}} . \tag{214} $$

The same analytic result can be directly derived by evaluating the integral (199).

Figure 12 shows the results of the total and separate wave contributions to the differential cross section for $1s$-$1s$ electron transfer in $p+\mathrm{H}$ collisions at intermediate and high velocities $v = 1, 2, 4, 8$ and 16 a.u. One can see that the plane wave contribution, shown by a dashed line, is dominant at small scattering angles $\theta_{c.m.} < 0.5\,\mathrm{mrad}$, whereas near the Thomas angle

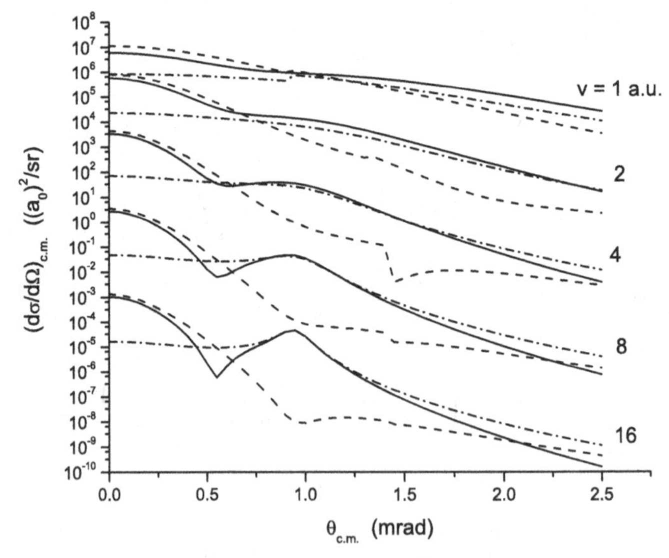

Fig. 12. Differential cross sections $(d\sigma/d\Omega)_{c.m.}$ (a_0^2/sr) for single-electron capture in the H^+–H collisions as a function of the scattering angle $\theta_{c.m.}$ (mrad) at intermediate and high velocities $v = 1$, 2, 4, 8, and 16 a.u. a_0 is the Bohr radius. Solid curve: the present CIS calculation including the coherent contributions from the plane and spherical waves. Dashed and dot-dashed curves: separate contributions from the plane and spherical waves, respectively.

$(\theta_T)_{c.m.} = \sqrt{3}/(m_p) = 9.4\,\text{mrad}$, where the proton mass $m_p = 1836$, the spherical wave contribution (dash-dotted line) becomes dominant at high enough velocities $v \gtrsim 4$ a.u. At $\theta_{c.m.} \approx 0.5\,\text{mrad}$ the plane- and spherical wave contributions to the amplitude are seen to be comparable in magnitude and, thus, they can effectively interfere. In the composite differential cross sections (solid line), this destructive interference results in a dip at $\theta_{c.m.} \approx 0.5\,\text{mrad}$. Observe that at large $\theta_{c.m.} > 1.5\,\text{mrad}$, on the right-hand-side wing of the Thomas peak, the spherical wave contribution to the amplitude remains bigger than the plane-wave one, although the difference between them is slowly decreasing with increasing $\theta_{c.m.}$. This Section 6 deals only with the CB and CIS methods.[37] More involved calculations are required in the Cheshire method,[34] which employs the CDW wavefunctions for both the initial and final scattering states.

7. Discussion and Conclusions

In this work we have overviewed a number of analytical methods that proved to be useful in theoretical studies of energetic ion-atom collisions. Our starting point is the known exact two-body Coulomb scattering state which can be split into the two parts that asymptotically at large separations between particles behave, respectively, as distorted plane and spherical waves. These plane and spherical waves are exact solutions to the Schrödinger equation that are irregular (logarithmically divergent) at origin. However, coefficients in their sum are chosen such that the logarithmic divergences cancel each other and the total scattering state is regular at origin. Besides the QM solutions, we give a detailed account of the corresponding solutions in the WKB and eikonal approximations. The WKB solutions are shown to reproduce well behavior of the QM ones, except in the regions near the caustics points where the WKB solutions become singular. At caustics, the WKB solutions should be replaced by more accurate and more complicated integral expressions.[75–78]

Further, we consider possible generalizations to three-body scattering states. Here, our primary concern is to construct approximate scattering states for the system of three charged particles into the continuum that satisfy appropriate asymptotic Coulomb boundary conditions. At large interparticle separations the scattering state in the eikonal approximation can be factorized into pairwise distortion factors F_{ij}^{eik}. Then, using correspondence (118) between eikonal, WKB and QM distortion factors, we define the CDW function as in equation (119) and investigate its normalization and asymptotic properties in detail. In the asymptotic region Ω_0, where all particles are well separated, the CDW function satisfies the Schrödinger equation up to the second order terms, while in the asymptotic region Ω_{ij}, where the pair of particles (i, j) is not well separated and the 3rd particle is far way, the residual term is of the first order. In order to get a scattering solution up to a second order residual term in Ω_{ij}, we introduce the concept of three-body *local* momenta that take into account effects of the long-range Coulomb field of a distant particle on the relative motion of a slightly separated pair of particles i and j. Within the WKB approximation, we first find approximate solutions including the local momenta that satisfy the Hamilton-Jakobi equation up to the second order terms in Ω_{ij}. If the local momenta were replaced by asymptotic ones, then after substitution this solution into the equation the residual terms would be of the first order. Having obtained an action function as our solution, we can readily calculate the WKB amplitude using an analytic expression (137), by evaluating the determinant of a matrix obtained as a result of the second-order differentiation of the action function with respect to momenta and coordinates. Applying this formula to a simplified action function, where the local momenta are replaced by asymptotic ones, we demonstrate that, in general, the amplitude cannot be reduced to a factorized form (148). From this result one can conclude that the CDW function (119) is not properly normalized even in the asymptotic regions. Taking advantage of equation (137) or its possible generalizations might greatly improve the normalization properties of the CDW function. As of now, its usefulness has not yet been tested in applications.

The bound state CDW function can be obtained by doing an analytic continuation from the two-particle continuum state to a bound state. The CDW function (153) satisfies the Schrödinger equation up to the first order terms. Similar to the case of three asymptotically free particles, we have to introduce the local momenta into collisional dynamics of two-body discrete states in order to get a solution up to the second order residual terms. In this way, we have derived equations in the basis of discrete states that describe, in general, non-unitary collisional dynamics induced by the *complex* local momenta. Solving this equation gives an approximate solution (154), (155), where $\psi_{p,s}^{(+)}$ are defined by equations (161), (165) or (166), that satisfies the Schrödinger equation up to the second order terms.

Thus, in the first half of the paper we have reviewed basic analytic methods that allowed us to derive approximate three-body continuum and bound state solutions that satisfy the Schrödinger equation up to the second order terms in asymptotic regions. It is of great interest to test these solutions in applications to energetic ion-atom collisions and, in particular, to reveal effects due to the local momenta. We restricted ourselves to the processes involving single-electron transitions. For possible generalizations of the Cheshire[34] CDW method to the processes involving two- or few-electron transitions, see, e.g., Refs. 20, 21 and references therein. In two- or more electron processes the dynamic correlations due to interelectron interactions are shown to play a significant role.

The numerical results presented in Section 4 clearly demonstrate that the local momenta in the CDW function (172) have a significant effect on the direct ionization DDCS. Moreover, separation of the scattering state into the plane and spherical parts proves to be helpful in understanding the underlying scattering mechanisms that form the observed electron spectra. For example, we find that on the high-energy wing of the ECC peak the spherical wave contribution turns out to be dominant over the plane wave. This means that electrons ejected in this part of the spectrum will predominantly be rescattered by the projectile ion. This numerical example shows that ionization models, including the CDW function with local momenta as scattering states, can be advantageous. It stimulates further exploration of more complicated expressions for the CDW functions with local momenta that have been derived in Section 3. However, the computational cost in evaluating the corresponding ionization amplitudes with such scattering states is expected to be higher.

In resonance ionization the CDW function with local momenta has been used in studying the PCI effects. In Section 5 we gave a detailed account of the strong PCI effects in the spectra of autoionization electrons, such as the Coulomb focusing effect and an inteference effect in the autoionization line profiles at small ejection angles that is caused by the interference between the plane and spherical wave contributions to the amplitude. There is another kind of interference effect that can be observed in autoionization profiles,[79–83] even under the kinematic conditions when PCI is weak and well described in the eikonal approximation. These effects are due to interference between resonance and direct ionization amplitudes. As a result of PCI, the Shore parametrization of a resonance lineshape is generally violated.

Using the CIS method, numerical evidence has been presented in Section 6 that strongly supports a view that the Thomas peak in the DCS for charge exchange at high velocities is formed as a result of interference between plane and spherical wave contributions to the amplitude. Here, the plane and spherical waves correspond to the electron-residual target ion continuum distortion factors that asymptotically behave as distorted plane and spherical waves. In the final CDW state within the illustrated CIS method, the electron is bound with respect to the projectile ion but is in a continuum state with respect to the target ion. Interference between these waves can be interpreted as an interference between electrons in the continuum states with and without rescattering by the residual target ion. This is a new look on the old physical scattering mechanism leading to the Thomas peak for the CDW function that emphasizes this interference rather than a two-step nature as originally suggested by Thomas in his classical picture for electron transfer. The role of local momenta in the bound CDW states and its effects on the electron transfer processes has not been presently investigated.

We can conclude that the more complicated three-body CDW functions with local momenta reviewed in this paper basically give a better description of scattering states in asymptotic regions. Therefore, its systematic application to the ion-atom collision processes is anticipated to produce better results than the standard CDW functions where the local momenta are replaced by asymptotic ones. However, numerical efforts associated with calculating the amplitudes are expected to be higher. The formal splitting of the two-body continuum distortion factors into plane and spherical wave parts can be very helpful in revealing characteristic features observed in the spectra of reaction products of various ion-atom collision processes. Finally, it should be noted that possible extensions of the present three-body CDW functions to four- or few-body problems can be implemented along the lines given in Refs. 20, 21, 84, 85.

Acknowledgments

This work has been done during my stay at the University of Rhode Island. I am grateful to D. Freeman for hospitality and help in work.

References

1. A. Z. Capri, *Non-relativistic Quantum Mechanics*, 3rd Ed. World Scientific Co. Pte. Ltd., Singapore, 2002.
2. D. S. F. Crothers, *J. At. Mol. Opt. Phys.*, **2010**, Article ID 604572, 3 pages (2010), doi 10.1155/2010/604572.
3. J. S. Briggs and J. Berakdar, *Phys. Rev. Lett.*, **72**, 3799–3802 (1996).
4. J. Berakdar, *Phys. Rev. A*, **53**, 2314–2326 (1999).
5. M. Abramowitz and I. Stegun, *Handbook of Mathematical Functions With Formulas, Graphs, and Mathematical Tables*, National Bureau of Standards, Washington, D.C., 1972.

6. L. D. Landau and E. M. Lifshitz, *Quantum Mechanics. Non-Relativistic Theory*, Vol. 3 Pergamon Press, Oxford, 1991.
7. L. C. Evans, *Partial Differential Equations. Graduate Studies in Mathematics*, Vol. 19 American Mathematical Society, 1997.
8. H. Goldstein, C. P. Poole, Jr., and J. L. Safko, *Classical Mechanics*, 3rd Ed., Addison Wesley, San Fransisco, 2000.
9. Sh. D. Kunikeev, *J. Phys. A: Math. Gen.*, **32**, 677–692 (1999).
10. Sh. D. Kunikeev, *J. Phys. A: Math. Gen.*, **33**, 5405–5428 (2000).
11. Sh. D. Kunikeev, in book *Many-Particle Spectroscopy of Atoms, Molecules, Clusters and Surfaces*, Eds J. Berakdar and J. Kirschner, Kluwer Academic/Plenum Publishers, New York, 2001, pp. 203–212.
12. S. P. Merkuriev, *Theor. Math. Phys.*, **32**, 680–694 (1977).
13. L. D. Faddeev and S. P. Merkuriev, *Quantum Scattering Theory for Several Particle Systems*, Kluwer Academic Pub., Dordrecht, 1993.
14. A. Galindo and P. Pascual, *Quantum Mechanics II*, Springer-Verlag, Berlin, 1991.
15. E. Guth and C. J. Mullin, *Phys. Rev.*, **83**, 667–668 (1951).
16. I. W. Herbst, *Commun. Math. Phys.*, **35**, 181–191 (1974).
17. J. D. Dollard, *J. Math. Phys.*, **5**, 729–738 (1964).
18. J. D. Dollard, *Commun. Math. Phys.*, **12**, 193–203 (1969).
19. A. M. Vesselova, *Theor. Math. Phys.*, **13**, 1200–1206 (1972).
20. Dž. Belkić, I. Mančev, and J. Hanssen, *Rev. Mod. Phys.*, **80**, 249–314 (2008).
21. Dž. Belkić, *Quantum Theory of High-Energy Ion-Atom Collisions*, CRC Press: Taylor and Francis Group, 2009.
22. L. Rosenberg, *Phys. Rev. D*, **8**, 1833–1843 (1973).
23. Dž. Belkić, *J. Phys. B: At. Mol. Phys.*, **11**, 3529–3552 (1978).
24. S. P. Merkuriev, *Theor. Math. Phys.*, **38**, 134–145 (1979).
25. D. S. F. Crothers and J. F. McCann, *J. Phys. B: At. Mol. Phys.*, **16**, 3229–3242 (1983).
26. A. L. Godunov, V. N. Mileev, and V. S. Senashenko, *Zh. Tekh. Fiz.*, **53**, 436 (1983).
27. A. L. Godunov, Sh. D. Kunikeev, and V. S. Senashenko, *Mosc. Univ. Phys. Bull.*, **38**, 95–97 (1983).
28. P. D. Fainstein, V. H. Ponce, and R. D. Rivarola, *J. Phys. B: At. Mol. Opt. Phys.*, **24**, 3091–3120 (1991).
29. M. Brauner, J. S. Briggs, and H. Klar, *J. Phys. B: At. Mol. Opt. Phys.*, **22**, 2265–2288 (1989).
30. D. H. Madison and O. Al-Hagan, *J. At. Mol. Opt. Phys.*, **2010**, Article ID 367180, 24 pages (2010), doi 10.1155/2010/267180.
31. J. H. Macek and S. Jones, *Rad. Phys. Chem.*, **74**, 7–11 (2005).
32. E. O. Alt and A. M. Mukhamedzhanov, *JETP Lett.*, **56**, 435–439 (1992).
33. E. O. Alt and A. M. Mukhamedzhanov, *Phys. Rev. A*, **47**, 2004–2022 (1993).
34. I. M. Cheshire, *Proc. Phys. Soc.*, **84**, 89–98 (1964).
35. L. Vainstein, L. Presnyakov, and I. Sobelman, *Sov. Phys. — JETP*, **18**, 1383–1395 (1964).
36. Sh. D. Kunikeev, *J. Phys. B: At. Mol. Opt. Phys.*, **31**, L849–L853 (1998).

37. Dž. Belkić, R. Gayet, and A. Salin, *Phys. Rep.*, **56**, 279–369 (1979).
38. J. E. Miraglia and J. Macek, *Phys. Rev. A*, **43**, 5919–5928 (1991).
39. Sh. D. Kunikeev, *J. Phys. B: At. Mol. Opt. Phys.*, **31**, 2649–2658 (1998).
40. Sh. D. Kunikeev, *Nucl. Instr. Meth. B*, **154**, 252–258 (1999).
41. G. Bernardi, S. Suarez, P. Fainstein, C. Garibotti, and W. Meckbach, *Phys. Rev. A*, **40**, 6863–6872 (1989).
42. Sh. D. Kunikeev and V. S. Senashenko, *Sov. Phys. — Tech. Phys.*, **35**, 778–780 (1990).
43. Sh. D. Kunikeev, V. S. Senashenko, and V. A. Sidorovich, *Lecture Notes in Physics*, Eds. D. Brerenyi and G. Hock, **376**, 112–116 Springer-Verlag, 1991.
44. Sh. D. Kunikeev, V. S. Senashenko, and V. A. Sidorovich, *Nucl. Instr. Meth. B*, **53**, 122–126 (1991).
45. M. Yu. Kuchiev and S. A. Sheinerman, *Sov. Phys. — JETP*, **63**, 986–990 (1986).
46. P. Van der Straten and R. Morgenstern, *J. Phys. B: At. Mol. Phys.*, **19**, 1361–1370 (1986).
47. M. Yu. Kuchiev and S. A. Sheinerman, *Sov. Phys. Usp.*, **32** 569–587 (1989).
48. Sh. D. Kunikeev and V. S. Senashenko, *XV ICPEAC, Abstr. of late papers,* p.17 Brighton, UK, 1987.
49. Sh. D. Kunikeev and V. S. Senashenko, *Sov. Phys. — Tech. Phys. Lett.*, **14**, 786–788 (1988).
50. M. Yu. Kuchiev and S. A. Sheinerman, *J. Phys. B: At. Mol. Phys.*, **21**, 2027–2038 (1988).
51. R. Barrachina and J. H. Macek, *J. Phys. B: At. Mol. Phys.*, **22**, 2151–2160 (1989).
52. J. K. Swenson, C. C. Havener, and N. Stolterfoht *et al.*, *Phys. Rev. Lett.*, **63**, 35–38 (1989).
53. J. K. Swenson, J. Burgdörfer, and P. H. Meyer *et al.*, *Phys. Rev. Lett.*, **66**, 417–420 (1991).
54. Sh. D. Kunikeev, *JETP*, **82**, 22–31 (1996).
55. Sh. D. Kunikeev and V. S. Senashenko, *JETP*, **82**, 839–852 (1996).
56. Sh. D. Kunikeev, *Workshop Autoionization Phenomena in Atoms, Inv. Talks*, 116–120, Moscow Univ. Press, Dubna, 1996.
57. Sh. D. Kunikeev and V. S. Senashenko, *Workshop Autoionization Phenomena in Atoms, Inv. Talks*, 121–125, Moscow Univ. Press, Dubna, 1996.
58. Sh. D. Kunikeev, *Nucl. Instr. Meth. B*, **124**, 350–353 (1997).
59. Sh. D. Kunikeev, *Nucl. Instr. Meth. B*, **124**, 354–357 (1997).
60. A. P. Prudnikov, Y. A. Brychkov, and O. I. Marichev, *Integrals and Series*, Vols. 1–3, Gordon and Breach, New York, 1986–1989.
61. S. Otranto and R. E. Olson, *Phys. Rev. A*, **72**, 022716 [11 pages] (2005).
62. Sh. D. Kunikeev and V. S. Senashenko, *Proc. XVII ICPEAC*, p. 419, Brisbane, Australia, 1991.
63. A. L. Godunov, Sh. D. Kunikeev, N. V. Novikov, V. S. Senashenko, and V. A. Shipakov, *Sov. Phys. — Tech. Phys.*, **38**, 541 (1993).
64. L. H. Thomas, *Proc. R. Soc. Lond., Ser. A,* **114**, 561–576 (1927).
65. R. M. Drisko, *thesis*, (Carnegie Inst. of Tech., 1955) (unpublished).

66. E. Horsdal-Pedersen, C. L. Cocke, and M. Stockli, *Phys. Rev. Lett.*, **50**, 1910–1913 (1983).
67. P. S. Vinitsky, Yu.V. Popov, and O. Chuluunbaatar, *Phys. Rev. A*, **71**, 012706 [9 pages] (2005).
68. J. Macek and S. Alston, *Phys. Rev. A*, **26**, 250–270 (1982).
69. J. S. Briggs, *J. Phys. B: At. Mol. Phys.*, **10**, 3075–3091 (1977).
70. P. A. Amundsen and D. Jakubassa, *J. Phys. B: At. Mol. Phys.*, **13**, L467–L472 (1980).
71. J. S. Briggs, P.T. Greenland, and L. Kocbach, *J. Phys. B: At. Mol. Phys.*, **15**, 3085–3102 (1982).
72. J. E. Miraglia, R. D. Piacentini, R. D. Rivarola, and A. Salin, *J. Phys. B: At. Mol. Phys.*, **14**, L197–L202 (1981).
73. A. L. Godunov, Sh. D. Kunikeev, and V. S. Senashenko, *Fiz. Plaz.*, **12**, 1355–1361 (1986) (in Russian).
74. P. J. Martin, *Phys. Rev. A*, **23**, 2858–2865 (1981).
75. S. Levit and U. Smilansky, *Ann. Phys., NY*, **108**, 165–197 (1977).
76. S. Levit, K. Mohring, U. Smilansky, and T. Dreyfus, *Ann. Phys., NY*, **114**, 223–242 (1978).
77. V. P. Maslov and M. V. Fedoriuk, *Semi-Classical Approximation in Quantum Mechanics* Dordrecht: Reidel 1981.
78. Yu. A. Kravtsov and Yu. I. Orlov, *Sov. Phys. Usp.*, **26**, 1038–1058 (1983).
79. Sh. D. Kunikeev and V. S. Senashenko, *Mosc. Univ. Phys. Bull.*, **44**, 18–20 (1989).
80. A. L. Godunov, Sh. D. Kunikeev, N. V. Novikov, and V. S. Senashenko, *Sov. Phys. — JETP*, **69**, 927–933 (1989).
81. P. Moretto-Capelle, D. Bordenave-Montesquieu, A. Bordenave-Montesquieu, A. L. Godunov, and V. A. Schipakov, *Phys. Rev. Lett.*, **79**, 5230–5233 (1997).
82. A. L. Godunov, V. A. Schipakov, P. Moretto-Capelle, D. Bordenave-Montesquieu, M. Benhenni and A. Bordenave-Montesquieu, *J. Phys. B: At. Mol. Opt. Phys.*, **30**, 5451–5478 (1997).
83. A. L. Godunov, P. B. Ivanov, V. A. Schipakov, P. Moretto-Capelle, D. Bordenave-Montesquieu, and A. Bordenave-Montesquieu, *J. Phys. B: At. Mol. Opt. Phys.*, **33**, 971–1000 (2000).
84. Dž. Belkić, *J. Math. Chem.*, **47**, 1366–1419 (2010).
85. Dž. Belkić, *J. Math. Chem.*, **47**, 1420–1467 (2010).

Chapter 6

Critical Assessment of Theoretical Methods for Li^{3+} Collisions with He at Intermediate and High Impact Energies

Dževad Belkić[a,*], Ivan Mančev[b], and Nenad Milojević[b]

[a]*Nobel Medical Institute, Karolinska Institute, Stockholm, Sweden*
dzevad.belkic@ki.se

[b]*Department of Physics, Faculty of Sciences and Mathematics, University of Niš, Niš, Serbia*

The total cross sections for the various processes for Li^{3+}–He collisions at intermediate-to-high impact energies are compared with the corresponding theories. The possible reasons for the discrepancies among various theoretical predictions are thoroughly discussed. Special attention has been paid to single and double electron capture, simultaneous transfer and ionization, as well as to single and double ionization.

1. Introduction

The Li^{3+}–He collisions are important in both fundamental and applied physics. For example, such studies can help test theories of few-body problems arising in atomic physics, specifically the interactions of nuclei with atomic electrons. For this purpose, the collision system Li^{3+}–He is well suited because electron-electron effects can be observed. Helium is the simplest neutral target that shows electron-electron effects, while lithium can serve as a fully stripped projectile.

From the perspective of applications in other areas of physics and other sciences as well as technologies, these few-electron collisional systems are also outstanding. The associated charge-changing cross sections constitute an important part of the input data bases for modeling of energy losses of

*The author to whom corrspondence should be addressed.

heavy particles in their passage through matter as encountered in e.g. high-temperature thermonuclear fusion,[1] hot plasmas,[2,3] heavy-ion therapy,[4] etc.

Available experimental data for various processes for Li^{3+}–He collisions at intermediate-to-high impact energies will be compared with the corresponding theories in order to assess the relative performance of the main two strategies, the semi-classical independent particle model (IPM) and the full quantum-mechanical four-body (4B) formalism. We shall thoroughly discuss possible reasons for the discrepancies among various theoretical predictions from these two frameworks and point out several potential paths for their improvement.

In Ref. of Sant'Anna *et al.*,[5] beside the measured data for various processes in Li^{3+}–He collisions at intermediate impact energies, the discussions of various theories for double-electron capture have been given. In the present review we report critical assessment of available theoretical methods not only for double capture (DC), but for other charge-changing collision channels, such as single capture (SC), transfer ionization (TI), double ionization (DI), and single ionization (SI):

$$\mathrm{Li}^{3+} + \mathrm{He} \longrightarrow \mathrm{Li}^{+} + \mathrm{He}^{2+}, \qquad \text{(DC)} \qquad (1)$$

$$\mathrm{Li}^{3+} + \mathrm{He} \longrightarrow \mathrm{Li}^{2+} + \mathrm{He}^{+}, \qquad \text{(SC)} \qquad (2)$$

$$\mathrm{Li}^{3+} + \mathrm{He} \longrightarrow \mathrm{Li}^{2+} + \mathrm{He}^{2+} + \mathrm{e}, \qquad \text{(TI)} \qquad (3)$$

$$\mathrm{Li}^{3+} + \mathrm{He} \longrightarrow \mathrm{Li}^{3+} + \mathrm{He}^{2+} + 2\mathrm{e}, \qquad \text{(DI)} \qquad (4)$$

$$\mathrm{Li}^{3+} + \mathrm{He} \longrightarrow \mathrm{Li}^{3+} + \mathrm{He}^{+} + \mathrm{e}. \qquad \text{(SI)} \qquad (5)$$

The presently studied collision system Li^{3+}–He is a pure four-body problem. Different measured data for this system should be suitable for assessing the adequacy of pure four-body theories that, in principle, are capable of describing both sequential and coherent mechanisms.

2. Double Electron Capture

For the DC in Li^{3+}–He collision, there are just three sets of experimental data. These sets, measured by Nikolaev *et al.*,[6] as well as by Shah and Gilbody[7] in the 1–2 MeV energy range, and very recently Sant'Anna *et al.*,[5] reported measured cross sections in the 2.0–5.6 MeV energy range (corresponding to the incident velocities in the 3.4–5.6 range in atomic units) not only for double capture, but also for the single capture, double ionization, single ionization, and transfer ionization.

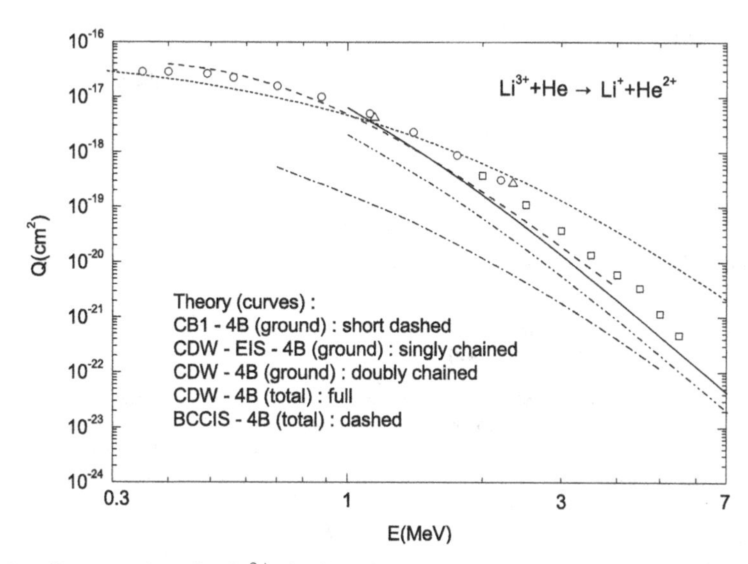

Fig. 1. Cross sections for Li^{3+} double electron capture on He as a function of the projectile energy. Experimental data: open squares (Sant'Anna *et al.*[5]); open triangles (Nikolaev *et al.*[6]) and open circles (Shah and Gilbody[7]); Theoretical models: singly chained curve, CDW-EIS-4B {final ground state} (Gayet *et al.*[12,13]); doubly chained curve, CDW-4B {final ground state} (Gayet *et al.*[11]); full curve, CDW-4B {final ground + excited states} (Gayet *et al.*[11]); dashed curve, BCCIS-4B {final ground plus excited states} (Gosh *et al.*[10]) and dotted curve, CB1-4B {final ground state} (Belkić[8,9]).

Regarding comparisons between experimental and theoretical findings for DC, to avoid clutter, the full quantum-mechanical four body (4B) methods and semi-classical independent particle models are depicted at two separate Figs. 1 and 2, respectively. The results from the 4B formalism shown in Fig. 1 are due to the boundary-corrected first Born (CB1-4B),[8,9] the boundary-corrected continuum intermediate states (BCCIS-4B),[10] the continuum distorted wave (CDW-4B)[11] and the continuum distorted wave initial state (CDW-EIS-4B)[12] methods. Figure 2 displays the cross sections from the corresponding IPM treatments within the continuum distorted waves (CDW-IPM) and its eikonalization (CDW-EIS-IPM)[14] and the continuum intermediate states (CIS-IPM),[15] as well as the second-order impulse approximation (IA2-IPM).[16]

All the presented 4B methods from Fig. 1 satisfy the correct boundary conditions, i.e. they possess the properly behaving wave functions at asymptotically large inter-particle separations in the entrance and exit channels. Importantly, these total scattering wave functions are correctly connected to the pertinent perturbation potentials that are responsible for the transitions

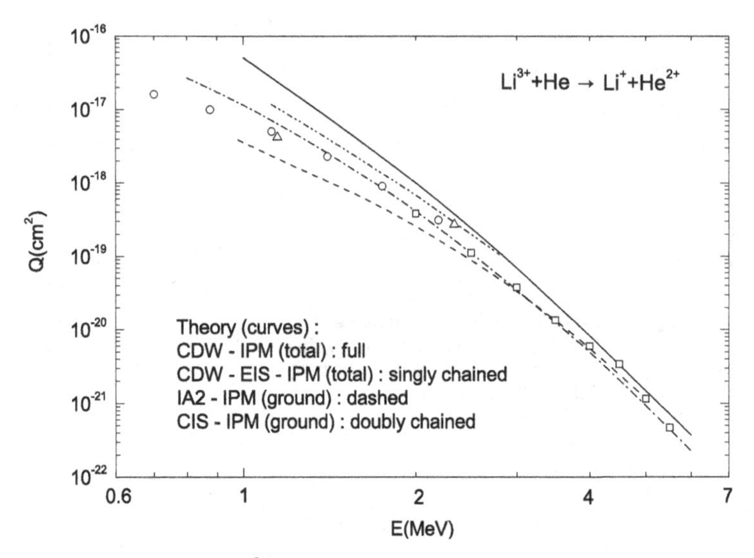

Fig. 2. Cross sections for Li^{3+} double electron capture on He as a function of the projectile energy. Experimental data: open squares (Sant'Anna *et al.*[5]); open triangles (Nikolaev *et al.*[6]) and open circles (Shah and Gilbody[7]); Theoretical models: dashed curve, IA2-IPM {final ground state} (Gravielle and Miraglia[16]); singly chained curve, CDW-EIS-IPM {final ground + excited states} (Martinez *et al.*[14]); doubly chained curve, CIS-IPM {final ground state} (Ghosh *et al.*[15]) and full curve, CDW-IPM {final ground + excited states} (Martinez *et al.*[14]).

from the initial to the final states of the whole system under study. By contrast, the CIS method[17] and the standard impulse approximation[18,19] disobey the correct boundary conditions for any rearranging collision within both IPM and 4B formalisms. A formal relationship of CIS and IA is that the former can be deduced from the latter by means of the usual peaking approximation applied to the momentum-space bound state vector from the total scattering wave function in the impulse approximation.[19] Total cross sections available from the past, abundant literature on SC show that performances of CIS and IA are similar at impact energies at which the classical Thomas double scattering is not dominant.

A common feature seen in Figs. 1 and 2 is a large spread among the presented theories and a varying degree of agreement with the experimental data. Such discrepancies within the 4B formalism in Fig. 1 are indeed enormous and attain two orders of magnitude when comparing the CB1-4B and CDW-EIS-4B methods. Differences within one order of magnitude are also observed between the CDW-IPM and IA2-IPM as evidenced in Fig. 2 concerning the IPM treatments. When Figs. 1 and 2 are juxtaposed to each other, it follows that even at high energies, there is a significant disaccord of the cross sections obtained by the IPM and 4B methods.

At first glance, and contrary to the common expectation, it appears that certain methods in their simpler, IPM variants outperform the more elaborate, 4B formulations, by exhibiting a better agreement especially with the high-energy experimental data in the 2–6 MeV range.[5] This is evidently clear for the CDW-EIS-4B and CDW-EIS-IPM from Figs. 1 and 2, respectively. Contrary to this, it appears that the same conclusion does not hold true for the CDW-4B and CDW-IPM, since the former/latter underestimate/overestimate the experimental data of (Sant'Anna *et al.*[5]), respectively, as obvious from Figs. 1–3. Furthermore, in the 3–5 MeV energy range, Fig. 2 indicates that IA2-IPM is in good agreement with measured cross sections.[5] However, this is deceiving due to a twofold circumstance: (i) the matrix element due to IA1-IPM neglected in IA2-IPM is far from being negligible, as is clear from CIS-IPM in Fig. 2 (as stated, based upon the experience with SC, it is expected that the first-order of IA and CIS, i.e. IA1-IPM and CIS-IPM ≡ CIS1-IPM for DC also give similar cross sections at non-asymptotic impact energies), and (ii) IA2-IPM from

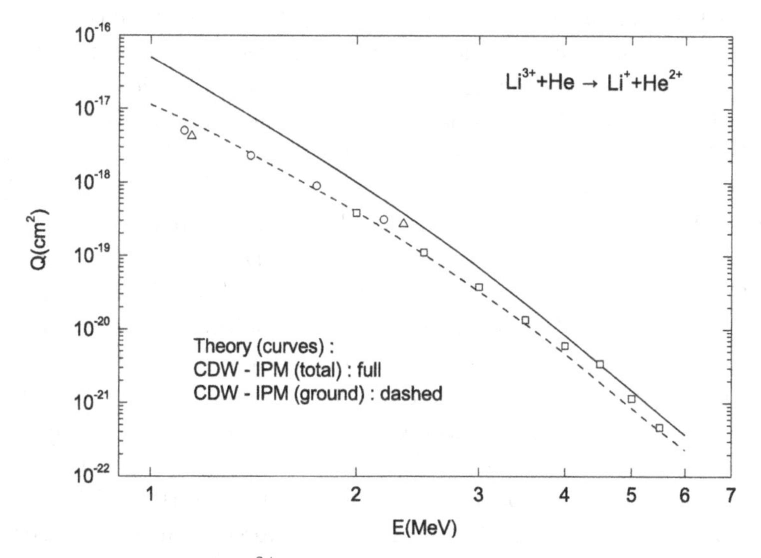

Fig. 3. Cross sections for Li^{3+} double-electron-capture on He as a function of the projectile energy. Experimental data: open squares (Sant'Anna *et al.*[5]); open triangles (Nikolaev *et al.*[6]) and open circles (Shah and Gilbody[7]); Theoretical models: dashed curve, CDW-IPM {final ground state} (Martinez *et al.*[14]) and full curve CDW-IPM{final ground plus excited states} (Martinez *et al.*[14]).

Ref. 16 includes only the non-resonant ground-to-ground state transition $He(1s^2) \to Li^+(1s^2)$ in process (1). In short, the apparent good agreement between IA2-IPM and the recently measured data[5] at energies 3–5 MeV would not persist in a more complete computation with the inclusion of the neglected contributions from IA1-IPM and the excited states of Li^+. Note that in all cited measurements,[5–7] the post-collisional state of Li^+ is not detected and, therefore, the theory should contain the sum of the dominantly contributing channels for the helium-like ion Li^+.

In IPM for DC each electron is captured independently due to the outright neglect of the inter-electron potential. These two independent encounters are modeled by the product $P_1(b)P_2(b)$ of the three-body (3B) impact parameter (b) dependent probabilities $P_1(b)$ and $P_2(b)$ for each event. The total cross section Q is obtained by integration of the composite probability $P_1(b)P_2(b)$ over all the impact parameters ($0 \leq b \leq \infty$). In principle, total cross sections from IPM should tend to those from the 4B method at high impact energies. However, at intermediate and lower energies, the concept of independent capture of each electron ceases to be applicable.[20–22] This can also be inferred from Figs. 1 and 2. Nevertheless, despite the inadequacy of IPM at intermediate and lower energies, the computations within CDW-IPM[13] were needed to direct attention towards the importance of the contributions from excited states. This was subsequently confirmed in the 4B formalism indicating that at intermediate and smaller energies, the main contributions to DC stems from single excited states, although the ground states of helium-like systems also provide a large contribution of about 40%.[10,11,14,23] Conversely, at high energies, DC to the ground states yields the dominant contributions.

Previous comparisons between e.g. the CDW-4B and CDW-IPM for DC revealed a varying degree of success of each of these two treatments.[11–14,20–22,24–32] Thus, for H^+–He collisions, excellent agreement between measurements and CDW-4B exists[20–22,26,27] already above 80 keV, as opposed to a significant overestimation by the CDW-IPM.[13] By contrast, for He^{2+}–He collision, CDW-IPM appears to be superior to CDW-4B.[13,20–22] Regarding Li^{3+}–He and B^{5+}–He collisions, large discrepancies were found between CDW-4B and CDW-IPM[11–14] even at relatively high energies in the MeV/amu range. Moreover, neither of the two variants of the same theory emerged with a clear advantage in comparisons with previous measurements for these two collisional systems.[7,33]

Other theoretical models were also used for DC in He^{2+}–He collisions such as the four-body boundary corrected continuum intermediate state

(BCIS-4B)[20,21,35] as well as the four-body Born distorted wave (BDW-4B) approximations[20–22,24,25] yielding a close mutual agreement and comparing favorably with the experimental data for total cross sections. Purkait *et al.*[23] and Ghosh *et al.*[10] used the acronym BCCIS-4B to refer to the variant of BCIS-4B with the full Coulomb waves for the relative motions of heavy nuclei instead of the corresponding logarithmic phase asymptotes employed by Belkić.[35] These latter two treatments with the said full and asymptotic Coulomb wave functions are equivalent for heavy particle collisions and, as such, are expected to give the same total cross sections in the eikonal approximation in which the mass of the nuclei is much larger than the electronic mass (for more details, see Refs. 20–22).

Computations in BCCIS-4B[10,23] show good agreement with the earlier experimental data[6,7] on DC in He^{2+}–He and Li^{3+}–He collisions. However, for the latter collision, it is seen in Fig. 1 that BCCIS-4B underestimates all measured cross sections of Sant'Anna *et al.*[5] Also applied to DC in the He^{2+}–He and Li^{3+}–He collisional channels were the CDW-EIS-4B and CDW-EIS-IPM.[12,14,32] Here, however, while CDW-EIS-IPM is reasonably acceptable for DC due to the reliance upon the corresponding successful 3B counterpart (CDW-EIS-3B), it was found in Refs. 12 and 14 that CDW-EIS-4B flagrantly underestimates (by orders of magnitude) the measurements on DC in He^{2+}–He and Li^{3+}–He collisions even at MeV/amu energies. This can also be seen in Fig. 1.

When compared to the experimental data, another four-body model is also known to dramatically break down at MeV/amu energies. This is the case with the CB1-4B approximation[8,9] for He^{2+}–He and Li^{3+}–He collisions, although relatively good success was achieved at intermediate impact energies according to Fig. 1 for the latter scattering (this agreement, which is based only upon the ground-to-ground state transition alone, would cease to exist if the excited states of Li^{+} are taken into account).

An approximate version of the IA2-IPM was tried by Gravielle and Miraglia[16] for DC in He^{2+}–He and Li^{3+}–He collisions with a reasonable success relative to measurements, as seen in Fig. 2 for the latter scattering. However, such an agreement is fortuitous for the reasons (i) and (ii) given earlier in this section. The model from Ref. 16 uses the second-order propagator for a two-step collision with the inclusion of the intermediate state on-shell Green function centered on the target and projectile nuclei. In IA2-IPM, as already emphasized, the contribution from the matrix element due to the associated first-order transition in this approximation (IA1-IPM, or simply, IA-IPM) was omitted. It was assumed[16] that IA1-IPM should yield small cross sections as in an earlier work of Crothers and McCarroll[36] who employed the independent event (IEV) model of CDW, as denoted by CDW-IEV, which is, in spirit, quite similar to CDW-IPM. However, as discussed, this assumption of Gravielle and Miraglia[16] is not supported by CIS-IPM. It is not supported by CDW-4B2 either, since CDW-4B2 contains a large contribution from CDW-4B1[32] for He^{2+}–He collisions.[32] To re-emphasize, as opposed to all the other mentioned 4B methods, IA always disregards the proper boundary conditions for DC for any scattering aggregates, similar to the corresponding well-known unsatisfactory status of this theory for SC. Therefore, any agreement of this method with experimental data should be taken with caution because of the incorrect description of the asymptotic region of scattering where all experiments are performed.

The CDW-4B2 improves significantly CDW-4B1 for He^{2+}–He collisions and yields a good agreement with experimental data.[32] Nevertheless, this improvement may be fortuitous, since the second-order propagator in the CDW-4B2 is from the ordinary distorted wave perturbation series with disconnected diagrams, as opposed to the Dodd-Greider expansion.[20,21,37] Moreover, the computations in CDW-4B2[32] include only the on-shell part of a model two-center Green function and retain merely the intermediate ground states centered at the projectile and target nucleus with no estimate on the contribution from the intermediate excited states and continuum. Further, CDW-4B2[32] is concerned only with the ground-to-ground state transition $He(1s^2) \to Li^+(1s^2)$, thus neglecting all the final excited states of Li^+ that yield a sizable contribution. As pointed out, all these drawbacks of CDW-4B2[32] are also present in IA2-IPM[16] due to the use of the same model Green function for intermediate states. Thus far, CDW-4B2 has not been applied to Li^{3+}–He collisions.

Faced with an unexpected failure of CDW-EIS-4B for DC, Martínez *et al.*[32] attempted to find an improvement in this model by following Gravielle and Miraglia[16] and thus employed the second-order transition operator via the two-center on-shell Green function with the one-electron intermediate ground states on the projectile and target nuclei. However, the ensuing CDW-EIS-4B2 has not met with success, since a huge overestimation of experimental data for DC in He^{2+}–He collisions was recorded.[32] To date, no computations have been reported using CDW-EIS-4B2 for Li^{3+}–He collisions.

To better comprehend the potential reasons for the varying performance of the discussed models when passing from one to another collisional system, it is necessary to highlight the key ingredient of the computations.

First of all, some of the cited references on theory dealt exclusively with the ground-to-ground state DC,[8,9,24–27] whereas there were studies that additionally included the contributions from a limited number of excited states.[10,11,13,14,16,23] The case of $H^+ + He(1s^2) \to H^-(1s^2) + He^{2+}$ collision is special, since the negative hydrogen ion H^- has no excited states.[34] This is DC to which the first order of CDW-4B, i.e. CDW-4B1, was originally applied by Belkić and Mančev[26,27] with the conclusion cohering to an empirically established fact: this 4B model is adequate at impact energies above 80 keV.

Approximately the same lowest limit of the applicability of CDW-3B and CDW-4B is expected, since this validity limit should be independent of the number of captured electrons. Other helium-like ions with nuclear charges larger than unity possess excited states and these need to be included in computations of cross sections for DC. Thus, Gayet *et al.*[11] used CDW-4B for DC with allowance for certain low-lying singly (1s, nl) $\{n \leq 3\,(0 \leq l \leq n-1)\}$ and the first doubly $(2l, 2l')\{l = 0,1\, l' = 0,1\}$ excited helium-like states in the exit channels of He^{2+} + He $\to$ He + He^{2+}, Li^{3+}+He $\to$ Li^+ – He^{2+} and B^{5+}+He $\to$ B^{3+}+He^{2+} collisions. These excited state helium-like wave functions can be reasonably well described by linear combinations of hydrogen-like bound states within the formalism of the configuration of interactions (CI), as exemplified by e.g. Bachau.[38] Precisely such excited state wave functions were used for DC in the studies by Gayet *et al.*,[11] as well as by Purkait *et al.*[23] and Ghosh *et al.*[10] It was found in Ref. 11 that excited states were not very important within CDW-4B for DC in He^{2+}–He collisions because of the dominant contribution from the resonant $1s^2 \to 1s^2$ transition. As mentioned, CDW-4B1 is not fully satisfactory for He^{2+}–He collisions, but the situation is conditionally improved by CDW-4B2. Further computations within CDW-4B2 are necessary for this collision to include the ignored off-shell matrix elements and the intermediate as well as final excited states. Intermediate excited states would check the convergence properties of the expansion of the two-center Green function. Final excited states are important, since the transition $He(1s^2) \to Li^+(1s^2)$ in process (1) does not provide the dominant contribution. This is clearly seen in Figs. 1 and 3 where the excited states of Li^+ included within CDW-4B and CDW-IPM, respectively, give a large contribution relative to the cross sections for the ground-to-ground state transition. In particular, Fig. 3 is a good illustration of the caution which one must exercize when comparing theory and experiment, since the good agreement found for the ground-to-ground state transition in process (1) disappears altogether when the excited states are taken into account.

For DC in Li^{3+}–He and B^{5+}–He collisions, singly excited states $(1s, nl)$ with $n \leq 3\,(l \leq n-1)$ were found to provide especially significant contributions.[11] At the same time, higher doubly excited states were shown to give cross sections that are at least an order of magnitude smaller than those due to singly excited states.[11] Overall, despite the inclusion of a restricted set of helium-like excited states, agreement between CDW-4B[11] and experimental data was still only qualitative. This could be due the replacement of the original CI orbitals $(1s, 3l)$ with the actual nuclear charges[38] by the single-configuration product of two hydrogen-like wave functions with certain effective (screened) nuclear charges.[11] Such a simplification, which was motivated by a reduction of computational demands, leads to non-orthogonality between the single-configuration orbitals and the remaining CI orbitals from Ref. 38. It is known that the excited state wave functions of Bachau[38] are not well adapted to describe the singly excited states above the $(1s, 2l)$ level and, moreover, the modification from Ref. 11 is non-unique. Since helium-like singly excited states are important for DC, they must be very accurate and should be included by using some other wave functions to verify the results obtained with the CI orbitals from Ref. 38. It remains to be seen whether this could improve the overall standing of CDW-4B. A more favorable situation was encountered when comparing BCCIS[10] with measurements for He^{2+}–He collisions, despite the the same ambiguous modification[11] of the CI wave functions from Ref. 38. However, for Li^{3+}–He collisions, BCCIS-4B underestimates measured data of Sant'Anna *et al.*[5] similarly to CDW-4B, as seen in Fig. 1.

Regarding the wave functions for the $1s^2$ ground state of helium target, it was found by Belkić and Mančev[27] that total cross sections for DC are weakly sensitive to the number of terms in the CI expansion. Specifically, quite simple 1–4 parameter orbitals of Hylleraas,[39] Löwdin,[40] Green *et al.*[41] and Silverman *et al.*[42] were shown to be sufficient.[27] Following this finding, Gayet *et al.*[11,13] employed the 4-parameter wave function of Löwdin[40] for the initial ground state of helium. Earlier, within IA2-IPM, Gravielle and Miraglia[16] also used a relatively simple Hartree-Fock wave function for $He(1s^2)$ in the form of the Slater-type orbitals given by Clementi and Roetti.[43] This type of simple two-electron wave functions were used not only in CDW, but also in CDW-EIS[12,13] and BCCIS.[10,23]

The most unanticipated conclusion from these previous investigations was that CDW-EIS-4B underestimates the experimental data on DC in

He^{2+}–He collisions by orders of magnitude at intermediate and MeV/amu energies.[32] This continues to be the case for Li^{3+}–He collisions at high impact energies 2–5.5 MeV from the measurement of Sant'Anna *et al.*,[5] as evidenced by Fig. 1. Such an occurrence is very surprising, since at these impact energies, CDW-EIS-3B is known to be remarkably successful. Thus, the above-mentioned conjecture that the number of captured electrons should not change the lower validity limit of CDW does not apply to CDW-EIS. This new unfavorable situation with CDW-EIS-4B casts serious doubts on the whole concept of eikonal electronic continuum intermediate states, since such states for two active electrons totally fail to describe DC at intermediate and high energies. This experience might indicate that with every additional electron to be captured from a multi-electron target, the validity of CDW-EIS could be pushed to ever higher energies and this, in turn, would severely limit the usefulness of the eikonal Coulomb multiple electronic states. As stated, in order to see whether the two-step collisional mechanism could improve the prospect for CDW-EIS-4B, Martinez *et al.*,[32] took into account the second-order in a perturbation expansion as in IA2-IPM.[16] However, such an attempt has not rescued the situation, since this time the resulting cross sections computed by means of CDW-EIS-4B2 were much larger than the experimental data for DC in He^{2+}–He collisions at MeV energies. Thus, CDW-EIS definitely appears as inadequate for DC, as pointed out earlier in Refs. 20, 21.

As to CB1-4B for DC in the mentioned pure four-body collisional systems, all the previous computations were carried out only for the ground-to-ground state capture using the one-parameter Hylleraas[39] wave function with the Slater-screened nuclear charge for the initial and final states.[8,9] The obtained total cross sections for H^{+}–He collisions largely overestimate measurements at all energies.[21] On the other hand, this model agrees well with experiments on DC in He^{2+}–He and Li^{3+}–He collisions at intermediate energies. However, due to the important, but neglected contributions from excited states, such an agreement between CB1-4B and measurements at intermediate energies seen in Fig. 1 for Li^{3+}–He collisions is fortuitous. Thus, it follows that CB1-4B is inadequate for DC.[20,21] This is expected, given that DC is a process with two active electrons for which a first-order model is unsuitable due to the exclusive reliance only upon the direct, separate interactions of the projectile with the two target electrons and upon the inclusion of the asymptotic eikonal relative motions of nuclei in the fields of screened nuclear charges in the entrance and exit channels. In other words, one reason for the failure of CB1-4B could be the neglect of continuum intermediate states of the two electrons. This is indeed true, as was demonstrated by Belkić[24,35] in BCIS-4B and BDW-4B that both include the twofold electronic continua in one channel, while in the other channel the descriptions coincide with those from CB1-4B. Computations show that the total cross sections from BCIS-4B and BDW-4B[35–25] are in good agreement with measurements on DC in He^{2+} – He collisions at higher MeV energies where CB1-4B overestimated the experimental data by orders of magnitude. Overall, despite its demonstrated inadequacy for DC, it is still convenient to have the cross sections from CB1-4B to highlight the potential significance of a missing effect due to double continuum intermediate states of the electrons. When CB1-4B is amended by the inclusion of these lacking Coulomb states, BCIS-4B emerges showing reasonable agreement with experimental data for He^{2+}–He collisions.[35] Likewise, in the case Li^{3+}–He collisions, Fig. 1 shows that BCCIS-4B[10] is in a fair agreement with our data at $E \leq 4\,\mathrm{MeV}$.

It can be concluded from the above comparisons between experiment and theory on DC that the situation is not clear-cut regarding consistency of performance of the various models. Therefore, more work is needed to better clarify the status of theory for two-electron transfer at high energies. All the existing theoretical models are restricted to sequential capture without any dynamic inter-electron correlations. Since static correlations in the bound helium-like systems were not found to be essential for DC, it would be important to clarify the role of dynamic, collisional correlations between two electrons. This has not been done thus far. It would also be of interest to examine the contributions from other mechanisms for DC, such as the coherence effect of the simultaneous capture of two electrons. Admittedly, it is very difficult to design a practical theory which would incorporate all these mechanisms and effects for DC, but it is certainly a goal which would be worthwhile to achieve in the near future. From the experimental view point, it would be significant to complement the data on total cross sections for DC by measuring angular distributions of scattered projectiles at sufficiently high energies. Here, the main goal would be to clearly detect the triple billiard-type collisions with the emergence of the three Thomas peaks predicted theoretically by Belkić *et al.*[25] within the 4B treatments, as opposed to only two such maximae stemming from IPM used by Martínez *et al.*[31] One of the triple Thomas peaks from the 4B formalism involves explicitly the dynamic electron-electron correlation effect and this is precisely the maximum which is missing from IPM. As such, differential cross sections measured at sufficiently high impact energies would offer an advantageous testing ground for a stringent validation of 4B theories versus IPM on the level of angular distributions, especially when the corresponding total cross sections cannot discriminate between these two approaches.

3. Single Electron Capture

In the next stage of investigation we shall analyze total cross sections for single capture. This reaction has previously been theoretically treated by a number of authors using various methods. For example, the IPM and Roothan-Hartree-Fock (RHF) target screening were adopted within the Corrected First Born (CB1-3B) theory of Belkić.[44] It has given good agreement with experimental data at intermediate and high impact energies. Belkić *et al.*[19] originally introduced the RHF method for single electron capture from multielectron target within CDW-3B approximation of Cheshire.[45] This RHF model and CDW-3B approach was used by Saha *et al.*[46] for charge changing reaction (2) at energy interval 0.2–4 MeV/amu.

For the same collision system, theoretical total cross sections were obtained by Busnengo *et al.*,[48] using the continuum distorted wave eikonal initial state (CDW-EIS) and the continuum distorted wave eikonal final state (CDW-EFS) as well as CDW-3B method. Although their models treated problem as the three-body, the CDW-EIS model correctly predicted the behaviour of the measurements even for lower energies, whereas all of three approximations were in good agreement in the high-energy region. Gravielle and Miraglia[49] studied (2) using the prior form of eikonal Impulse approximation and assumed the IPM for the helium target. Sidorovich *et al.*[50] have applied the approximation of Bessel and Gerjuoy[51] to calculate the charge-changing cross sections in collisions of H^+, He^{2+} and Li^{3+} ions with He atoms at energy region of 0.025–4 MeV/amu in the independent-electron approximation. The IPM and the Unitarized distorted wave approximation (UDWA), which is based on the atomic-orbital expansion, is used by Suzuki *et al.*[52] for description the reaction (2). However, in their method the correlation term is simplified, according to $Z_T/r_1 - Z_T/r_2 + 1/r_{12} = -\alpha/r_1 - \alpha/r_2$ where α is the effective charge, so that their model cannot yield any information about the correlation effect.

All these studies[19,44,46,48,49] consider the process (2) as an equivalent three-body problem ignoring correlation effects from the outset.

The contribution from electron-electron interaction during the collision in the Li^{3+}+He scattering has been assessed for the first time by means of the CDW-4B theory.[53] The CDW-4B model for SC is introduced by Belkić *et al.*,[54] and explicitly includes the dynamic electronic correlations through the dielectronic interactions $1/r_{12}$ in the transition T operator via term $V_{\rm Corr} = 1/r_{12} - 1/x_1$, where $\vec{x}_1$ is the position vector of captured electron relative (say e_1) to target nucleus $Z_{\rm T}$.

The potential $1/r_{12}$ in $V_{\rm Corr} = 1/r_{12} - 1/x_1$ represents the direct Coulomb interaction between e_1 and e_2, whereas $1/x_1$ is the asymptotic tail of the $1/r_{12}$, since $r_{12} \to x_1$ as $x_1 \to \infty$. Hence, the term $V_{\rm Corr}$ is precisely the difference between the finite and asymptotic values of the *same* potential. As such, $V_{\rm Corr}$ is a short-range interaction in accordance with the correct boundary conditions.[19,55,56] Using the relation $r_{12} = |\vec{x}_1 - \vec{x}_2|$, where x_2 is the position vector of non-captured electron relative to $Z_{\rm T}$, we can develop $1/x_1 = 1/|\vec{r}_{12} - \vec{x}_2|$ in a power series around $\vec{x}_2$ according to $1/x_1 = 1/|\vec{r}_{12} - \vec{x}_2| = 1/r_{12} - \vec{r}_{12}\cdot\vec{x}_2/r_{12}^3 + \cdots$ so that $V_{\rm Corr} = 1/r_{12} - 1/x_1 = \vec{r}_{12}\cdot\vec{x}_2/r_{12}^3 + \cdots$. This is justified by the small value of the x_2 coordinate (of the order of Bohr radius a_0), since electron e_2 always remains bound in the projectile. From here we can see that the potential $V_{\rm Corr}$ contains information on the dielectronic correlation $e_1 - e_2$. Therefore, $V_{\rm Corr}$ can be interpreted as a correlation term. When the potential $V_{\rm Corr}$ is placed in the T matrix, it plays the role of a perturbation which causes capture of electron e_1.

In Fig. 4, the theoretical post CDW-4B total cross sections[53] for SC to the ground state in Li^{3+} + He collisions are plotted (full line) together with the experimental data of Shah and Gilbody,[7] Dmitriev *et al.*,[59] Woitke *et al.*[60] as well as Sant'Anna *et al.*[5] As can be seen, the computed cross sections[53] are in satisfactory agreement with the experimental measurements. The CDW-4B curve lies slightly below the experimental findings, due to the fact that the displayed CDW-4B results include capture only in the ground state, while the contribution from the exited states is accounted by a factor 1.202 which additionally multiplies the total cross sections. When we dropped the relevant term for dynamic correlations $V_{\rm Corr} = 1/r_{12} - 1/x_1$ in the transition amplitude, we obtained the cross sections, which grossly underestimate the experimental data (see dashed curve in Fig. 1). This is evidence that the dynamic correlations play an important role in the electron capture in the ground state, especially at higher impact energies.

By increasing the charge of the projectile, the contribution from capture into excited states becomes more important. This has been examined in Ref. 53 via the extension of the CDW-4B method to encompass capture to the final excited states: $1s$, $2s$, $2p$, $3s$, $3p$ and $3d$, with the

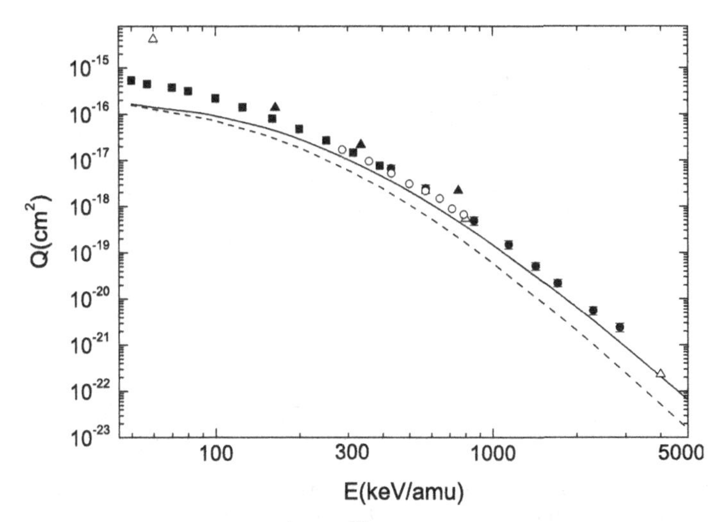

Fig. 4. The total cross sections (in cm^2) as a function of the laboratory incident energy E(keV/amu) for the reaction: $Li^{3+} + He \longrightarrow Li^{2+} + He^{+}$. The full and the dashed lines represent the post cross sections of the CDW-4B method[53] with the complete perturbation potential and without the potential V_{Corr}, respectively. Both curves correspond to the capture in ground state while the contribution from the excited states is accounted for by a factor 1.202 which additionally multiplies the total cross sections. The symbol ($\triangle$) refers to theoretical post total cross sections of the CDW-4B method which includes capture to the final excited states: $1s$, $2s$, $2p$, $3s$, $3p$ and $3d$. The displayed theoretical results are obtained by means of the orbital of Hylleraas[39] for the ground state of the helium target atom. Experimental data: ■ Shah and Gilbody[7]; • Woitke *et al.*[60]; ∘ Sant'Anna *et al.*[5]; ▲ Dmitriev *et al.*[59]

purpose of determining whether electronic correlations remain important. As an illustration, Mančev[53] gave the results at energies 60, 800 and 4000 keV/amu. The values of these total cross sections are denoted by symbol ($\triangle$) in Fig. 4. As expected, the contribution from the excited states becomes less important as the impact energy increases. However, at lower energies, in our case 60 keV/amu, total cross sections notably overestimate the experimental data.

The four-body first Born approximation with correct boundary conditions has been carried out for single electron capture[57] in p − He and He^{2+}−He collisions. In the present work the CB1-4B model[57] is applied to the process (2) and obtained results are presented in Fig. 5, together with CDW-4B theory.[53] As can be seen from Fig. 5, the CB1-4B results are in very good agreement above 300 keV/amu. It should be noted that both theoretical results, CDW-4B and CB1-4B cross sections presented in

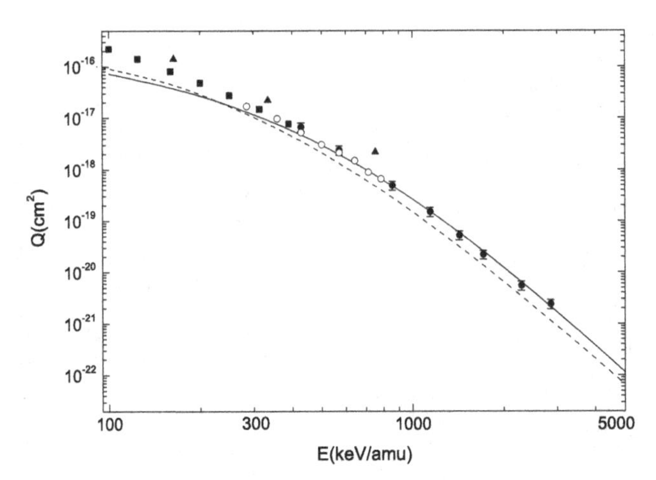

Fig. 5. The total cross sections (in cm^2) as a function of the laboratory incident energy E(keV/amu) for the reaction: $Li^{3+} + He \longrightarrow Li^{2+} + He^{+}$. The full line represents the post total cross sections of the CB1-4B method (present results), whereas dashed line relates to the post CDW-4B method.[53] Both curves are obtained with the complete perturbation potential and correspond to the capture in ground state while the contribution from the excited states is accounted for by a factor 1.202 which additionally multiplies the total cross sections. The displayed results are obtained by means of the orbital of Hylleraas[39] for the ground state of the helium target atom. Experimental data: ■ Shah and Gilbody[7]; • Woitke *et al.*[60]; ∘ Sant'Anna *et al.*[5]; ▲ Dmitriev *et al.*[59]

Fig. 5, are obtained with the complete perturbation potentials which include correlation term V_{Corr}.

Samanta *et al.*[61] used the BCCIS-4B model and obtained good agreement with experimental data on SC. It is of interest in the future to adopt the BCIS-4B model of Belkić[35] (which was originally formulated for DC) to SC in order to check the values for total cross sections from Ref. 61.

4. Transfer Ionization

Several research groups have presented theoretical calculations[50,53,62–66] of the total cross sections for transfer ionization in reaction (3). Most of these investigations[50,62,63,65] were based on the independent particle model.

Using a couplet channel semiclassical impact parameter model with the traveling atomic orbital expansion and employing the IPM, Singhal and Lin[62] obtained results which overestimated measured cross sections and exposed the inadequacy of simple uncorrelated models to describe this process. A simple model within IPM based on the Bohr-Lindhard model and classical statistical model has been developed for transfer ionization of helium by ions $A^{q+}(q = 1-3)$ in Ref. 64. As known, the IPM model completely ignores the dynamic correlations and computes the net probability for the transfer ionization as a product of the individual probabilities for transfer of one electron and independent ionization of the other electron: $Q = 2\pi \int \mathrm{d}\rho\rho P_C(\rho)P_I(\rho)$, where $P_C(\rho)$ and $P_I(\rho)$ are probabilities of capture and ionization, respectively, for a single electron in a collision with impact parameter ρ. In Ref. 65, for the $P_C(\rho)$, the CDW-3B model was used, whereas the ionization probability was calculated using two different approaches: the Born approximation and the multipole expansion defined on one centre (MEDOC). However, for Li^{3+}–He collisions, Ref. 65 reported the total cross section at only one impact energy 0.4MeV/amu, so that one could not conclude about the validity of the MEDOC model for the case of reaction (3).

The four-body continuum distorted wave (CDW-4B) model for the TI has been formulated and implemented by Belkić *et al.*[54] for He^{2+}–He collisions and good agreement with measurements has been obtained. Later, the CDW-4B method used by Mančev[53] has found to be in good agreement with the experimental data for Li^{3+}–He collisions.

The post and prior total cross sections of the CDW-4B model for TI in reaction (3), derived with the full perturbations[53] are plotted in Fig. 6, where the experimental findings from Refs. 5, 7, 60 are also displayed. The CDW-4B theory is found to be in good agreement with the experimental data. The post cross sections lie below the prior ones at impact energies between 100 and 3000 keV/amu, with the reverse behavior outside this energy interval. These computations have been performed for electron transfer to the ground state. The theoretical results of Bhattacharyya *et al.*,[63] are also depicted in Fig. 6. Their cross sections were computed within a relativistically covariant field approach using the second-order Feynman diagrams. As can be seen from Fig. 6, their results greatly overestimate the experimental measurements.

Using the second Born approximation, Godunov *et al.*[66] found that allowance for (static) electron correlation directly determines how closely theoretical calculations agree with experimental data. Their second-order calculation with uncorrelated functions differ considerably from the experimental data, while including correlations gives better agreement. In their calculation,[66] an additional approximation is used, such that the electron-electron interaction $1/|\vec{r}_1 - \vec{r}_2|$ from perturbation potential in exit channel is replaced according to: $-Z_{\mathrm{T}}/r_1 + 1/|\vec{r}_1 - \vec{r}_2| \simeq -(Z_{\mathrm{T}} - 1)/r_1$, where $\vec{r}_{1,2}$

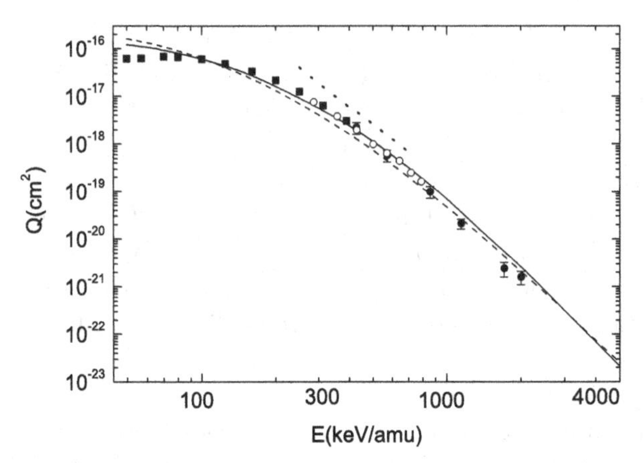

Fig. 6. The total cross sections (in cm^2) as a function of the laboratory incident energy E(keV/amu) for the reaction: $Li^{3+} + He \longrightarrow Li^{2+} + He^{2+} + e$. The full and dashed lines represent, respectively, the prior and post cross sections of the CDW-4B approximation[53] with the complete perturbation potentials. The dotted curve refers to the theoretical results of Bhattacharyya *et al.*[63] Experimental data: ■ Shah and Gilbody[7]; • Woitke *et al.*[60]; ○ Sant'Anna *et al.*[5]

are the position vectors for the electrons relative to the target nucleus Z_T. As known, a unique Born expansion of the transition amplitude does not exist. Godunov *et al.*[66] used the propagator $V_f G_0^+ V_i$ of the second order term in the Born expansions with the free-particle Green's function G_0^+, where $V_{i,f}$ are interaction potentials in the incoming and outgoing channel.

In Fig. 7, the transfer-ionization cross sections for Li^{3+}−He collisions are presented. It is clear that the first Born calculations[66] are much lower than the experimental data. Inclusion of the second-order terms (on- and off-shell) considerably improves the agreement with experimental data. As the collision energy increases, the effect of the second-order terms decreases but remains noticeable. The off-shell term relates to the Cauchy principal value integral, whereas on-shell relates to the term $\pm i\pi f$ in the Sokhatsky–Weierstrass theorem: $\lim_{\epsilon\to 0} \int \mathrm{d}x f(x)/[x - x_0 \pm \epsilon] = P \int \mathrm{d}x f(x)/(x - x_0) \pm i\pi f(x_0)$ which was used in Ref. 66, where P stands for the Cauchy principal value integral. The off-shell term may lead to time correlation between electrons.[66]

4.1. *BDW-4B Model*

The BDW-4B model for transfer ionization has been formulated by Belkić and Mančev.[67] This model is a fully quantum mechanical four-body model,

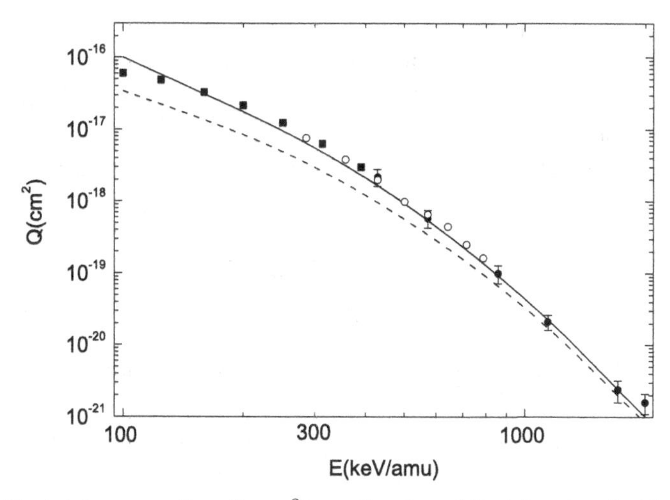

Fig. 7. The total cross sections (in cm^2) as a function of the laboratory incident energy E(keV/amu) for the reaction: $Li^{3+} + He \longrightarrow Li^{2+} + He^{2+} + e$. The full line represents the total cross sections of the second Born approximation which includes both on- and off-shell terms,[66] whereas dashed line relates to the first Born calculations.[66] The displayed results are obtained by means of the multiconfigurational Hartree-Fock approximation (MCHF)[71] for the ground state of the helium target atom. Experimental data: ■ Shah and Gilbody[7]; • Woitke *et al.*;[60] ∘ Sant'Anna *et al.*[5]

and it explicitly takes into account each particle and interaction in the collision, allowing for a systematic study of the dynamics of the process. The BDW-4B method preserves the correct boundary conditions in both scattering channels according to the principles of scattering theory.[56] The BDW-4B method takes full account of the Coulomb continuum intermediate states of the captured electron *only* in the exit channel, such that it coincides with the CDW-4B[54] method in the exit channel and with the CB1-4B in the entrance channel.

The transition amplitude in the BDW-4B model can be written as[67]:

$$T_{if}^{+} = \langle \chi_f^{-} | U_f | \Phi_i^{+} \rangle. \tag{6}$$

The initial wave function Φ_i^+ with asymptotically correct boundary condition can be written as follows:

$$\Phi_i^{+} = \varphi_i(\vec{x}_1, \vec{x}_2) \mathrm{e}^{i\vec{k}_i \cdot \vec{r}_i + i\nu_i \ln(vR - \vec{v}\cdot\vec{R})}, \tag{7}$$

where $\vec{r}_i$ is the relative vector of Z_P with respect to the center of mass of $(Z_T; e_1, e_2)_i$. Function $\varphi_i(\vec{x}_1, \vec{x}_2)$ represents the two-electron bound state wave function of the atomic system $(Z_T; e_1, e_2)_i$, whereas $\vec{k}_i$ is the initial wave-vector, v is the incident velocity, and $\nu_i = \mathrm{Z_P}(\mathrm{Z_T}-2)/v$. The initial state is distorted even at infinity, due to the presence of the asymptotic Coulomb repulsive potential, $V_i^{\infty} = \mathrm{Z_P}(\mathrm{Z_T}-2)/R$, between the projectile and the screened target nucleus.

In the BDW-4B model the distorted potential U_f and distorted wave χ_f^- are chosen according to:[54]

$$U_f = V_\mathrm{P}(R, s_2) + V(r_{12}, x_1) - \vec{\nabla}_{s_1}\varphi_f \cdot \vec{\nabla}_{x_1} 1/\varphi_f, \tag{8}$$

$$\chi_f^- = N^-(\zeta)N^-(\nu_\mathrm{T})\phi_f\varphi_f(\vec{s}_1)\,{}_1F_1(-i\zeta, 1, -ipx_2 - i\vec{p}\cdot\vec{x}_2) \\ \times {}_1F_1(-i\nu_\mathrm{T}, 1, -ivx_1 - i\vec{v}\cdot\vec{x}_1)\mathrm{e}^{-i\vec{k}_f\cdot\vec{r}_f - i\nu\ln(vR+\vec{v}\cdot\vec{R})}, \tag{9}$$

where the function ϕ_f is defined by $\phi_f = (2\pi)^{-3/2}\mathrm{e}^{-i\vec{k}_f\cdot\vec{r}_f + i\vec{\kappa}\cdot\vec{x}_2}$, whereas $\vec{r}_f$ is the position vector of T with respect to the center of mass of the system $(\mathrm{Z_P}, e_1)_f + e_2$ in the exit channel, $\vec{k}_f$ is the final wave-vector, and $\nu = \mathrm{Z_P}(\mathrm{Z_T}-1)/v$. The quantities $N^-(\zeta)$ and $N^-(\nu_\mathrm{T})$ are defined by:

$$N^-(\zeta) = \Gamma(1+i\zeta)\mathrm{e}^{\pi\zeta/2}, \quad N^-(\nu_\mathrm{T}) = \Gamma(1+i\nu_\mathrm{T})\mathrm{e}^{\pi\nu_\mathrm{T}/2},$$
$$\nu_\mathrm{T} = \frac{\mathrm{Z_T}-1}{v}, \quad \zeta = \frac{\mathrm{Z_T}}{p}, \quad \vec{p} = \vec{v} + \vec{\kappa}.$$

The wave function χ_f^- obeys the correct boundary conditions.

Using equations (7), (8) and (9), the expression for the transition amplitude in the BDW-4B model, T_{if}^+, becomes:

$$T_{if}^+ = \mathcal{M}\iiint \mathrm{d}\vec{R}\mathrm{d}\vec{x}_1\mathrm{d}\vec{x}_2\mathrm{e}^{i\vec{\alpha}\cdot\vec{s}_1 + i\vec{\beta}\cdot\vec{x}_1 - i\vec{\kappa}\cdot\vec{x}_2}\mathcal{L}(R)\varphi_i(\vec{x}_1, \vec{x}_2) \\ \times {}_1F_1(i\zeta, 1, ipx_2 + i\vec{p}\cdot\vec{x}_2)\{[V_\mathrm{P}(R, s_2) + V(r_{12}, x_1)] \\ \times {}_1F_1(i\nu_\mathrm{T}, 1, ivx_1 + i\vec{v}\cdot\vec{x}_1)\varphi_f^*(\vec{s}_1) \\ - \vec{\nabla}_{s_1}\varphi_f^*(\vec{s}_1)\cdot\vec{\nabla}_{x_1}{}_1F_1(i\nu_\mathrm{T}, 1, ivx_1 + i\vec{v}\cdot\vec{x}_1)\}, \tag{10}$$

where

$$\mathcal{M} = (2\pi)^{-3/2}N^{-*}(\nu_\mathrm{T})N^{-*}(\zeta).$$

The product of logarithmic Coulomb factors is denoted by the auxiliary function $\mathcal{L}(R)$ which can be reduced to the single term:

$$\mathcal{L}(R) = \mathrm{e}^{i\nu_i\ln(vR - \vec{v}\cdot\vec{R}) + i\nu\ln(vR+\vec{v}\cdot\vec{R})} = (\rho v)^{2i\nu_i}(vR + \vec{v}\cdot\vec{R})^{i\xi}, \tag{11}$$

where $\xi = Z_P/v$. The multiplying term $(\rho v)^{2i\nu_i}$ does not contribute to the total cross section and can be dropped from the transition amplitudes.

Hence, such a formulated BDW-4B method exactly coincides with the CDW-4B method[54] in the exit channel and with the CB1-4B method in the entrance channel. Therefore, the BDW-4B method satisfies the correct boundary conditions in both scattering channels.

4.2. *Calculation of the Matrix Elements*

In this study, the analytical calculation will be limited to a helium-like target in the ground sate which is described by means of one-parameter wave function of Hylleraas-type: $\varphi_i(\vec{x}_1, \vec{x}_2) = \varphi_\lambda(\vec{x}_1)\varphi_\lambda(\vec{x}_1)$, $\varphi_\lambda(\vec{r}) = N_\lambda \exp(-\lambda r)$, $N_\lambda = (\lambda^3/\pi)^{1/2}$ where effective charge is $\lambda = 1.6875$.

Using the inverse Fourier transform:

$$\frac{1}{\omega} = \frac{1}{2\pi^2}\int \frac{d\vec{\tau}}{\tau^2} e^{-i\vec{\tau}\cdot\vec{\omega}}, \quad \text{for } \vec{\omega} \in \{\vec{s}_2, \vec{r}_{12}\}, \tag{12}$$

the transition amplitude T_{if}^{+} can be written as:

$$T_{if}^{+} = T_{if}^{(1)} + T_{if}^{(2)} - T_{if}^{(3)}, \tag{13}$$

$$T_{if}^{(1)} = Z_P \mathcal{M}_{PT} \int d\vec{R}(vR + \vec{v}\cdot\vec{R})^{i\xi} e^{-i\vec{\alpha}\cdot\vec{R}} \Delta T_{Rs} \equiv T'_R - T'_{s_2}, \tag{14}$$

$$T_{if}^{(2)} = \mathcal{M}_{PT} \int d\vec{R}(vR + \vec{v}\cdot\vec{R})^{i\xi} e^{-i\vec{\alpha}\cdot\vec{R}} \Delta T_{12} \equiv T'_{12} - T'_{x_1}, \tag{15}$$

$$T_{if}^{(3)} = \mathcal{M}_{PT} \int d\vec{R}(vR + \vec{v}\cdot\vec{R})^{i\xi} e^{-i\vec{\alpha}\cdot\vec{R}} T_{\nabla}, \tag{16}$$

where $\mathcal{M}_{PT} = \mathcal{M} Z_P^{3/2} N_\lambda^2/\sqrt{\pi}$,

$$\Delta T_{12} = T_{12} - T_{x_1}, \quad \Delta T_{Rs} = T_R - T_{s_2}, \tag{17}$$

$$T_{12} = \frac{1}{2\pi^2}\int \frac{d\vec{\tau}}{\tau^2} T_{x_1} \mathcal{D}_{x_2}^{(+)}, \quad T_{s_2} = \frac{1}{2\pi^2}\int \frac{d\vec{\tau}}{\tau^2} e^{i\vec{\tau}\cdot\vec{R}} \mathcal{L}_{x_1} \mathcal{D}_{x_2}^{(-)}, \tag{18}$$

$$T_{\nabla} = \mathcal{D}_{x_1}\mathcal{D}_{x_2}^{(0)}, \quad T_{x_1} = \mathcal{K}_{x_1}\mathcal{D}_{x_2}^{(0)}, \quad T_R = \frac{1}{R}\mathcal{L}_{x_1}\mathcal{D}_{x_2}^{(0)}, \tag{19}$$

$$\mathcal{D}_{x_1} = \int d\vec{x}_1 e^{-i\vec{v}\cdot\vec{x}_1 - \lambda x_1}[\vec{\nabla}_{s_1} e^{-Z_P s_1}] \cdot [\vec{\nabla}_{x_1} {}_1F_1(i\nu_T, 1, ivx_1 + i\vec{v}\cdot\vec{x}_1)], \tag{20}$$

$$\mathcal{D}_{x_2}^{(0)} = \int \mathrm{d}\vec{x}_2 \mathrm{e}^{-i\vec{\kappa}\cdot\vec{x}_2-\lambda x_2} {}_1F_1(i\zeta, 1, ipx_2 + i\vec{p}\cdot\vec{x}_2), \tag{21}$$

$$\mathcal{D}_{x_2}^{(\pm)} = \int \mathrm{d}\vec{x}_2 \mathrm{e}^{i(\pm\vec{\tau}-\vec{\kappa})\cdot\vec{x}_2-\lambda x_2} {}_1F_1(i\zeta, 1, ipx_2 + i\vec{p}\cdot\vec{x}_2), \tag{22}$$

$$\mathcal{T}_{x_1} = \int \mathrm{d}\vec{x}_1 \mathrm{e}^{-i(\vec{\tau}+\vec{v})\cdot\vec{x}_1-\lambda x_1} {}_1F_1(i\nu_T, 1, ivx_1 + i\vec{v}\cdot\vec{x}_1)\mathrm{e}^{-Z_\mathrm{P}s_1}, \tag{23}$$

$$\mathcal{L}_{x_1} = \int \mathrm{d}\vec{x}_1 \mathrm{e}^{-i\vec{v}\cdot\vec{x}_1-\lambda x_1} {}_1F_1(i\nu_T, 1, ivx_1 + i\vec{v}\cdot\vec{x}_1)\mathrm{e}^{-Z_\mathrm{P}s_1}, \tag{24}$$

$$\mathcal{K}_{x_1} = \int \mathrm{d}\vec{x}_1 \mathrm{e}^{-i\vec{v}\cdot\vec{x}_1-\lambda x_1} {}_1F_1(i\nu_T, 1, ivx_1 + i\vec{v}\cdot\vec{x}_1)\frac{\mathrm{e}^{-Z_\mathrm{P}s_1}}{x_1}. \tag{25}$$

Using following relation which is obtained by means of the complex contour technique of Nordsieck:[68]

$$\int \mathrm{d}\vec{r}\mathrm{e}^{i\vec{q}\cdot\vec{r}-\lambda r} {}_1F_1(i\nu, 1, i\omega r + i\vec{\omega}\cdot\vec{r}) = \frac{8\pi}{q^2+\lambda^2}\left(1 + 2\frac{\vec{q}\cdot\vec{\omega} - i\lambda\omega}{q^2+\lambda^2}\right)^{-i\nu}$$
$$\times\left(\lambda\frac{1-i\nu}{q^2+\lambda^2} + i\nu\frac{\lambda - iv}{q^2+\lambda^2+2\vec{q}\cdot\vec{\omega}-2i\lambda\omega}\right), \tag{26}$$

the integrals $\mathcal{D}_{x_2}^{(0)}$ and $\mathcal{D}_{x_2}^{(\pm)}$ can be analytically solved with results:

$$\mathcal{D}_{x_2}^{(0)} = 8\pi\frac{S_0^{i\zeta}\mathcal{L}_0}{\kappa_0} \equiv 8\pi W_0, \quad \mathcal{D}_{x_2}^{(\pm)} = \frac{8\pi}{q_\pm^2+\lambda^2}T_\pm^{i\zeta}\mathcal{R}_\pm \equiv 8\pi W_\pm, \tag{27}$$

with

$$S_0^{-1} = 1 - 2\frac{\vec{\kappa}\cdot\vec{p} + i\lambda p}{\kappa_0}, \qquad T_\pm^{-1} = 1 + 2\frac{\vec{q}_\pm\cdot\vec{p} - i\lambda p}{q_\pm^2+\lambda^2}, \tag{28}$$

$$\mathcal{L}_0 = [\lambda(1-i\zeta) + i\zeta(\lambda - ip)S_0]/\kappa_0, \tag{29}$$

$$\mathcal{R}_\pm = \lambda\frac{1-i\zeta}{q_\pm^2+\lambda^2} + i\zeta\frac{\lambda - ip}{q_\pm^2+\lambda^2+2\vec{q}_\pm\cdot\vec{p}-2i\lambda p}, \tag{30}$$

where $\vec{q}_\pm = \pm\vec{\tau} - \kappa$, $\kappa_0 = \kappa^2 + \lambda^2$.

Employing integral representation of the confluent hypergeometric function:

$${}_1F_1(i\nu_T, 1, ivx_1 + i\vec{v}\cdot\vec{x}_1)] = \frac{1}{\Gamma(i\nu_T)\Gamma(1-i\nu_T)}\int_0^1 \mathrm{d}t_1 f_T(t_1)\mathrm{e}^{i(vx_1+\vec{v}\cdot\vec{x}_1)t_1}, \tag{31}$$

where $f_T(t_1) = t_1^{i\nu_T - 1}(1 - t_1)^{-i\nu_T}$, integral $\mathcal{D}_{x_1}$ becomes

$$\mathcal{D}_{x_1} = \frac{1}{\Gamma(i\nu_T)\Gamma(1 - i\nu_T)} \int_0^1 \mathrm{d}t_1 f_T(t_1)\mathcal{F}_1(\vec{R}), \tag{32}$$

$$\mathcal{F}_1(\vec{R}) = \int \mathrm{d}\vec{x}_1 \mathrm{e}^{-i\vec{v}\cdot\vec{x}_1 - \lambda x_1}[\vec{\nabla}_{s_1}\mathrm{e}^{-Z_{\mathrm{P}}s_1}] \cdot [\vec{\nabla}_{x_1}\Omega(\vec{x}_1)], \tag{33}$$

with $\Omega(\vec{x}_1) = \exp[i(vx_1 + \vec{v}\cdot\vec{x}_1)t_1]$.

Integral $\mathcal{F}_1(\vec{R})$ can be analytically reduced to a one-dimensional integral. Following Ref. 69, after long calculations, we have the result:

$$\mathcal{F}_1(\vec{R}) = 2\pi Z_{\mathrm{P}} t_1 \omega \mathrm{e}^{i\vec{\alpha}\cdot\vec{R}} \int_0^1 \mathrm{d}t \frac{t(1-t)\mathrm{e}^{-i\vec{Q}_1\cdot\vec{R} - \Delta_1 R}}{\Delta_1^5} \times [a_1 R^2 - b_1 R - c_1 - d_1(1 + \Delta_1 R)\vec{R}\cdot\hat{v}], \tag{34}$$

where

$$\vec{Q}_1 = \vec{\alpha}t - \vec{\beta}_1(1 - t), \qquad \vec{\beta}_1 = \vec{\beta} + \vec{v}t_1, \tag{35}$$

$$\Delta_1^2 = v_1^2 t(1 - t) + Z_{\mathrm{P}}^2 t + \lambda_1^2(1 - t), \tag{36}$$

$$\vec{v}_1 = \vec{v}(1 - t_1), \quad \lambda_1 = \lambda - \omega t_1, \quad a_1 = \Delta_1^2\delta_1^+, \quad b_1 = c_1\Delta_1, \tag{37}$$

$$c_1 = 3\delta_1^-, \quad \omega = iv, \quad \delta_1 = \lambda_1 - \gamma_1, \quad \gamma_1 = \omega t(1 - t_1), \tag{38}$$

$$\delta_1^{\pm} = \Delta_1^2 \pm \gamma_1'\delta_1, \quad d_1 = -\Delta_1^2(\gamma_1' + \delta_1), \quad \gamma_1' = \omega(1 - t)(1 - t_1). \tag{39}$$

Now, the expression for the term $T_{if}^{(3)}$ from equation (16) in the transition amplitude becomes:

$$T_{if}^{(3)} = \frac{16\mathcal{M}Z_{\mathrm{P}}^{5/2}N_\lambda^2\pi^{3/2}\omega}{\Gamma(i\nu_T)\Gamma(1 - i\nu_T)} W_0 \int_0^1 \mathrm{d}t_1 f_T(t_1)t_1 \int_0^1 \mathrm{d}t \frac{t(1-t)}{\Delta_1^5} J(t_1), \tag{40}$$

$$J(t_1) = a_1 I_3 - b_1 I_2 - c_1 I_1 + \frac{d_1}{\omega}(I_1' + \Delta_1 I_2'), \tag{41}$$

where

$$I_n = \int \mathrm{d}\vec{R} R^{n-1}(vR + \vec{v}\cdot\vec{R})^{i\xi}\mathrm{e}^{-\vec{Q}_1\cdot\vec{R} - \Delta_1 R}, \tag{42}$$

$$I_n' = \vec{v}\cdot\vec{\nabla}_{Q_1} I_n \quad (n = 0, 1, 2, \ldots). \tag{43}$$

Integrals I_n and I'_n can be analytically calculated along the lines in Ref. 69 and the following result is obtained for $J(t_1)$:

$$J(t_1) = 8\pi\Gamma(1+i\xi)\mathcal{T}D(\mathcal{A}'' - i\xi\mathcal{B}''), \tag{44}$$

where

$$\mathcal{A}'' = -2a_1\frac{D}{\Delta_1}A_\beta + \frac{b_1}{\Delta_1}A_\alpha - c_1 - 2\frac{d_1 D}{\omega\Delta_1}(A_1 - A_2), \tag{45}$$

$$\mathcal{B}'' = 2a_1\frac{D}{\Delta_1}B_\beta - \frac{b_1}{\Delta_1}B_\alpha - c_1 C + 2\frac{d_1 D}{\omega\Delta_1}(B_1 - B_2), \tag{46}$$

$$\mathcal{F} = \frac{B^{i\xi}}{Q_1^2 + \Delta_1^2}, \quad B = \frac{2(v\Delta_1 - i\vec{Q}_1\cdot\vec{v})}{Q_1^2 + \Delta_1^2}, \tag{47}$$

$$C = \frac{v}{B\Delta_1} - 1, \quad A = \frac{\Delta_1^2}{Q_1^2 + \Delta_1^2}, \quad D = \frac{A}{\Delta_1}, \tag{48}$$

$$A_\alpha = 1 - 4A, \quad B_\alpha = 1 + 2AC_\alpha, \quad C_\alpha = C[4 + (1-i\xi)C], \tag{49}$$

$$A_\beta = 6(1-2A), \quad B_\beta = 2AC_\beta + 3D_\beta, \tag{50}$$

$$C_\beta = C[18 + 9(1-i\xi)C + (1-i\xi)(2-i\xi)C^2], \tag{51}$$

$$D_\beta = 2 - (1+i\xi)C, \tag{52}$$

$$A_1 = \{2 + i\xi[2 - (1+i\xi)C]\}(\vec{Q}_1\cdot\vec{v}), \quad B_1 = u\omega[2 + (1-i\xi)C], \tag{53}$$

$$A_2 = \{2(1-6A) + i\xi[2AC_2 + (3+i\xi)]\}(\vec{Q}_1\cdot\vec{v}), \tag{54}$$

$$C_2 = (1+i\xi)C[6 + (1-i\xi)C] - 6, \quad u = (1+C)\Delta_1, \tag{55}$$

$$B_2 = u\omega[(1+i\xi) - 2AC_2'], \quad C_2' = (1-i\xi)C[6 + (2-i\xi)C] + 6. \tag{56}$$

In that way, the nine-dimensional integral for term $T_{if}^{(3)}$ is reduced to a two-dimensional real quadrature.

Using a similar technique for the quantities $\mathcal{L}_{x_1}$, $\mathcal{K}_{x_1}$ and $\mathcal{T}_{x_1}$, introduced by equation (19), we can obtain following expressions:

$$\mathcal{L}_{x_1} = \mathcal{M}_{x_1}\mathrm{e}^{i\vec{\alpha}\cdot\vec{R}}\int_0^1 \mathrm{d}t_1 f_T(t_1)\lambda_1\int_0^1 \mathrm{d}t\frac{t(1-t)\mathrm{e}^{-i\vec{Q}_1\cdot\vec{R}-\Delta_1 R}}{\Delta_1^5} \times (3 + 3\Delta_1 R + \Delta_1^2 R^2), \tag{57}$$

$$\mathcal{K}_{x_1} = \mathcal{M}_{x_1} \mathrm{e}^{i\vec{\alpha}\cdot\vec{R}} \int_0^1 \mathrm{d}t_1 f_T(t_1) \int_0^1 \mathrm{d}t \frac{t\mathrm{e}^{-i\vec{Q}_1\cdot\vec{R}-\Delta_1 R}}{\Delta_1^3}(1+\Delta_1 R), \quad (58)$$

$$\mathcal{T}_{x_1} = \mathcal{M}_{x_1} \mathrm{e}^{i\vec{\alpha}\cdot\vec{R}} \int_0^1 \mathrm{d}t_1 f_T(t_1)\lambda_1$$

$$\times \int_0^1 \mathrm{d}t \frac{t(1-t)\mathrm{e}^{-i\vec{Q}_{12}\cdot\vec{R}-\Delta_{12}R}}{\Delta_{12}^5}(3+3\Delta_{12}R+\Delta_{12}^2R^2), \quad (59)$$

where

$$\mathcal{M}_{x_1} = 2\pi Z_{\mathrm{P}}/[\Gamma(i\nu_T)\Gamma(1-i\nu_T)], \quad \vec{Q}_{12} = \vec{Q}_1 + \vec{\tau}, \quad (60)$$

$$\Delta_{12}^2 = (\vec{v}_1+\vec{\tau})^2 t(1-t) + Z_{\mathrm{P}}^2 t + \lambda_1^2(1-t). \quad (61)$$

Inserting equations (57) and (27) in quantity T_R', introduced by equation (14), we obtain:

$$T_R' = 2\pi^2 N_c Z_{\mathrm{P}} W_0 \int_0^1 \mathrm{d}t_1 f_T(t_1)\lambda_1 \int_0^1 \mathrm{d}t \frac{t(1-t)}{\Delta_1^5}(3I_0 + 3\Delta_1 I_1 + \Delta_1^2 I_2)$$

$$= 16\pi^3\Gamma(1+i\xi)N_c Z_{\mathrm{P}} W_0 \int_0^1 \mathrm{d}t_1 f_T(t_1)\lambda_1$$

$$\times \int_0^1 \mathrm{d}t \frac{t(1-t)\mathcal{F}}{\Delta_1^5}(\mathcal{A}_R - i\xi\mathcal{B}_R), \quad (62)$$

where $N_c = 8\pi\mathcal{M}_{PT}\mathcal{M}_{x_1}/(2\pi^2) = 4\mathcal{M}_{PT}\mathcal{M}_{x_1}/\pi$. The analytical results for integrals $I_{0,1,2}$ defined by equation (42) is also utilized. The quantities $\mathcal{A}_R$ and $\mathcal{B}_R$ are introduced via

$$\mathcal{A}_R = 3/2 + 3\Delta_1 D - \Delta_1 D A_\alpha, \quad \mathcal{B}_R = 3\Delta_1 DC + \Delta_1 D B_\alpha. \quad (63)$$

Inserting equations (58) and (27) in quantity T_{x_1}' introduced by equation (15) we have:

$$T_{x_1}' = 2\pi^2 N_c W_0 \int_0^1 \mathrm{d}t_1 f_T(t_1) \int_0^1 \mathrm{d}t \frac{t}{\Delta_1^3}(I_1 + \Delta_1 I_2)$$

$$= 16\pi^3\Gamma(1+i\xi)N_c W_0 \int_0^1 \mathrm{d}t_1 f_T(t_1) \int_0^1 \mathrm{d}t \frac{t\mathcal{F}D}{\Delta_1^3}(\mathcal{A}_3 - i\xi\mathcal{B}_3), \quad (64)$$

where

$$\mathcal{A}_3 = 1 - A_\alpha, \quad \mathcal{B}_3 = C + B_\alpha. \quad (65)$$

Employing equations (57), (59) and (27), the expressions for the remaining quantities T'_{12} and T'_{s_2} from equations (15) and (14) become:

$$T'_{12} = N_c \int \frac{\mathrm{d}\vec{\tau}}{\tau^2} \int_0^1 \mathrm{d}t_1 f_T(t_1)\lambda_1 W_+ \int_0^1 \mathrm{d}t \frac{t(1-t)}{\Delta_{12}^5}(3K_1 + 3\Delta_{12}K_2 + \Delta_{12}^2 K_3)$$

$$= 8\pi\Gamma(1+i\xi)N_c \int_0^1 \mathrm{d}t_1 f_T(t_1)\lambda_1 \int_0^1 dtt(1-t)$$

$$\times \int \frac{\mathrm{d}\vec{\tau}}{\tau^2} W_+ \frac{D_{12}\mathcal{F}_{12}}{\Delta_{12}^5}(\mathcal{A}_{12} - i\xi\mathcal{B}_{12}), \tag{66}$$

where

$$K_n = \int \mathrm{d}\vec{R}R^{n-1}(vR + \vec{v}\cdot\vec{R})^{i\xi}\mathrm{e}^{-\vec{Q}_{12}\cdot\vec{R} - \Delta_{12}R}, \quad (n = 1,2,3) \tag{67}$$

$$\mathcal{A}_{12} = 3 - 3A''_\alpha - 2A_{12}A''_\beta, \quad \mathcal{B}_{12} = 3C_{12} + 3B''_\alpha + 2A_{12}B''_\beta, \tag{68}$$

$$T'_{s_2} = Z_\mathrm{P}N_c \int \frac{\mathrm{d}\vec{\tau}}{\tau^2} \int_0^1 \mathrm{d}t_1 f_T(t_1)\lambda_1 W_- \int_0^1 \mathrm{d}t \frac{t(1-t)}{\Delta_1^5}(3J_1 + 3\Delta_1 J_2 + \Delta_1^2 J_3)$$

$$= 8\pi\Gamma(1+i\xi)Z_\mathrm{P}N_c \int_0^1 \mathrm{d}t_1 f_T(t_1)\lambda_1 \int_0^1 dtt(1-t)$$

$$\times \int \frac{\mathrm{d}\vec{\tau}}{\tau^2} W_- \frac{D_\tau\mathcal{F}_\tau}{\Delta_1^5}(\mathcal{A}_{s_2} - i\xi\mathcal{B}_{s_2}), \tag{69}$$

where

$$\mathcal{A}_{s_2} = 3 - 3A'_\alpha - 2A_\tau A'_\beta, \quad \mathcal{B}_{s_2} = 3C_\tau + 3B'_\alpha + 2A_\tau B'_\beta. \tag{70}$$

All quantities which appear in equations (68) and (70) are listed in the Appendix. Finally, for transition amplitude we have

$$T_{if}^+ = 8\pi\Gamma(1+i\xi)N_c \int_0^1 \mathrm{d}t_1 f_T(t_1) \int_0^1 \mathrm{d}tt(1-t)\left\{\frac{1}{\Delta_1^5}\mathcal{H} + \Phi(t_1,t)\right\}, \tag{71}$$

with

$$\Phi(t_1,t) = \int \frac{\mathrm{d}\vec{\tau}}{\tau^2}F(\vec{\tau}) = \int_0^\infty \mathrm{d}\tau \int_0^\pi \mathrm{d}\theta_\tau \sin\theta_\tau \int_0^{2\pi} \mathrm{d}\phi_\tau F(\vec{\tau}), \tag{72}$$

$$\mathcal{H} = 2\pi^2[\lambda_1 Z_\mathrm{P}\mathcal{P}_R - \Delta_1^2\mathcal{P}_{x_1} - \omega t_1\mathcal{P}_\nabla], \tag{73}$$

$$F(\vec{\tau}) = \lambda_1[\mathcal{P}_{12} - Z_\mathrm{P}\mathcal{P}_{s_2}], \quad \mathcal{P}_R = \mathcal{F}(\mathcal{A}_R - i\xi\mathcal{B}_R)W_0, \tag{74}$$

$$\mathcal{P}_\nabla = \mathcal{F}D(\mathcal{A}'' - i\xi\mathcal{B}'')W_0, \quad \mathcal{P}_{x_1} = \mathcal{F}D(\mathcal{A}_3 - i\xi\mathcal{B}_3)W_0, \tag{75}$$

$$\mathcal{P}_{12} = \mathcal{F}_{12}D_{12}\frac{(\mathcal{A}_{12} - i\xi\mathcal{B}_{12})}{\Delta_{12}^5}W_+, \quad \mathcal{P}_{s_2} = \mathcal{F}_\tau D_\tau\frac{(\mathcal{A}_{s_2} - i\xi\mathcal{B}_{s_2})}{\Delta_1^5}W_-. \tag{76}$$

The triple differential cross sections for simultaneous transfer and ionization can be obtained from the expressions:

$$Q_{if}(\vec{\kappa}) \equiv \frac{\mathrm{d}^3 Q_{if}}{\mathrm{d}\vec{\kappa}} = \frac{1}{2\pi^2 v^2}\int_0^\infty \mathrm{d}\eta\eta|T_{if}^+(\vec{\eta})|^2. \tag{77}$$

The last step to get a value for corresponding total cross sections is a triple integration over $\vec{\kappa} = (\kappa\sin\theta_\kappa\cos\phi_\kappa, \kappa\sin\theta_\kappa\sin\phi_\kappa, \kappa\cos\theta_\kappa)$:

$$Q_{if}(\pi a_0^2) = \int_0^\infty \mathrm{d}\kappa\kappa^2 \int_0^\pi \mathrm{d}\theta_\kappa \sin\theta_\kappa \int_0^{2\pi} \mathrm{d}\phi_\kappa\, Q_{if}(\vec{\kappa}). \tag{78}$$

Hence, according to the present BDW-4B model the total cross section is derived in terms of *nine*-dimensional numerical quadratures.

In order to apply the Gauss-Legendre numerical quadrature for the integration over κ, τ, θ_κ, θ_τ and η, it is convenient to introduce the change of variables according to: $\kappa = (1+x)/(1-x)$, $x \in [-1,+1]$; $\tau = (1+y)/(1-y)$, $y \in [-1,+1]$; $\cos\theta_\kappa = u$, $u \in [-1,+1]$; $\cos\theta_\tau = v$, $v \in [-1,+1]$; $\eta = \sqrt{2(1+z)/(1-z)}$; $z \in [-1,+1]$. The last change of variable is important since it concentrates the integration points near the forward cone,[70] which gives the dominant contributions because of the eikonal nature of scattering for heavy projectiles. The singularities at the point $x = 1$, $y = 1$ and $z = 1$ disappear altogether after analytical scaling of the integrand. The integrations over ϕ_κ and ϕ_τ are performed by Gauss-Mehler quadratures. It should be pointed out that special attention has been paid to integration over t_1. Namely, this integration requires a Cauchy-type procedure for simultaneous regularization of the branch-points singularities at $t_1 = 0$ and $t_1 = 1$.

5. Double Ionization

The idea that the single ionization cross section can be extended to double ionization mechanisms is frequently used via various models. The most basic models which treated Li^{3+}−He collisions are reviewed here.

5.1. *Independent Particle Model*

In the IPM it is assumed that either the two electrons are removed simultaneously, or that the conditions for the ionization of the second electron remain unchainged even though the first electron has been removed, i.e. the ionization probability is independent of the state of other electron active in the collision. This assumption may have a degree of validity in the collision where the charge of the target is large, however it would almost certainly be false when considering the double ionization of the helium, where the energy required to ionize the second electron would be mush greater than that to remove the first electron.

In this model the total cross section is given by $Q = 2\pi \int \mathrm{d}\rho\rho[P_{Z_{\mathrm{T}}^*}(\rho)]^2$ where $P_{Z_{\mathrm{T}}^*}(\rho)$ is the probability of single ionization at impact parameter ρ as $P(\rho) = \int \mathrm{d}\vec{\kappa}|a_{if}(\rho,\vec{\kappa})|^2$, where $a_{if}(\rho,\kappa)$ is the transition amplitude. The momentum of the ionized electron with respect to the residual target atom is given by $\vec{\kappa}$. Interaction between the ionized electron and residual target ion can be modelled by a Coulomb potential. This has been done following Belkić *et al.*,[19] by assuming that an ejected electron, ionized from an orbital of Roothan-Hartree-Fock energy E_i and principal quantum numbers n, moves in the potential of the form $V(r_T) = -Z_{\mathrm{T}}^*/r_T$ where effective charge $Z_{\mathrm{T}}^* = \sqrt{-2n^2E_i}$ (for helium $Z_{\mathrm{T}}^* = 1.35495$), and r_T is the position of the electron with respect to the target nucleus.

Another way to choose the Z_{T}^* has been done by Shingal and Lin.[62] In the context of the close-coupling travelling atomic orbital impact-parameter model, they[62] assumed that the bonding energy of each electron is half of the double ionization energy of 2.9 a.u. and the ejected electrons move in Coulombic potential with $Z_{\mathrm{T}}^* = 1.703$.

5.2. *Independent Event Model*

The independent event model, as denoted earlier by IEV, originally introduced by Crothers and McCarroll,[72] considers that the second electron is emitted from the single ionized atom, after the first electron makes a transition i.e. DI would be therefore be considered as a two stage process: $\mathrm{P} + \mathrm{T} \to \mathrm{P} + \mathrm{T}^+ + \mathrm{e}$ with probability $P_1(\rho)$, $\mathrm{P} + \mathrm{T}^+ \to \mathrm{P} + \mathrm{T}^{2+} + \mathrm{e}$ with probability $P_2(\rho)$. In this model the total cross section is given by $Q = 2\pi \int \mathrm{d}\rho\rho P_1(\rho)P_2(\rho)$.

5.3. *Coulomb Density of States Correlation Model*

The IPM and IEV models do not account for the electron-electron correlation as well as the exclusion principle, which forbids two electrons to be emitted with the same momentum. McCartney[73] introduced these constraints by multiplying the single probabilities by the continuum Coulomb density of states factor, also denoted as the Gamow factor $C(k_{12})$, for the electron-electron interaction in the final state. With this factor, the IPM and IEV models were called the CDS-IPM and CDS-IEV model, respectively. The total cross section in this approximation is now defined as $Q = 2\pi \int d\rho\rho \int d\vec{\kappa}_1 \int d\kappa_2 C(\kappa_{12}) W_1(\rho, \vec{\kappa}_1) W_2(\rho, \vec{\kappa}_2)$, where $W_i(\rho, \vec{\kappa}_i) = |a(\rho, \vec{\kappa}_i)|^2$, $i = 1, 2$. The new factor $C(k_{12})$ is given by[73] $C(\kappa_{12}) = 2\pi/[\kappa_{12}(e^{2\pi/\kappa_{12}} - 1)]$, where $\kappa_{12} = |\vec{\kappa}_1 - \vec{\kappa}_2|$. This implies that $C(\kappa_{12}) \to 0$ when $\kappa_{12} \to 0$, and $C(\kappa_{12}) \to 1$ as $\kappa_{12} \to \infty$. Thus if the electrons are well separated in the momentum space, the CDS factor allows the electrons to move independently, but the probability of the electrons being in the same region of momentum space is exponentially damped.

Bradley *et al.*[75] published calculations of the multiple ionization cross sections of helium, lithium and neon atoms by using Roothan-Hartree-Fock[43] atomic functions for the initial state, and a final state given by hydrogenic Coulomb continuum waves with an effective charge Z_T^*, derived from the binding energies. They compared results obtained with the IPM and IEV approaches, with and without the CDS factor, and found that the CDS-IEV agrees with experiment at intermediate energies. Subsequently, Fiori *et al.*[76,77] calculated the total double ionization cross sections for helium and lithium atoms by the impact of H^+, He^{2+}, and Li^{3+} in the context of the two-step CDS-IEV model. This takes into account the final state electronic correlation as well as the relaxation of the atom after the first ionization event. The individual probabilities are calculated within the continuum distorted wave eikonal initial state approximation of Crothers and McCann[78] with numerically evaluated atomic wave functions for the bound and continuum states. They solved the wave equation with the optimized potential model (OPM) of Talman and Shadwick[79] avoiding, in this way, the use of parametric charges, and discussed the influence of these improved wave functions on the cross sections.

As can be seen from Fig. 8, the results in the IPM of McCartney[73] are very sensitive to the value of the effective charge of the target. Both effective charges (the dotted curve with effective charge $Z_T^* = 1.703$ and dashed curve with $Z_T^* = 1.35495$) considerably overestimate experimental data. Since the CDS factor suppresses events where both electrons are emitted to the same region of momentum space, in other words in the CDS model the weighted probability profile shifts to lower ρ, where the electrons have a greater likelihood of being emitted to more widely separated regions of

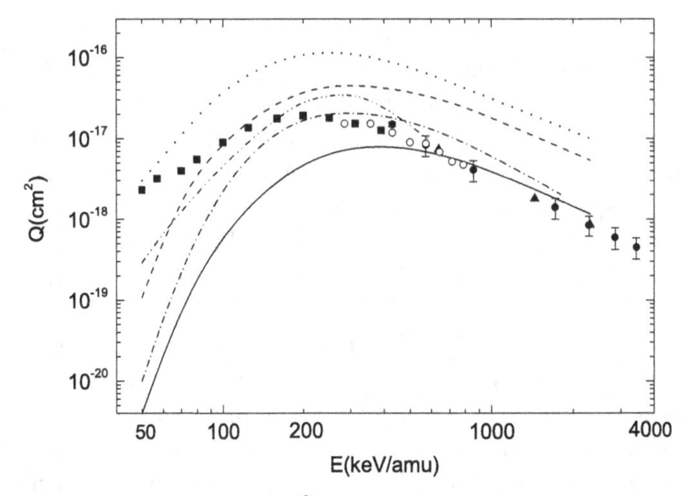

Fig. 8. The total cross sections (in cm^2) as a function of the laboratory incident energy E(keV/amu) for double ionization in the reaction: $Li^{3+} + He \longrightarrow Li^{3+} + He^{2+} + 2e$. The full curve: the CDS model[73]; Dotted curve: IPM model with effective charge $Z_T^* = 1.703$[73]; Dashed curve: IPM model with effective charge $Z_T^* = 1.35495$[73]; Dashed-dot curve: CDS-IEV results with OPM wave functions for the initial and final orbitals of Fiori *et al.*[76]; Dashed-dot-dot curve: close-coupling results of Shingal and Lin[62] (their model B). Experimental data: ■ Shah and Gilbody[7]; ▲ Knudsen *et al.*[58]; • Woitke *et al.*[60]; ○ Sant'Anna *et al.*[5]

momentum space. As such, the CDS model gives satisfactory agreement with experimental data for energy range 200–2000 keV/amu (see full line in Fig. 8).

The optimized potential model allows for a more accurate representation of one-electron orbitals than does the rather crude effective charge model. Fiori *et al.*[76,77] showed that the OPM approach leads to better agreement than the effective charge model when compared to the experimental results. They also noted that differences between cross sections obtained with the OPM and the effective charge model are larger for Li than for He targets. The OPM approach gives a better description of the spatial electronic distribution of each orbital and is more appropriate for dealing with complex atoms.

It should be noted that the theoretical CDS-IEV results of Fiori *et al.*[76] with OPM wave functions for the initial and final orbitals (the dashed-dot curve in Fig. 8) are similar to the CDS-IEV results[74,75] obtained with a RHF initial state and a final hydrogenic Coulomb wave (not shown in Fig. 8 to avoid clutter). At higher impact velocities the CDS-IEV results overestimate experimental measurements and this may be associated with a shake-off mechanism where an electron is emitted and the second electron relaxes from an ionic bound state to continuum. For higher impact energies, the shake-off processes, not considered in an independent electron description, may contribute significantly. The contribution of the shake-off was roughly estimated by McGuire[80] within the Born approximation. In this approximation and for the double ionization of He by protons the contribution of shake-off is only important for energies larger than 2 MeV.

The two-electron coupled-channel method (close-coupling method) was successfully used by Barna *et al.*[81,82] to calculate single- and double-ionization total cross sections for fast and highly charged ion collisions. The main idea of their method is the discretization of the electron continuum by Coulomb wave packets. The double-electron continuum is approximated by a large number of symmetrized products of single-particle Coulomb packets and includes a high degree of correlation.

5.4. *The Classical Trajectory Monte Carlo (CTMC) Method*

The CTMC method is an alternative classical theory to calculate the cross sections for the ejected electrons in bare ion and He collisions. Differences between the classical models result from different models for the underlying mean field by which the electron-electron interaction is taken into account. Barna *et al.*[81] refer to them as non-equivalent electron (NEE) CTMC and equivalent electron (EE) CTMC models, respectively. In both versions of CTMC approaches, Newton's classical non-relativistic equations of motion have been solved numerically[81] for a large number of trajectories for given initial conditions.

5.4.1. *Non-equivalent electron CTMC model*

In this model the four structureless particles are characterized by their masses and charges. The forces acting among the four bodies are taken to be purely Coulombic. The interaction between the two active electrons of the helium atom is, however neglected during the collision. The two electrons are treated as nonequivalent. They are represented by micro-canonical ensembles with energies corresponding to the first and second ionization potentials, respectively. Barna *et al.*[81] noted that this type of CTMC model is the classical analogue of the quantum-mechanical treatment of He atom when the two electron wavefunction is built as a product of the two *different* single-particle wavefunctions. The total cross section for a specific event i is calculated[81] from $Q_i = \pi\rho_{max}^2 N_i/N$, where N is the total number of trajectories calculated for the impact parameters less than ρ_{max}, and N_i is the number of trajectories that satisfy the criteria for the process under consideration.

5.4.2. *Equivalent electron CTMC model*

In this model three particles are explicitly treated: the projectile, the active atomic electron and the remaining helium ion (He^+). The interaction between the active target electron and the projectile is purely Coulombic. For the description of the interaction between the projectile and the helium core and between the active electron and the helium core, a model potential is used which is based on Hartree-Fock calculations $V(r) = -[Z_T - 1)\Omega(r) + 1]/r$ where $\Omega(r) = [Hd(\exp(r/d) - 1) + 1]^{-1}$ with values of parameters[83] $H = 1.77\,$a.u and $d = 0.381\,$a.u for helium atom. This type of CTMC model is the classical analogue of the quantum-mechanical model in which the time-dependent two-electron wavefunction is built as a product of two *identical* single-particle wavefunctions.

It should be noted that both CTMC models completely ignore dynamical correlation effects (i.e. the ionization of the second electron by the first ionized electron via the electron-electron repulsion).

In Fig. 9, we compare the coupled-channel (CC) and CTMC calculations[81] together with experimental data.[5,7,58,60] As shown in Fig. 9, the coupled-channel method gives better results than the CTMC model. Thanks to the large number of double-ionized channels, even for strong perturbation the results agree well with experimental data. The differences between the EE-CTMC and NEE-CTMC models are significant. In contrast to the case of single ionization, here the results of NEE-CTMC model are closer to the experiment than the results of EE-CTMC. As pointed out by Barna *et al.*[81] the reason is most likely that in the non-equivalent electron approximation the binding energies of the two electrons in the atom can be independently specified according to the first and second ionization threshold. At the same time CTMC models, using different independent-particle approximations, show somewhat larger discrepancies compared to the experimental cross sections. The discrepancies between coupled-channel and CTMC calculations may be due to the incomplete treatment of the electron-electron interaction within the classical simulations. However, the coupled-channel and CTMC calculations[81] have been performed only at three impact energies (0.64, 1.44 and 2.31 MeV/amu) in a relatively small energy interval and it is difficult to make general conclusions about the validity of these models.

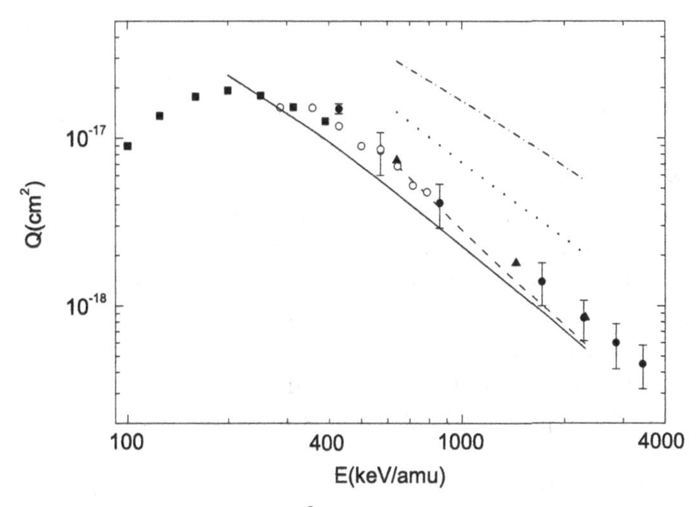

Fig. 9. The total cross sections (in cm^2) as a function of the laboratory incident energy E(keV/amu) for double ionization in the reaction: $Li^{3+} + He \longrightarrow Li^{3+} + He^{2+} + 2e$. The full curve: IPM results of Deb and Crothers[84]; Dotted curve: NEE-CTMC model[81]; Dashed curve: close-coupling results of Barna *et al.*;[81,82] Dashed-dot curve: EE-CTMC model.[81] Experimental data: ■ Shah and Gilbody;[7] ▲ Knudsen *et al.*[58]; • Woitke *et al.*[60]; ○ Sant'Anna *et al.*[5]

In Fig. 9 also shown are the results of Deb and Crothers[84] obtained by the independent-event model (full line). Ionization of each of the two electrons from helium is considered to be an independent event and probabilities for these events are calculated individually by using the continuum-distorted-wave approximation. The CDW-3B for single ionization was introduced by Belkić.[86] Static electron correlations in Ref. 84 have been accounted for through the explicitly correlated two-electron wavefunction of Pluvinage.[85] Good agreement between the results of Deb and Crothers[84] and measurements has been attributed to this static correlation.

Unlike single ionization, electron correlation plays an important role and the inclusion of its effects is very difficult. This is compounded by the fact that three separate mechanisms: shake off effects, dynamic correlation and projectile-electron correlation, contribute to the double ionization process. We conclude that the IPM and IEV models do not have the level of sophistication required to predict accurate cross sections for double ionization.

In the two step models (IPM and IEV), the value of double ionization cross sections is supposed to depend strongly on the quality of the model used in the description of the single ionization cross sections.

6. Single Ionization

For the ionization of atoms, the complication arises that the electron in the final state moves under the combined influences of the projectile and the residual target. So one has to solve at least a three-body problem in which three particles interact via Coulomb potentials. Due to the long-range nature of the Coulomb potential, the free electronic state cannot be represented by only a plane wave. An exact solution for the problem is difficult to obtain; however, its correct asymptotic form is available.[86,87]

Assuming an independent-electron model, Fainstein *et al.*[88] considered that one electron is ionized while the other remains "frozen" during the collision. In this way, the problem is reduced to a one-active-electron system, and they[88] presented calculations using the continuum distorted wave eikonal-initial-state model.

In Fig. 10, the total cross sections for ionization of He by Li^{3+} calculated with the CDW-EIS model[88] are compared with calculations using the first Born approximation (FBA)[89] and with experimental data. The CDW-EIS results[88] are in good agreement with experimental data and give the correct behavior at the cross-section maximum. As can be seen from Fig. 10 the CDW-EIS results agree much better than do the FBA results which fail to reproduce the experimental data at lower impact energies.

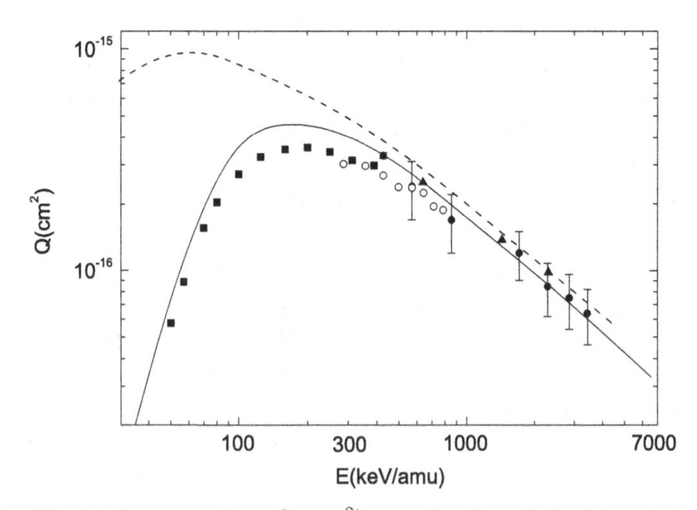

Fig. 10. The total cross sections (in cm^2) as a function of the laboratory incident energy E(keV/amu) for single ionization in the reaction: $Li^{3+} + He \longrightarrow Li^{3+} + He^+ + e$. The full curve: CDW-EIS results of Fainstein *et al.*;[88] Dashed curve: FBA results.[88,89] Experimental data: ■ Shah and Gilbody;[7] ▲ Knudsen *et al.*;[58] • Woitke *et al.*;[60] ○ Sant'Anna *et al.*[5]

Sahoo *et al.*[90] have applied the perturbative approach to study ionization of helium atoms by Li^{3+} ions. The targets were treated as one-electron atoms, where the interaction of the active electron with the rest of the target is represented by a model potential. The final electronic state is described by the product of two Coulomb waves and thus the ejected electron is considered to be moving under the combined influence of the projectile and the residual target. Their results[90] (not shown in Fig. 10 to avoid clutter) are close to that of Fainstein *et al.*,[88] and show good accord with the predictions of the experimental data.

The calculation of the total cross sections for ionization of He by Li^{3+} has also been the subject of investigation in ref.[62] The calculated results[62] within the close-coupling travelling atomic orbital impact-parameter model for SI have been in reasonable agreement with experimental data at intermediate energies.

Figure 11 shows the measured single ionization total cross sections, together with the theoretical results of coupled-channel and CTMC calculations[81] at impact energies 0.164–2.31 MeV/amu. For single-ionization cross sections the coupled-channel method gives better agreement with experiments than the CTMC methods. Barna *et al.*[81,82] found that the equivalent electron CTMC results are typically closer to the experimental

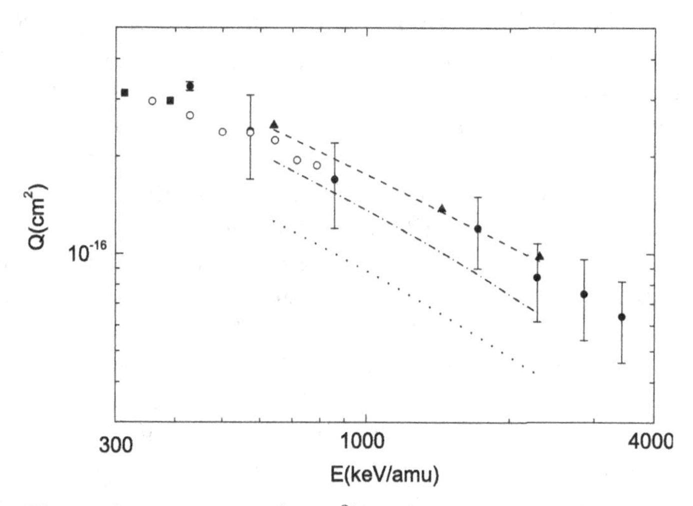

Fig. 11. The total cross sections (in cm^2) as a function of the laboratory incident energy E(keV/amu) for single ionization in the reaction: $Li^{3+} + He \longrightarrow Li^{3+} + He^{+} + e$. The dotted curve: NEE-CTMC model;[81] Dashed-dot curve: EE-CTMC model;[81] Dashed curve: coupled channel calculations of Barna *et al.*[81,82] Experimental data: ■ Shah and Gilbody;[7] ▲ Knudsen *et al.*;[58] • Woitke *et al.*;[60] o Sant'Anna *et al.*[5]

cross sections than the non-equivalent electron CTMC. This is not surprising as the NEE model underestimates the multiplicity of the available electrons relatively easily ionized by an energy transfer of the order of the first ionization potential. Their experience[81,82] shows that in order to obtain the converged results, a particular choice of an optimized basis set for each collision system is required. As a pragmatic solution to find an optimized basis, they always started with the first Born approximation including all channels.

7. Conclusions

The two major strategies from the literature on collision system Li^{3+}–He, the semi-classical independent particle model and the quantum-mechanical four-body formalism are found to exhibit a varying degree of success when passing from one concrete method to another within which the said strategies were implemented. Certainly, the most frequently used theoretical methodologies with correct boundary conditions are from the realm of the continuum distorted waves and continuum distorted wave eikonal initial states. As it is well-known, the CDW method employs the full Coulomb waves for initial and final continuum intermediate states for electronic motions in the field of the projectile and target nuclei. These Coulomb wave functions are simplified in CDW-EIS in the entrance channel by means of the corresponding spatially asymptotic logarithmic eikonal phase factors for the active electrons in the projectile field in the initial state. This simplification was initially introduced in the case of SI and SC for the reason of having normalized total scattering wave functions for the initial state of the whole system. Such a simplifying approximation of CDW proved fruitful in practice, since CDW-EIS was systematically found to produce the needed bending of the curves for total cross sections in SI and SC near and below the Massey peak (i.e. in the resonant energy region where the incident velocity matches the classical Bohr velocity in the target orbit from which one active electron is ionized or captured). By contrast, total cross sections in CDW for SI and SC keep on rising with decreased impact energies and this is due to an overabundant density of the continuum intermediate states manifested by the presence of the Coulomb normalization factors. These latter factors disappear altogether from the asymptotic forms of Coulomb waves, or equivalently, from the eikonal initial states, and it is precisely for this reason of a drastically reduced density of continuum intermediate states in SI and SC that CDW-EIS is capable of producing the Massey peaks in accordance with the adiabatic hypothesis. Such a reduction happens to be in the right direction for SI and SC, thus yielding systematically excellent agreements of CDW-EIS with the corresponding experimental data virtually at all energies. This is the case even at lower impact energies that lie far below the expected limit of validity of applicability of first-order terms from perturbation expansion of transition amplitudes.

The classical trajectory Monte Carlo method is an alternative classical theory to calculate the cross sections for the ejected electrons in Li^{3+} and He collisions. In the present work two different variants of CTMC model are used: non-equivalent electron NEE-CTMC and equivalent electron EE-CTMC models. The CTMC models ignore the dynamical correlation effects because the electron-electron interaction is neglected during the entire collision. In contrast to the case of single ionization, for double ionization the results of NEE-CTMC model are closer to the experimental than the results of EE-CTMC.

The independent-particle close-coupling methods in the semi-classical impact parameter treatment where the electron wave function is expanded either around the target or around target and projectile was successfully used by different authors to calculate ionization cross sections. The CTMC calculations describe the general trend of the experimental data correctly. However, the agreement between experiments and calculations is, in general, superior for the close coupling method.

Acknowledgements

Dž. Belkić acknowledges support from Cancerfonden, Radiumhemmet Research Fund and the Karolinska Institute Fund.

I. Mančev and N. Milojević acknowledge support from Ministry of Science of the Republic of Serbia through Project No. 171020.

Appendix A

$$\mathcal{F}_{12} = \frac{B_{12}^{i\xi}}{Q_{12}^2 + \Delta_{12}^2}, \quad B_{12} = \frac{2(v\Delta_{12} - i\vec{Q}_{12}\cdot\vec{v})}{Q_{12}^2 + \Delta_{12}^2}, \quad \vec{Q}_{12} = \vec{Q}_1 + \vec{\tau}, \tag{A.1}$$

$$C_{12} = \frac{v}{B_{12}\Delta_{12}} - 1, \quad A_{12} = \frac{\Delta_{12}^2}{Q_{12}^2 + \Delta_{12}^2}, \quad D_{12} = \frac{A_{12}}{\Delta_{12}}, \tag{A.2}$$

$$A''_\alpha = 1 - 4A_{12}, \quad B''_\alpha = 1 + 2A_{12}C''_\alpha, \quad C''_\alpha = C_{12}[4 + (1 - i\xi)C_{12}], \tag{A.3}$$

$$A''_\beta = 6(1 - 2A_{12}), \quad B''_\beta = 2A_{12}C''_\beta + 3D''_\beta,$$

$$D''_\beta = 2 - (1 + i\xi)C_{12}, \tag{A.4}$$

$$C''_\beta = C_{12}[18 + 9(1 - i\xi)C_{12} + (1 - i\xi)(2 - i\xi)C_{12}^2], \tag{A.5}$$

$$\mathcal{F}_\tau = \frac{B_\tau^{i\xi}}{Q_\tau^2 + \Delta_1^2}, \quad B_\tau = \frac{2(v\Delta_1 - i\vec{Q}_\tau \cdot \vec{v})}{Q_\tau^2 + \Delta_1^2}, \quad \vec{Q}_\tau = \vec{Q}_1 - \vec{\tau}, \tag{A.6}$$

$$C_\tau = \frac{v}{B_\tau \Delta_1} - 1, \quad A_\tau = \frac{\Delta_1^2}{Q_\tau^2 + \Delta_1^2}, \quad D_\tau = \frac{A_\tau}{\Delta_1}, \tag{A.7}$$

$$A'_\alpha = 1 - 4A_\tau, \quad B'_\alpha = 1 + 2A_\tau C'_\alpha, \quad C'_\alpha = C_\tau[4 + (1 - i\xi)C_\tau], \tag{A.8}$$

$$A'_\beta = 6(1 - 2A_\tau), \; B'_\beta = 2A_\tau C'_\beta + 3D'_\beta, \quad D'_\beta = 2 - (1 + i\xi)C_\tau, \tag{A.9}$$

$$C'_\beta = C_\tau[18 + 9(1 - i\xi)C_\tau + (1 - i\xi)(2 - i\xi)C_\tau^2]. \tag{A.10}$$

References

1. S. S. Yu, B. G. Logan, J. J. Barnard, F. M. Bieniosek, R. J. Briggs, R. H. Cohen, J. E. Coleman, R. C. Davidson, A. Friedman, E. P. Gilson, L. R. Grisham, D. P. Grote, E. Henestroza, I. D. Kaganovich, M. Kireeff Covo, R. A. Kishek, J. W. Kwan, E. P. Lee, M. A. Leitner, S. M. Lund, A. W. Molvik, C. L. Olson, H. Qin, P. K. Roy, A. Sefkow, P. A. Seidl, E. A. Startsev, J.-L. Vay, W. L. Waldron, and D. R. Welch, *Nucl. Fusion*, **47**, 721–727 (2007).
2. M. Sasao, K. Sato, A. Matsumoto, A. Nishizawa, S. Takagi, S. Amemiya, T. Masuda, Y. Tsurita, F. Fukuzawa, Y. Haruyama, and Y. Kanamori, *J. Phys. Soc. Japan*, **55**, 102–105 (1986).
3. I. Murakami, J. Yan, H. Sato, M. Kimura, R. K. Janev, and T. Kato, *At. Data Nucl. Data Tables*, **94**, 161–222 (2008).
4. U. Amaldi and G. Kraft, *Rep. Prog. Phys.*, **68**, 1861–1882 (2005).
5. M. M. Sant'Anna, A. C. Santos, L. F. Coelho, G. Jalbert, N. V. de Castro Faria, F. Zappa, P. Focke, and Dž. Belkić, *Phys. Rev. A*, **80**, 042707 (2009).
6. V. S. Nikolaev, V. M. Tubaev, A. N. Teplova, and M. T. Novikov, *Sov. Phys. JETP*, **14**, 20–24, (1962) [*Zh. Eks. Teor. Fiz.* **14**, 201–205, (1962)].
7. M. B. Shah and H. B. Gilbody, *J. Phys. B: At. Mol. Phys.*, **18**, 899–913 (1985).
8. Dž. Belkić, *Phys. Rev. A*, **47**, 189–200 (1993).
9. Dž. Belkić, *J. Phys. B*, **26**, 497–508 (1993).
10. S. Ghosh, A. Dhara, C. R. Mandal, and M. Purkait, *Phys. Rev. A*, **78**, 042708 (2008).
11. R. Gayet, J. Hanssen, L. Jacqui, A. Martínez, and R. Rivarola, *Phys. Scr.*, **53**, 549–556 (1996).

12. R. Gayet, J. Hanssen, A. Martínez, and R. Rivarola, *Comments At. Mol. Phys.*, **30**, 231–248 (1994).
13. R. Gayet, J. Hanssen, A. Martínez, and R. Rivarola, *Nucl. Instr. Meth. B*, **86**, 158–160 (1994).
14. A. E. Martínez, H. F. Busnengo, R. Gayet, J. Hanssen, and R. D. Rivarola, *Nucl. Instr. Meth. B*, **132**, 344–349 (1997).
15. M. Ghosh, C. R. Mandal, and S. C. Mukherjee, *Phys. Rev. A*, **35**, 5259–5261 (1987).
16. M. S. Gravielle and J. E. Miraglia, *Phys. Rev. A*, **45**, 2965–2973 (1992).
17. Dž. Belkić, *J. Phys. B*, **10**, 3491–3510 (1977).
18. J. P. Coleman, in *Case studies in atomic physics*, Ch. 3, Eds. E. W. McDaniel and M.R.C. McDowell, North-Holland, The Netherlands, 1969.
19. Dž. Belkić, R. Gayet, and Salin, *Phys. Rep.*, **56**, 279–369 (1979).
20. Dž. Belkić, I. Mančev, and J. Hanssen, *Rev. Mod. Phys.*, **80**, 249–314 (2008).
21. Dž. Belkić, *Quantum theory of high-energy ion-atom collisions*, Taylor & Francis, London, 2008.
22. Dž. Belkić, *J. Math. Chem.*, **47**, 1420–1467 (2010).
23. M. Purkait, S. Sounda, A. Dhara, and C. R. Mandal, *Phys. Rev. A*, **74**, 042723 (2006).
24. Dž. Belkić, *Nucl. Inst. Meth. B*, **86**, 62–81 (1994).
25. Dž. Belkić, I. Mančev, and M. Mudrinić, *Phys. Rev. A*, **49**, 3646–3658 (1994).
26. Dž. Belkić and I. Mančev, *Phys. Scr.*, **45**, 35–42 (1992).
27. Dž. Belkić and I. Mančev, *Phys. Scr.*, **46**, 18–23 (1993).
28. R. Gayet, R. D. Rivarola, and A. Salin, *J. Phys. B*, **14**, 2421–2427 (1981).
29. G. Deco and N. Grün, *Z. Phys. D*, **18**, 339–343 (1991).
30. R. Gayet, J. Hanssen, A. Martínez, and R. Rivarola, *Z. Phys. D*, **18**, 345–350 (1991).
31. A. E. Martínez, R. Gayet, J. Hanssen, and R. D. Rivarola, *J. Phys. B*, **27**, L375–L382 (1994).
32. A. E. Martínez, R.D. Rivarola, R. Gayet, and J. Hanssen, *Phys. Scr.*, **T80**, 124–127 (1999).
33. R. Hippler, S. Datz, P. Miller, P. Pepmiler, and P. Dittner, *Phys. Rev. A*, **35**, 585–590 (1987).
34. R. N. Hill, *Phys. Rev. Lett.*, **38**, 643–646 (1977).
35. Dž. Belkić, *Phys. Rev. A*, **47**, 3824–3844 (1993).
36. D. S. F. Crothers and R. McCarroll, *J. Phys. B*, **20**, 2835–3842 (1987).
37. L. D. Dodd and K. R. Greider, *Phys. Rev.*, **146**, 675–686 (1966).
38. H. Bachau, *J. Phys. B*, **17**, 1771–1784 (1984).
39. E. A. Hylleraas, *Z. Phys.*, **54**, 347–366 (1929).
40. P. O. Löwdin, *Phys. Rev.*, **90**, 120–125 (1953).
41. L. C. Green, M. M. Mulder, M. N. Lewis, and J. W. Jr. Woll, *Phys. Rev.*, **93**, 757–761 (1954).
42. J. Silverman, O. Platas, and F. A. Matsen, *J. Chem. Phys.*, **32**, 1402–1406 (1960).
43. E. Clementi and C. Roetti, *At. Data Nucl. Data Tables*, **14**, 177–478 (1974).
44. Dž. Belkić, *Phys. Scr.*, **40**, 610–624 (1989).

45. I. M. Cheshire, *Proc. Phys. Soc.* **84**, 89–98 (1964).
46. G. S. Saha, S. Data, and S. C. Mukherjee, *Phys. Rev.* **A34**, 2809–2821 (1986).
47. D. S. F. Crothers and K. M. Dunseath, *J. Phys. B: At. Mol. Phys.*, **15**, 2061–2074 (1982).
48. H. F. Busnengo, A. E. Martinez, and R. D. Rivarola, *J. Phys. B: At. Mol. Opt. Phys.*, **29**, 4193–4205 (1996).
49. M. S. Gravielle and J. E. Miraglia, *Phys. Rev. A*, **51**, 2131–2139 (1995).
50. V. A. Sidorovich, V. S. Nikolaev, and J. H. McGuire, *Phys. Rev. A*, **31**, 2193–2201 (1985).
51. R. H. Bassel and E. Gerjuoy, *Phys. Rev.* **117**, 749–756 (1960).
52. H. Suzuki, Y. Kajikawa, N. Toshima, H. Ryufuku, and T. Watanabe, *Phys. Rev. A*, **29**, 525–528 (1984).
53. I. Mančev, *Phys. Rev. A*, **64**, 012708 (2001).
54. Dž. Belkić, R. Gayet, J. Hanssen, I. Mančev, and A. Nuñez, *Phys. Rev. A*, **56**, 3675–3681 (1997).
55. J. D. Dollard, *J. Math. Phys.*, **5**, 729–738 (1964); J. D. Dollard, Ph. D. thesis (University of Michigan) (1963).
56. Dž. Belkić, *Principles of Quantum Scattering Theory*, Institute of Physics, Bristol 2004.
57. I. Mančev and N. Milojević, *Phys. Rev. A*, 81, 022710 (2010).
58. H. Knudsen, L. H. Andersen, P. Hvelplund, G. Astner, H. Cederquist, H. Danared, L. Liljeby, and K.-G. Rensfelt *J. Phys. B: At. Mol. Phys.* **17**, 3545–3564 (1984).
59. I. S. Dmitriev, Ya. A. Teplova, Ya. A. Belkova, N. V. Novikov, and Yu. A. Fainberg, *Atomic Data and Nuclear Data Tables*, **96**, 85–121 (2010) [V. S. Nikolaev, I. S. Dmitriev, L. N. Fateyeva, Ya. A. Teplova, *Zh. Eksp. Teor. Fiz.* **40**, 989–1000 (1961): *JETP* **13**, 695–702 (1961)].
60. O. Woitke, P. A. Závodszky, S. M. Ferguson, J. H. Houck, and J. A. Tanis, *Phys. Rev. A*, **57**, 2692–2700 (1998).
61. R. Samanta, M. Purkait, and C. R. Mandal, *Phys. Rev. A*, **83**, 032706 (2011).
62. R. Shingal and C. D. Lin, *J. Phys. B: At. Mol. Phys.*, **24**, 251–264 (1991).
63. S. Bhattacharyya, K. Rinn, E. Salzborn, and L. Chatterjee, *J. Phys. B: At. Mol. Phys.* **21**, 111–118 (1988).
64. B. W. Ding, D. Y. Yu, and X. M. Chen, *Nucl. Instr. Meth. Phys. Res. B*, **266**, 886–888 (2008).
65. R. Gayet and A. Salin, *J. Phys. B*, **20**, L571–L576 (1987).
66. A. L. Godunov, J. H. McGuire, V. S. Schipakov, H. R. J. Walters, and Colm T. Whelan, *J. Phys. B*, **39**, 987–996 (2006).
67. Dž. Belkić and I. Mančev, *Phys. Rev. A*, **83**, 012703 (2011).
68. A. Nordsieck, *Phys. Rev.* **93**, 785–787 (1954).
69. Dž. Belkić, *Nucl. Instr. Meth. Phys. Res. B*, **86**, 62–81 (1994).
70. Dž. Belkić and H. Taylor, *Phys. Rev. A* **35**, 1991–2006 (1987).
71. C. Froese Fisher *Atomic, Molecular and Optical Physics References Book* ed. G. W. F. Drake (New York: AIP) (chapter 21) (1996).
72. D.S. F. Crothers and J. F. McCarrol, *J. Phys. B*, **20**, 2835–2842 (1987).
73. M. McCartney, *J. Phys. B*, **30**, L155–L160 (1997).

74. M. McCartney, *Nucl. Instr. Meth. Phys. Res. B*, **155**, 343–348 (1999).
75. J. Bradley, R. J. Lee, M. McCartney, and D. S. F. Crothers, *J. Phys. B*, **37**, 3723–3734 (2004).
76. Marcelo Fiori, A. B. Rocha, C. E. Bielschowsky, Ginette Jalbert, and C. R. Garibotti, *J. Phys. B*, **39**, 1751–1762 (2006).
77. Marcelo Fiori, Ginette Jalbert, and C.R. Garibotti, *Journal Electron Spectroscopy and Related Phenomena*, **161**, 191–193 (2007).
78. D. S. F. Crothers and J. F. McCann, *J. Phys. B*, **16**, 3229–3242 (1983).
79. J. D. Talman and W. F. Shadwick, *Phys. Rev. A*, **14**, 36–40 (1975); J. D. Talman, *Comput. Phys. Commun.*, **54**, 85–94 (1989).
80. J. H. McGuire, *Phys. Rev. Lett.*, **49**, 1153–1156 (1982).
81. I. F. Barna, K. Tökési, and J. Burgdörfer, *J. Phys. B*, **38**, 1001–1014 (2005).
82. I. F. Barna, N. Grün, and W. Scheid, *Eur. Phys. J. D*, **25**, 239–246 (2003); Imre F. Barna, Ph. D. thesis, University of Giessen (2002).
83. R. H. Garvey, C. H. Jackman, and A. E. S. Green, *Phys. Rev. A*, **12**, 1144–1152 (1975).
84. N. C. Deb and D. S. F. Crothers, *J. Phys. B*, **24**, 2359–2366 (1991).
85. P. Pluvinage, *Ann. Phys. NY*, **5**, 145–152 (1950).
86. Dž. Belkić, *J. Phys. B*, **11**, 3529–3552 (1978).
87. Dž. Belkić, *J. Math. Chem.* **47**, 1366–1419 (2010).
88. P. D. Fainstein, V. H. Ponce, and R. D. Rivarola, *Phys. Rev. A*, **36**, 3639–3641 (1987).
89. K. L. Bell and A. E. Kingston, *J. Phys. B*, **2**, 653–661 (1969).
90. S. Sahoo, R. Das, N. C. Sil, S. C. Mukherjee, and K. Roy, *Phys. Rev. A*, **62**, 022716 (2000).

Chapter 7

Study of Inelastic Processes in Ion-H_2O Collisions Using Classical Trajectory Monte Carlo and Semiclassical Methods

L. F. Errea*, Clara Illescas, P. M. M. Gabás, L. Méndez,
I. Rabadán, and A. Riera
Laboratorio Asociado al CIEMAT de Física Atómica y Molecular en Plasmas de Fusión, Departamento de Química, módulo 13, Universidad Autónoma de Madrid, Cantoblanco 28049-Madrid, Spain
lf.errea@uam.es

B. Pons
CELIA, Université de Bordeaux I-CNRS-CEA, 351 Cours de la Libération, 33405-Talence, France

We present calculations of cross sections for one- and two-electron processes in collisions of H^+, He^{2+} and C^{6+} with water molecules. We employ two kind of methods: a classical trajectory Monte Carlo approach and a semiclassical treatment with expansions in terms of molecular wavefunctions. Anisotropy effects related to the structure of the target are explicitly incorporated by using a three-center model potential to describe the electron-H_2O^+ interaction. This is compared with a simple approach that employs a screened Coulomb potential. At high energies we derive scaling laws with respect to the projectile charge. We also estimate cross sections for molecular fragmentation subsequent to electron removal.

1. Introduction

Hadron therapy is a valuable alternative to X- or γ-ray radiotherapy (see Refs. 1, 2 for reviews). The use of this technique started in 1954 at the Lawrence Berkeley Laboratory (USA)[3]; since then, several thousands

*The author to whom corrspondence should be addressed.

of patients have been treated with proton beams in several installations and with carbon ions at Chiba (Japan)[4] and Darmstadt (Germany).[5] These techniques have several advantages when compared to conventional photon radiation: they allow access to deeply seated tumors, and the lethal tumor dose is raised while the surrounding healthy tissue remains unaffected. Physicists and biologists can measure and/or compute the intensity, penetration depth and lethal dose of the ion beams.[6–9] The mechanisms responsible for single and double strand breaks of the DNA, which subsequently lead to cell death, are complex, and several works (see Ref. 10 and references therein) have aimed to elucidate them. The interest of this subject has motivated experiments on collisions of multicharged ions with DNA bases to shed light on the DNA fragmentation processes (see, e.g., Ref. 11), but theoretical works are scarce.[12,13]

On the other hand, e-DNA experiments[14] have shown that collisions of relatively slow electrons (with energy of about 10 eV) can lead to the breakdown of DNA through a mechanism that involves the formation of intermediate resonant states. Therefore, processes that produce electrons are also relevant in the understanding of biological damage and ion therapy. In this respect, electron emission in collisions of ions with water provides the most significant source of electrons in the interaction of ion beams with the cell. Although several experiments have provided detailed information on ionizing proton-water collisions, data are scarce for multicharged ion impact. Furthermore, in addition to target ionization, other processes such as excitation and charge-exchange in ion-H_2O collisions can lead to molecular fragmentation, and the fragments formed in these reactions are also responsible for DNA breakdown. Therefore, these processes must also be explicitly considered to reliably simulate the passage of charged particles in biological environments.[15,16]

Experiments on ion-water collisions include early measurements of total cross sections for electron capture and electron loss of Dagnac et al.[17] Rudd et al. measured total cross sections for production of electrons and positive charges in collisions of protons[18–20] and alpha particles[21] with water vapour, at collision energies from a few keV/amu to several MeV/amu. Measurements of total electron capture cross sections at collision energies of about 1 keV/amu have been reported in Refs. 22–24; these results are particularly relevant in cometary X-ray emission. Recently, the experiments have focused on the molecular fragmentation,[25–31] and some works[32,33] have measured both target and projectile Balmer emission after ion-H_2O collisions.

Previous theoretical works aimed at filling in the collisional database of interest for radiation damage. Nevertheless, most of these works focused on H^++H_2O collisions and employed perturbative methods,[34] such as the continuum distorted wave-eikonal initial state (CDW-EIS)[35–37] and first Born (FB) approximations,[38] which are in general useful at impact energies E greater than 100 keV/amu. Recently, Lüdde et al.[39] have applied the basis set generator method (BGM) to H^++H_2O collisions, beyond the isotropic electron-core approximation; they have reported electron production and net capture cross sections in good agreement with experiment.

With respect to collisions of multicharged ions, He^{2+}+H_2O have been studied in Refs. 40 and 41, in the frameworks of FB and classical models, respectively. Comparison with experimental data showed acceptable, but not very satisfactory, agreement. In spite of the interest of C^{6+}+H_2O collisions in ion-based cancer therapy, only CDW-EIS[42] and FB[43] calculations of the total ionization cross section have been reported so far, together with a single experimental point at 6 MeV/amu.[42]

In this work, we consider the use of the classical trajectory Monte Carlo (CTMC) method to evaluate cross sections for single ionization, single capture, and two-electron processes (transfer ionization, double capture and double ionization) in H^+, He^{2+}, and C^{6+}+H_2O collisions in the impact energy range $20 \leq E \leq 10000$ keV/amu, which covers the intermediate and high impact energy regimes of interest for therapy applications. This chapter is mainly based on the work published in Refs. 44 and 45. Since 1966, when it was first applied to ion-atom collisions,[46] the CTMC method has been widely employed to treat ion collisions with one-electron targets. In this respect, previous works of our group, which aimed at providing data sets for plasma diagnostics, considered multicharged ions–H collisions (see Refs. 47–51). With the exception of a few works[52–54] that dealt with the classical treatment of two-electron systems, the application of the CTMC method to many-electron targets is in general carried out by employing the Independent Particle Model (IPM),[55–57] which assumes that each electron moves in an effective field created by the nuclei and the remaining electrons. In practice, this assumption reduces the solution of the many-electron problem to a set of non-interacting one-electron problems for each electron of the system. In general, the effective potentials are screened-Coulomb potentials with effective charges obtained by fitting the energies of the target electrons. More sophisticated model potentials have been employed in Ref. 58.

A first difference between the application of the CTMC method to ion–molecule and ion–atom collisions is the existence of the target internal degrees of freedom; however, CTMC is usually applied at high collision energies ($E > 25\,\text{keV/amu}$), where one can assume that the target nuclei remain fixed during the collision. The main difference with respect to ion–atom collisions is therefore the anisotropy of the target. Previous CTMC calculations for ion-molecule collisions are scarce; they include some calculations of electron capture and ionization cross sections in ion–H_2,[59–61] using one-center potentials. Collisions with H_2O have been considered in the CTMC calculations of Otranto et al.,[62,63] who have evaluated state-selective electron capture cross sections at relatively low collision energies, motivated by the application of these data to model cometary X-ray emission. As in practically all available calculations, our treatment employs the IPM, which involves the use of electron-core effective potentials. In a first approach[44] we used one-center (isotropic) potentials. Afterwards, we have developed a technique that uses a three-center model potential to describe the interaction of the active electron with the H_2O^+ core; this allows us to explicitly consider the anisotropy of the molecular target. In this respect, it is noteworthy that relevant anisotropy effects have been detected in calculations of both H^+– and e–H_2O collisions (see, e.g., Refs. 39, 64). The evaluation of inelastic probabilities for the physical many-electron system, in terms of the monoelectronic ones resulting from the model potential calculations, is usually carried out by means of the IPM. In our work we have also considered the application of the so-called independent event (IEV) model,[65–68] which has been found[69,70] to be more adequate than the usual IPM for multielectronic targets.

As mentioned above, the application of the CTMC method is restricted to collision energies above 25 keV/amu. At lower energies, one must employ a semiclassical formalism, where the electronic motion is treated quantum-mechanically. Some calculations[71,72] have used this methodology, expanding the collision wavefunction in terms of the electronic wavefunctions of the supermolecule formed by the ion and the target molecule. Although *ab initio* electronic structure calculations can be performed with high accuracy for relatively small systems like H_2O[73] and H_3O^+,[74] dynamical calculations require simplified methods to avoid the need of regularizing conical intersections where dynamical couplings diverge (see, e.g. Ref. 75 and references therein), as these regions are the loci where non-adiabatic transitions can predominantly take place at low energies. In this work we describe and apply two relatively simple semiclassical *ab initio* methods to evaluate charge transfer and ionization cross sections, with the idea that they can be applied to study larger biomolecules. As in the CTMC approach, the two techniques are based on the use of a multi-center pseudopotential to describe the interaction of the active electron with the target polyatomic core, and the IPM to define transition probabilities.

The chapter is organized as follows: in Section 2, we summarize the basic assumptions of our CTMC approach, emphasizing the description of the initial electron densities associated to the molecular orbitals. The use of semiclassical methods at relatively high collision energies is considered in Section 3. We present in Section 4 our results for H^++H_2O collisions and compare them to both experimental and theoretical data. We also compare in this section the results obtained from applying the methods explained in Sections 2 and 3. Sections 5 and 6 contain our results for He^{2+} and C^{6+} ion impacts, respectively; useful scaling relations with respect to the projectile charge and the impact energy are presented in Section 7. Heavy fragments formed in the dissociation that takes place after molecular ionization are important sources of radiation damage, and we have evaluated fragmentation cross sections for H^++H_2O collisions; they are displayed in Section 7. Finally, conclusions and perspectives issued from the present work are given in Section 8. Atomic units are used throughout unless otherwise indicated.

2. Impact Parameter-CTMC Approach

Since we consider relatively high collision energies, we apply the Franck-Condon approximation where the nuclei of the H_2O molecule remain at their equilibrium positions during the collision. Furthermore, we employ the impact parameter approximation[76] in which the relative ion-molecule motion is described by straight-line trajectories $\mathbf{R}(t) = \mathbf{b} + \mathbf{v}t$, where $\mathbf{R}$ is the ion position vector with respect to the origin of coordinates, placed on the oxygen nucleus of the target molecule, $\mathbf{b}$ is the impact parameter and $\mathbf{v}$ the constant relative velocity.

2.1. *Motion of the Active Electron; Initial Distributions and Monoelectronic Probabilities*

2.1.1. *Model potentials*

We assume that each of the ten electrons involved in bare ion-H_2O collisions evolves independently, subject to the mean field created by the nuclei and the other nine electrons. Furthermore, we assume that the two electrons occupying the $1a_1$ molecular orbital (MO) of H_2O are too deeply bound to the O nucleus to significantly participate in the collision dynamics. We thus explicitly consider the dynamics of the eight electrons initially located in the four outermost MOs of H_2O, of symmetries $2a_1$, $1b_2$, $3a_1$ and $1b_1$.

In the CTMC calculation (see Ref. 46), the motion of the active electron, which initially occupies the MO ϕ_k of the water molecule, is described by means of a classical distribution function, $\rho_k(\mathbf{r}, \mathbf{p}, t)$, which is discretized by using a set of N independent trajectories $\{\mathbf{r}_j(t), \mathbf{p}_j(t)\}$, according to

$$\rho_k(\mathbf{r}, \mathbf{p}, t) = \frac{1}{N} \sum_{j=1}^{N} \delta(\mathbf{r} - \mathbf{r}_j(t))\delta(\mathbf{p} - \mathbf{p}_j(t)). \quad (1)$$

This discretized $\rho_k(\mathbf{r}, \mathbf{p}, t)$ is substituted in the Liouville equation $\partial \rho_k / \partial t = -\{\rho_k, h_k\}$, where $h_k(\mathbf{r}, \mathbf{p})$ is the one-electron Hamiltonian for the motion of the electron initially described by ϕ_k; this yields the Hamilton equations:

$$\begin{aligned} \dot{\mathbf{r}}_j &= \mathbf{p}_j, \\ \dot{\mathbf{p}}_j &= -\nabla_{\mathbf{r}_j}[h_k(\mathbf{r}_j, \mathbf{p}_j)], \end{aligned} \quad (2)$$

which monitor the temporal evolution of the jth trajectory among the set of N independent ones. The one-electron Hamiltonian is expressed as:

$$h_k = \frac{p^2}{2} - \frac{Z_{\mathrm{P}}}{r_{\mathrm{P}}} + V_k(\mathbf{r}_{\mathrm{t}}), \quad (3)$$

where $\mathbf{p}$ is the electronic momentum and $\mathbf{r}_{\mathrm{P}}$, $\mathbf{r}_{\mathrm{t}}$ are the electronic position vectors with respect to the projectile and the target, respectively.

In the one-center treatment, the electron in the MO ϕ_k moves in the screened Coulomb potential:

$$V_k = -\frac{Z_k}{r_{\mathrm{O}}}, \quad (4)$$

where r_{O} is the electron distance to the oxygen nucleus and Z_k are the effective charges that fulfill:

$$Z_k = \sqrt{-2n^2 \epsilon_k}, \quad (5)$$

with ϵ_k the energy of ϕ_k and, in the present case, $n = 2$.

In the CTMC three-center treatment we have used a model potential[69] of the form:

$$V(\mathbf{r}) = V_{\rm O}(r_{\rm O}) + V_{\rm H}(r_{\rm H1}) + V_{\rm H}(r_{\rm H2}), \tag{6}$$

with

$$\begin{aligned} V_{\rm O}(r_{\rm O}) &= -\frac{8-N_{\rm O}}{r_{\rm O}} - \frac{N_{\rm O}}{r_{\rm O}}(1+\alpha_{\rm O}r_{\rm O})\exp(-2\alpha_{\rm O}r_{\rm O}), \\ V_{\rm H}(r_{\rm H}) &= -\frac{1-N_{\rm H}}{r_{\rm H}} - \frac{N_{\rm H}}{r_{\rm H}}(1+\alpha_{\rm H}r_{\rm H})\exp(-2\alpha_{\rm H}r_{\rm H}), \end{aligned} \tag{7}$$

which is applied to describe the motion of the electron in the four MOs of the valence shell. In these expressions $r_{\rm O}$, $r_{\rm H1}$ and $r_{\rm H2}$ are the electron distances to the three target nuclei. To determine the parameters ($N_{\rm O} = 7.1$, $N_{\rm H} = (9 - N_{\rm O})/2$, $\alpha_{\rm O} = 1.500\,a_0$, $\alpha_{\rm H} = 0.665\,a_0$) we have solved the Schrödinger equation $H_0\phi_i = [p^2/2 + V(\mathbf{r})]\phi_i = \epsilon_i\phi_i$ by expanding ϕ_i in a large Gaussian-type orbital (GTO) basis (see Ref. 77). The values of the parameters are then obtained by minimizing the differences $|\epsilon_i - \epsilon_i^{\rm SCF}|$, where $\epsilon_i^{\rm SCF}$ are the self-consistent field (SCF) energies of the valence MOs of H_2O obtained in the same GTO basis set. In practice, we obtained $\max\{|\epsilon_i - \epsilon_i^{\rm SCF}|, i = 1,\ldots,4\} < 10^{-3}$ a.u..

2.1.2. *Initial distributions*

In the one-center treatment, the initial electron distribution $\rho_k(\mathbf{r},\mathbf{p},t = -\infty)$ can be constructed following the usual procedure for ion-H collisions. The standard method[46] employs a microcanonical distribution, $\rho_k^k(\mathbf{r},\mathbf{p};\epsilon_k)$, which is defined as

$$\rho_k^j(\mathbf{r},\mathbf{p};E_j) = \delta(h_k - E_j). \tag{8}$$

It is well-known (see e.g. Refs. 78, 79) that the microcanonical distribution shows important limitations to describe target ionization at low collision energies; they are related to the sharp cut-off of the spatial initial distribution in the classically forbidden region (see Fig. 1). To overcome this limitation, several improved distributions have been proposed.[78,80–82] Such improved initial conditions involve an electron energy distribution that spreads over the entire energy-bin associated to the entry channel[82] and therefore allow mimicking, to some extent, under-barrier transitions which are specially relevant for projectile charges $Z_{\rm P} > 1$. In this work, we have employed a hydrogenic distribution[78] that in practice is a linear combination of 7 microcanonical distributions, ρ_k^j, with different energies E_j:

$$\rho_k(\mathbf{r}, \mathbf{p}, t \to -\infty) = \sum_{j=1}^{7} a_k^j \rho_k^j(\mathbf{r}, \mathbf{p}; E_j), \tag{9}$$

where the coefficients a_k^j have been chosen by fitting the quantal position and momentum distributions, for the same one-center potential, and checked that $\sum_j a_k^j E_j \simeq \epsilon_k$. When using the hydrogenic distribution, the total ionization cross section is compatible with the Bethe limit,[47] although this cross section is in general underestimated.[83]

In the three-center calculations, we have limited our study to the use of microcanonical distributions, and, for simplicity, we will employ in the following a single-index notation to name them: $\rho_k(\mathbf{r}, \mathbf{p}, t = -\infty)$. To build up these distributions, we have worked by analogy with the central potential case,[84] using five random parameters. In practice, the $t \to -\infty$ condition is approached by setting $Z_P = 0$ in (2) and the initial phase-space conditions are established in spherical coordinates assuming that the electron is located at the perihelion of an elliptic orbit with $\mathbf{r}_j \perp \mathbf{p}_j$ at fixed t. Thus, the five random parameters consist of (i) $\cos\theta_r$ and (ii) φ_r that define the orientation of the orbit in coordinate space; (iii) the value of β, with $0 \leq \beta \leq 1$, which yields the perihelion radius $r = \frac{r_{\max}}{2}(1 + \sqrt{1-\beta})$, where $r_{\max}$ is the maximum value of r that is obtained by solving numerically the nonlinear equation $V_{\text{mod}}(r_{\max}, \theta_r, \varphi_r) - \epsilon_i = 0$; (iv) the azimuthal angle φ_p of the momentum vector; and (v) Δt, the time interval over which the Hamilton equations (2), with $Z_P = 0$, have to be propagated until the distribution becomes time-independent, given that the trajectories are not elliptical in the three-center V_{mod} potential.

As an illustration of the initial distributions generated by our methods, we compare in Figs. 1 and 2 the classical, $\rho_k^{\text{C}}(r)$, and the quantal radial distributions, $\rho_k^{\text{Q}}(r)$, obtained by using both one- and three-center model potentials of equations (4) and (6) for the four MOs considered in this work. The classical radial distributions are defined as:

$$\rho_k^{\text{C}}(r) = \int \text{d}\mathbf{p} \int_0^{\pi} \text{d}\theta_r \, \sin\theta_r \int_0^{2\pi} \text{d}\varphi_r \, r^2 \rho_k(\mathbf{r}, \mathbf{p}), \tag{10}$$

and the quantal radial distributions have been obtained as:

$$\rho_k^{\text{Q}}(r) = \int_0^{\pi} \text{d}\theta_r \sin\theta_r \int_0^{2\pi} \text{d}\varphi_r \, r^2 |\phi_k(\mathbf{r})|^2. \tag{11}$$

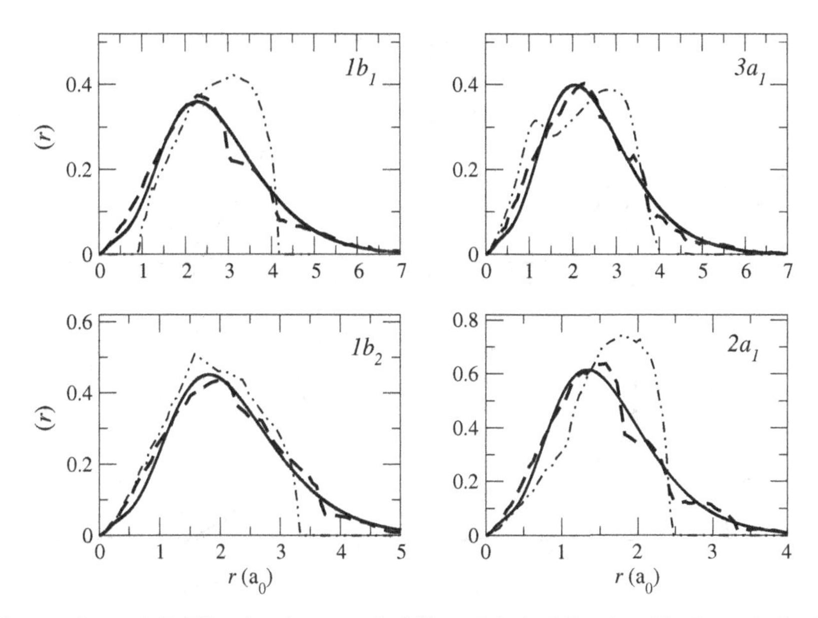

Fig. 1. Quantal (full lines), microcanonical (dotted-dashed lines) and hydrogenic distributions (dashed lines), as functions of r, for the four valence MOs of the H_2O molecule, obtained using the one-center model potential of equation (4) [see equation (12)].

We have checked that the plots are identical when integrating in (11) the probability densities obtained from either the SCF MOs or the MOs that are solutions or the one-electron Schrödinger equation with the model potential (6).

The shape of the microcanonical one-center radial distributions (Fig. 1) is easily understood by taking into account that the classical motion is restricted to $r < -Z_k/\epsilon_k$, which limits the extension of these distributions. To build the hydrogenic distributions we carry out a linear combination of microcanonical distributions with components of energy higher than ϵ_k, which, as already mentioned, leads to a distribution broader than the microcanonical one. The hydrogenic distributions were obtained by fitting quantal distributions, derived from the corresponding screened Coulomb potentials. In practice, we have fitted, for each value of Z_k, the combination of the quantal radial distributions for the corresponding 2s and 2p orbitals of the form:

$$\rho^{\mathrm{Q}}_{n=2}(r) = \frac{1}{4}\rho^{\mathrm{Q}}_{2s}(r) + \frac{3}{4}\rho^{\mathrm{Q}}_{2p}(r). \tag{12}$$

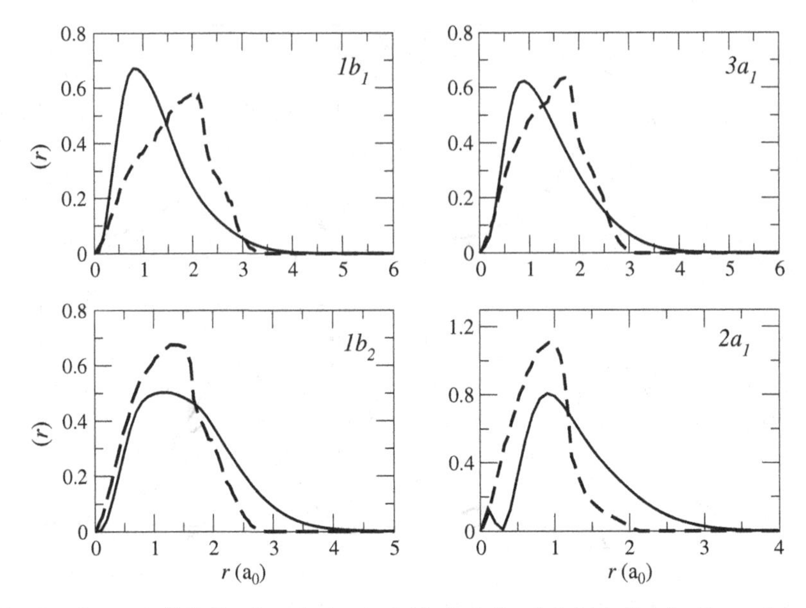

Fig. 2. Quantal (full lines) and classical (dashed lines) initial distributions $\rho(r)$ as functions of r, for the four valence MOs of the H_2O molecule, obtained using the three-center model potential of equation (6).

The comparison of the one-center distributions of Fig. 1 with both quantal and classical distributions of Fig. 2 shows that the latter are more diffuse for all MOs. This result is consistent with the fact that the potential that defines the electron motion in the MOs asymptotically behaves as $-1/r$, while $Z_k > 2$. It can be observed in Fig. 2 that the agreement between classical and quantal three-centre distributions is satisfactory over the whole r range; the classical densities are, in general, spread out in position space similarly to the quantal ones, although some discrepancies are noticeable for the innermost orbitals.

In Fig. 3 we compare the classical and quantal spatial densities associated with the $2a_1$, $3a_1$ and $1b_2$ MOs in the XZ plane of H_2O. $\hat{\mathbf{x}}$ corresponds to the direction between the two H atoms and $\hat{\mathbf{z}} \perp \hat{\mathbf{x}}$. In fact, the planar classical densities correspond to the fraction of electrons which initially lie within the slab $|\mathbf{y}| < 0.15\,a_0$. In order to gauge the improvement inherent to the $V_{\rm mod}$ description of H_2O, we also report in Fig. 3 the classical densities issued from a monocentric hydrogenic classical

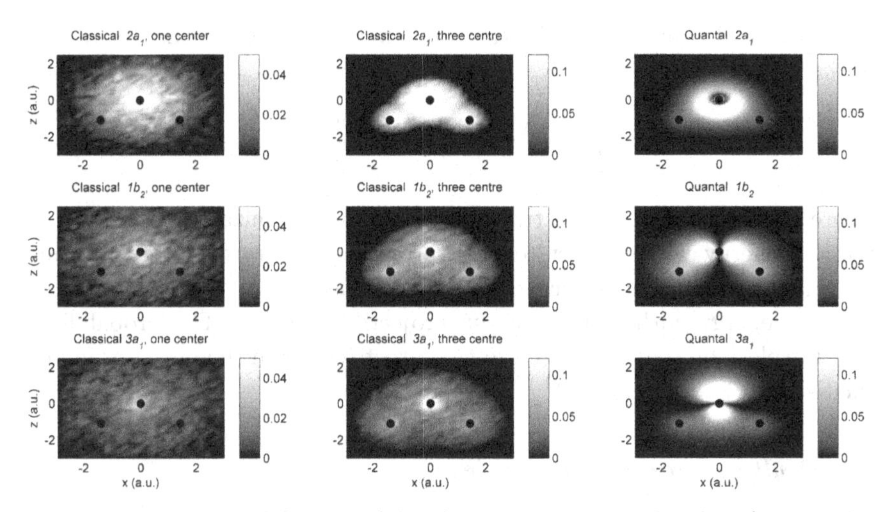

Fig. 3. Contour plots of the quantal distributions on the molecular plane ($y = 0$, right panels) and the classical distributions for those trajectories with $|y| < 0.15\,a_0$ (left panels, monocentric potential; middle panels, tricentric potential) for the $2a_1$, $1b_2$ and $3a_1$ MOs. The grey scale of each panel is also shown.

distribution. In spite of the fact that any classical description inevitably fails to reproduce the nodal structure of the quantal densities, we see in Fig. 3 that the three-center model approach allows us to better describe the electron delocalization over the three nuclei. Moreover, the three-center classical densities concentrate on the molecular plane, while those obtained by means of the one-center treatment spread out (note the different scale employed for the plots of the one-centre distributions), as in the quantal description. We notice in Fig. 3 that the electron density of the $3a_1$ MO is well described by means of the classical $V_{\rm mod}$ treatment. The same happens for the $1b_1$ MO, which is not displayed in Fig. 3 because it vanishes in the molecular plane. These are positive features since the electrons should be preferentially pulled out, at low and intermediate E, from these two MOs, with the smallest ionization potentials. We shall verify this point in the next section.

2.1.3. *One-electron transition probabilities*

Once the initial distributions are generated, the Hamilton equations (2) are integrated up to the time $t_{\rm max} = 500/v$, when the one-electron ionization and capture probabilities are defined, for each of the four MOs of the valence shell ($k = 1, \ldots, 4$), according to:

$$p_k^{\mathrm{ion}} = \frac{N_k^{\mathrm{ion}}}{N}; \quad p_k^{\mathrm{cap}} = \frac{N_k^{\mathrm{cap}}}{N}, \tag{13}$$

where N_k^{ion} is the number of trajectories leading to ionization (those with positive energy with respect to both projectile and target at t_{max}) and N_k^{cap} is the number of trajectories leading to electron capture (those with negative energy with respect to the projectile at t_{max}). The probability that the electron remains bound to the target (either in the initial or in an excited state) is therefore: $p_k^{\mathrm{el}} = 1 - p_k^{\mathrm{ion}} - p_k^{\mathrm{cap}}$.

2.2. *Multielectronic Probabilities*

In order to relate the one-electron probabilities of equation (13) with those of the physical many-electron system, one can apply the independent particle model. Assuming that the electrons occupying the same MO are equivalent, the probabilities for single ionization (SI) and single capture (SC) take the forms:

$$P^{\mathrm{SI}} = \sum_{k=1}^{4} P_k^{\mathrm{SI}} = 2\sum_{k=1}^{4} p_k^{\mathrm{ion}} p_k^{\mathrm{el}} \prod_{j \neq k} (p_j^{\mathrm{el}})^2,$$

$$P^{\mathrm{SC}} = \sum_{k=1}^{4} P_k^{\mathrm{SC}} = 2\sum_{k=1}^{4} p_k^{\mathrm{cap}} p_k^{\mathrm{el}} \prod_{j \neq k} (p_j^{\mathrm{el}})^2. \tag{14}$$

One alternative to the standard IPM is the independent event model, suggested in Ref. 65 and used in several works (e.g. Refs. 66–68) for ion collisions with He. Following the interpretation of Janev et al.,[66] the IEV assumes that the many-electron removal takes place sequentially. For ion-H_2O collisions, this means that the second electron is removed from H_2O^+. Therefore, the implementation of the IEV consists of substituting in (14) the elastic probabilities p_j^{el} by $\mathcal{P}_j^{\mathrm{el}}$, where $\mathcal{P}_j^{\mathrm{el}} = 1 - \mathcal{P}_j^{\mathrm{ion}} - \mathcal{P}_j^{\mathrm{cap}}$, and $\mathcal{P}_j^{\mathrm{ion,\,cap}}$ are one-electron probabilities calculated for ion-H_2O^+ collisions. However, previous calculations for ion-H_2 collisions[85] indicate that the electrons on the same shell are equivalent and, accordingly, we have not distinguished between ionization and capture probabilities for electrons on the same shell;

this leads to:

$$P^{\mathrm{SI}} = \sum_{k=1}^{4} P_k^{\mathrm{SI}} = 2\sum_{k=1}^{4} p_k^{\mathrm{ion}} p_k^{\mathrm{el}} \prod_{j\neq k} (\mathcal{P}_j^{\mathrm{el}})^2,$$

$$P^{\mathrm{SC}} = \sum_{k=1}^{4} P_k^{\mathrm{SC}} = 2\sum_{k=1}^{4} p_k^{\mathrm{cap}} p_k^{\mathrm{el}} \prod_{j\neq k} (\mathcal{P}_j^{\mathrm{el}})^2. \tag{15}$$

These formulae can be further simplified by taking into account that $\mathcal{P}_j^{\mathrm{ion,\,cap}}$ are small, because they involve electron removal from a positive ion, so that $\mathcal{P}_j^{\mathrm{el}} \approx 1$.

We have employed this simplification in the present calculations, yielding:

$$P^{\mathrm{SI}} = 2\sum_{k=1}^{4} p_k^{\mathrm{ion}} p_k^{\mathrm{el}},$$

$$P^{\mathrm{SC}} = 2\sum_{k=1}^{4} p_k^{\mathrm{cap}} p_k^{\mathrm{el}}. \tag{16}$$

It is well-known that processes involving the removal of more than one electron are, in general, not well described in the framework of the IPM. In this respect, we have found, taking experimental data on H^++H_2O collisions as references, that both single and double electron processes are better described by our implementation of the IEV (see however Refs. 68, 86). The probabilities for transfer ionization (TI), double ionization (DI) and double capture (DC) are obtained following arguments similar to those leading to (16) as

$$P^{\mathrm{TI}} = 2\sum_{k=1}^{4} p_k^{\mathrm{ion}} p_k^{\mathrm{cap}}; \quad P^{\mathrm{DI}} = \sum_{k=1}^{4} (p_k^{\mathrm{ion}})^2; \quad P^{\mathrm{DC}} = \sum_{k=1}^{4} (p_k^{\mathrm{cap}})^2. \tag{17}$$

Finally, in order to compare our ionization cross sections with the available experimental data, as well as to other theoretical predictions, it is useful to define the probability of electron production (EP) (or *net* ionization), P^{EP}, as

$$P^{\mathrm{EP}} = P^{\mathrm{SI}} + 2P^{\mathrm{DI}} + P^{\mathrm{TI}} = 2\sum_{k=1}^{4} p_k^{\mathrm{ion}}, \tag{18}$$

and, in a similar way, the *net* capture probability

$$P_{\rm net}^{\rm C} = P^{\rm SC} + 2P^{\rm DC} + P^{\rm TI} = 2\sum_{k=1}^{4} p_k^{\rm cap}, \tag{19}$$

where we have neglected the probabilities for three-electron processes.

2.3. *Anisotropy and Orientation Averaged Cross Sections*

The available experimental data correspond to ion collisions with gas phase targets, so that it is necessary to average the calculated cross sections over the molecule orientation to compare with them. The orientation average can be carried out in an equivalent way by averaging over the direction of the projectile velocity, keeping the molecule orientation fixed in the laboratory reference frame. Explicitly:

$$\sigma^{\rm X}(v) = \frac{1}{4\pi}\int {\rm d}\mathbf{b}\int {\rm d}\Omega\, P^{\rm X}(\mathbf{b}, v, \Omega), \tag{20}$$

where Ω is the solid angle that defines the direction of $\mathbf{v}$, and X=SI, SC, TI, DI or EP. The integration has been performed numerically by applying the method of Ref. 87, where the integral over $\mathrm{d}\Omega$ is obtained by means of a 6-point Newton-Côtes formula. The integration over $\mathrm{d}\hat{\mathbf{b}}$ is then carried out with the restriction $\mathbf{b} \perp \mathbf{v}$, and we have considered 4 orientations of the impact parameter $\mathbf{b}$ in the plane perpendicular to each orientation of $\mathbf{v}$. Finally, taking into account the molecular symmetry, the orientation averaged cross section is given by a combination of cross sections calculated for the 10 projectile trajectories illustrated in Fig. 4:

$$\sigma^{\rm X}(v) = \frac{1}{12}\left(\sum_{m=1}^{10} \sigma_m^{\rm X} + \sigma_4^{\rm X} + \sigma_8^{\rm X}\right), \tag{21}$$

with

$$\sigma_m^{\rm X} = 2\pi\int_0^\infty bP_m^{\rm X}(b, v){\rm d}b. \tag{22}$$

The probabilities $P_m^{\rm X}$ have been evaluated for nuclear trajectories with the orientation t_m (see Fig. 4).

In the case of employing a one-center model potential, the calculations for different trajectory orientations obviously yield the same cross section, and in order to incorporate the anisotropy effect in the total cross sections we have extended to the present case the method proposed in Ref. 47.

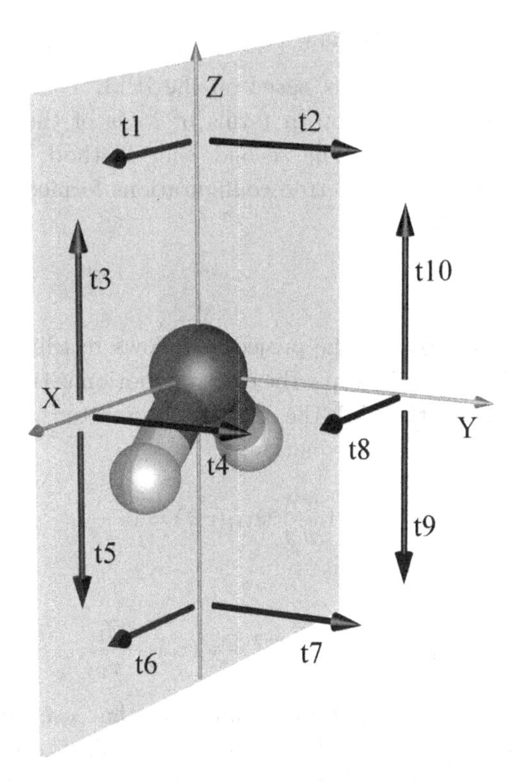

Fig. 4. Ion trajectories employed in the orientation average.

In this model, we assume that the minimum value of the impact parameter is equal to the dimension of the target nuclear skeleton, d, in the direction of $\hat{\mathbf{b}}$. We place the O nucleus at the origin; the H nuclei lie on the XZ half plane with $Z < 0$ and the molecular symmetry axis is in the Z direction (Fig. 4). We consider the set of ten trajectory orientations of Fig. 4; this set includes two trajectories (t_3 and t_5) with $\hat{\mathbf{v}}$ parallel to the symmetry axis and $\hat{\mathbf{b}} = \hat{\mathbf{X}}$, where we have taken $d = 2.873/2\,\mathrm{a}_0$ (the equilibrium H-H distance in the ground state of the water molecule is $2.873\,\mathrm{a}_0$). Two other trajectories (t_6 and t_7) have $\hat{\mathbf{b}} = -\hat{\mathbf{Z}}$, where we have taken for d the distance from the O nucleus to the H-H line ($1.102\,\mathrm{a}_0$). Assuming, as in Ref. 47, that the transition probabilities are independent of the trajectory orientation, we obtain the simple expression:

$$\sigma^{\mathrm{X}}(E) = 2\pi \int_0^\infty \left[b + \frac{1}{6}(1.436 + 1.102) \right] P^X(b, E)\mathrm{d}b. \qquad (23)$$

3. Semiclassical Calculations

We have employed two methods based on the IPM. The first one, called Method I (MI), is an expansion in terms of MOs of the supermolecule formed during the collision. The second one, Method II (MII), is an expansion in terms of many-electron configurations formed as products of the above-mentioned MOs.

3.1. *Method I*

In the semiclassical approach the projectile follows rectilinear trajectories with respect to the molecule, while the electron motion is treated quantum-mechanically. In the method MI, the one-electron scattering wave function, $\Psi_{\rm MI}$, is solution of the eikonal equation:

$$\left(h_{\rm MI} - {\rm i}\frac{\partial}{\partial t}\right)\Psi_{\rm MI}(\mathbf{r},t) = 0, \tag{24}$$

with

$$h_{\rm MI}(\mathbf{r};\mathbf{R}) = -\frac{1}{2}\nabla^2 + V(r_{\rm t}) - \frac{Z_{\rm P}}{r_{\rm P}}, \tag{25}$$

where we have employed a notation similar to that used in the CTMC method [see equation (3)]. To add additional flexibility to the description of the system, we have employed for the effective potential, $V(r_{\rm t})$, a pseudo-potential of the form:

$$V(r_{\rm O}, r_{\rm H_1}, r_{\rm H_2}) = V_{\rm O}(r_{\rm O}) + V_{\rm H}(r_{\rm H_1}) + V_{\rm H}(r_{\rm H_2}), \tag{26}$$

with

$$\begin{aligned} V_{\rm O} = &-\frac{8 - N_{\rm O}}{r_{\rm O}} - \frac{N_{\rm O}}{r_{\rm O}}(1 + \alpha_{\rm O} r_{\rm O}){\rm e}^{-2\alpha_{\rm O} r_{\rm O}}|s><s|, \\ &-\frac{N_{\rm O}}{r_{\rm O}}(1 + \beta_{\rm O} r_{\rm O})\,{\rm e}^{-2\beta_{\rm O} r_{\rm O}}[1 - |s><s|], \end{aligned} \tag{27}$$

$$V_{\rm H} = -\frac{1 - N_{\rm H}}{r_{\rm H}} - \frac{N_{\rm H}}{r_{\rm H}}(1 + \alpha_{\rm H} r_{\rm H})\,{\rm e}^{-2\alpha_{\rm H} r_{\rm H}}. \tag{28}$$

As in the CTMC calculations, these expressions contain free parameters that are fitted so that the eigenvalues of the target Hamiltonian, $[h_{\rm MI}(\mathbf{r};\mathbf{R}) + Z_{\rm P}/r_{\rm P}]$ differ in less than 4×10^{-3} a.u. from the energies of the SCF MOs of water. We obtain: $N_{\rm O} = 7.162$, $\alpha_{\rm O} = 1.48\,{\rm a}_0^{-1}$, $\beta_{\rm O} = 1.60\,{\rm a}_0^{-1}$ and $\alpha_{\rm H} = 0.665\,{\rm a}_0^{-1}$, with $N_{\rm H} = (9 - N_{\rm O})/2$.

At large values of R ($R_a = 1000\,a_0$ in our calculations), we solve the model-potential eigenvalue equation $h_{\mathrm{MI}}\Phi_i = \epsilon_i\Phi_i$ in a GTO basis set $\{\xi\}$; the eigenvectors are the asymptotic molecular orbitals:

$$\Phi_i(R_a) = \sum_k c_{ik}(R_a)\xi_k(R_a). \tag{29}$$

We, then, define the frozen molecular orbitals along the projectile trajectory by using for all values of R the asymptotic matrix of MO coefficients $\mathbf{C}(R_a)$:

$$\Phi_i(R) = \sum_k c_{ik}(R_a)\xi_k(R). \tag{30}$$

It must be noted that we use the asymptotic forms of the MOs instead of an adiabatic basis of eigenfunctions of h_{MI} at each R, since in the latter case the potential energy surfaces display conical intersections, where the dynamical coupling diverge.

The one-electron wave function Ψ_{MI} is then expanded in the molecular basis set $\{\Phi\}$ as:

$$\Psi_{\mathrm{MI}}(\mathbf{r};t) = D(\mathbf{r},t)\sum_j a_j(t)\Phi_j, \tag{31}$$

where $D(\mathbf{r},t)$ is a common translation factor (see[88] and references therein), which ensures that the initial conditions are fulfilled for a finite basis set. Substitution of expansion (31) into the semiclassical equation (24) leads, for each nuclear trajectory, to a system of first order differential equations:

$$\mathrm{i}\frac{\mathrm{d}a_i}{\mathrm{d}t} = \sum_k (\mathbf{s}^{-1}\mathbf{M})_{ik}a_k, \tag{32}$$

where

$$M_{ik} = \left\langle \Phi_i D \left| h_{\mathrm{MI}} - \mathrm{i}\frac{\partial}{\partial t} \right| \Phi_k D \right\rangle \tag{33}$$

and

$$s_{ik} = \langle \Phi_i D | \Phi_k D \rangle \tag{34}$$

are the coupling and overlap matrices in the basis $\{\Phi\}$. In this work, the basis set contains 108 MOs: 85 MOs, which have asymptotically positive energy, are SI channels, 7 are SC channels, and the rest are asymptotically target MOs, including the entrance channels (the occupied orbitals of the molecule) and excitation channels. In practice, the system of differential equations (32) is solved for each of the four initial conditions: $a_j(t=0)=\delta_{ij}$, with $j=2,3,4,5$, which correspond to the electron initially located on one of the four valence MOs of H_2O. The one-electron probability for the transition from Φ_j to Φ_i along a given nuclear trajectory is:

$$p_{ij} = |\langle \Psi_{\mathrm{MI}}(R_a)|\Phi_i(R_a)D\rangle|^2 = |a_i(R_a)|^2. \tag{35}$$

In practice, given the large basis employed, we have not included the translation factor and placed the origin of electronic coordinates on the oxygen nucleus, which allows us to evaluate total electron capture transition p_j^{cap} and ionization probabilities p_j^{ion}. Explicitly:

$$p_j^{\mathrm{sc}} = \sum_i p_{ij} \quad \text{if } \epsilon_i(R_a) < 0 \quad \text{and orbital } i \text{ is on the projectile,} \tag{36}$$

$$p_j^{\mathrm{ion}} = \sum_i p_{ij} \quad \text{if } \epsilon_i(R_a) > 0. \tag{37}$$

The many-electron probabilities are obtained from these (pseudo-potential) probabilities by using the IEVM [see equation (16)]:

$$P^{\mathrm{SC}} = 2\sum_{j=2}^{5} p_j^{\mathrm{sc}}(1-p_j^{\mathrm{ion}}-p_j^{\mathrm{sc}});$$

$$P^{\mathrm{ION}} = 2\sum_{j=2}^{5} p_j^{\mathrm{ion}}(1-p_j^{\mathrm{ion}}-p_j^{\mathrm{sc}}). \tag{38}$$

The orientation-averaged total cross sections are then evaluated by substituting the transition probabilities (38) into equations (21) and (22). Thus, for each impact velocity, the calculation involves the solution of the system of differential equations (32), with four initial conditions, for each of the ten trajectory orientations of Fig. 4. Moreover, the numerical integration of the transition probabilities of (22) requires repeating the calculation for an appropriate set of values of b.

3.2. *Method II*

In this alternative multi-electronic method, we develop the scattering wavefunction in terms of configurations, which are antisymetrized products of the asymptotic MOs Φ_i of equation (30). In the present calculation, the basis $\{\psi_i\}$ includes one configuration to represent the entrance channel $X^{q+} + H_2O$, with the target in the ground electronic state, where the lowest lying water orbitals Φ_i $(i = 1, \ldots, 5)$ are doubly occupied. The exit channel configurations are obtained as single excitations from the ground configuration:

$$\psi_1 = ||\Phi_1\bar{\Phi}_1 \ldots \Phi_5\bar{\Phi}_5|| \quad \text{entrance channel,} \tag{39}$$

$$\psi_j = ||\Phi_1\bar{\Phi}_1 \ldots \Phi_m\bar{\Phi}_l|| + ||\Phi_1\bar{\Phi}_1 \ldots \Phi_l\bar{\Phi}_m|| \quad \text{exit channels,} \tag{40}$$

where $m = 2, 3, 4$ and 5 correspond to one of the MOs of the valence shell Φ_l, with $l \geq 6$, are the unoccupied orbitals. They can be target orbitals, and transitions to these configurations are interpreted as excitation; projectile orbitals, and the populations of these configurations are interpreted as electron capture; or, finally, when the energies of these orbitals are positive, the transitions to the corresponding configurations lead to ionization.

The collisional wave function is:

$$\Psi_{\mathrm{MII}}(\mathbf{r};t) = D(\mathbf{r},t)\sum_j d_j(t)\psi_j \exp\left[-\mathrm{i}\int_0^t \mathrm{d}t'\,\Lambda_j\right], \tag{41}$$

where now $\mathbf{r}$ denotes the coordinates of all electrons; $\Lambda_j = (\mathbf{S}^{-1}\mathbf{h}_{\mathrm{MII}})_{jj}$, and $\mathbf{S}$ and $\mathbf{h}_{\mathrm{MII}}$ are the overlap and MII Hamiltonian matrices in the basis $\{\psi\}$ (see below). This leads to the transition probabilities from state ψ_1 to state ψ_i, along trajectory k:

$$P_i^k = |\langle \Psi_{\mathrm{MII}}(R_a)|\psi_i(R_a)D\rangle|^2 = |d_i(R_a)|^2. \tag{42}$$

The SC probability is:

$$P^{k,\mathrm{SC}} = \sum_i P_i^k, \tag{43}$$

where the sum extends over the configurations representing electron capture. Analogously, SI probabilities are obtained, when the exit configurations represent ionization channels. Since the method employs the eikonal equation for all electrons, it is not necessary to use a many-electron interpretation of the one-electron transition probabilities as in the other methods employed in this work, but it requires evaluating many two-electron integrals. To overcome this difficulty, we have employed in the present calculation a simplification, first applied in Ref. 89, where the IPM is used to evaluate the Hamiltonian matrix in the configuration basis set.

In practice:

$$\langle \psi_i | H | \psi_j \rangle \approx \langle \psi_i | h_{\mathrm{MII}} | \psi_j \rangle = \left\langle \psi_i \left| \sum_{j=1}^{2n} h_{\mathrm{MI}} \right| \psi_j \right\rangle . \tag{44}$$

As in the other methods, the orientation-averaged cross sections are obtained integrating the transition probabilities, as explained in equations (21) and (22).

4. $H^+ + H_2O$ Collisions

4.1. *CTMC Calculations*

In this section, we compare our total cross sections to different experimental and theoretical results for the $H^+ + H_2O$ system, in the energy range $20\,\mathrm{keV} \leq E \leq 10\,\mathrm{MeV}$. In Fig. 5(a), we compare our cross sections for single ionization (SI) with perturbative CDW[36,37] and FB[38] calculations and experiments.[25,27,31] Our calculations, using the three-center model potential and the IEV, lead to cross sections in better agreement with the experimental ones than the one-center calculations, reported in Ref. 44, which employ a Z_{eff} description of the target and the IPM. Moreover, at high energies our SI cross section shows an energy dependence similar to that of the CDW-EIS[36,37] and FB[38] calculations. However, it is worth pointing out the reasonable agreement of the cross section evaluated by means of the simple one-center treatment with experiments. This fact can be understood as a partial compensation of the two main differences between both one- and three-center CTMC calculations. First, the one-center model potential leads to a diffuse initial distribution (Figs. 1 and 3), which yields one-electron SI probabilities larger than those obtained in the three-center calculation. Second, the use of the conventional IPM from Eq. (14) reduces the multielectronic SI probabilities with respect to those obtained by means of the IEV from Eq. (16). This is further illustrated in Fig. 5(a), where we include the results obtained by employing the one-center model potential together with the IEV, which clearly overestimates the SI total cross section.

We compare in Fig. 5(b) the total cross sections for electron production with the experimental data of Refs. 19,20 and the theoretical data of Ref. 39, issued from BGM calculations. We have also included in this figure the data of Ref. 31 for $E < 100$ keV/amu, which, as explained by the authors, correspond to the sum of SI and TI cross sections; indeed, double ionization

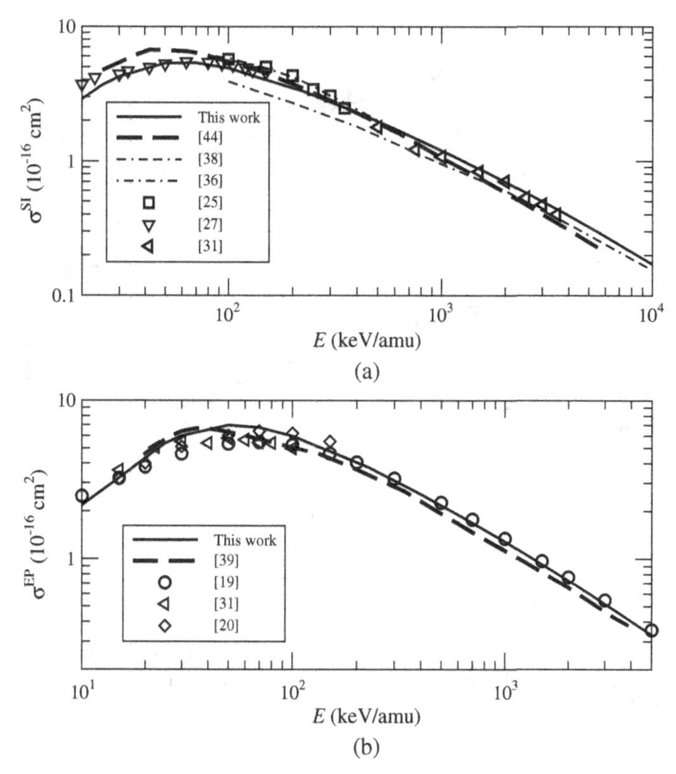

Fig. 5. Total cross sections for (a) single ionization, σ^{SI}, and (b) electron production, σ^{EP}, in proton collisions with water as functions of the collision energy. Present calculations: solid lines, CTMC with a three-center model potential and IEV: dashed line,[44] CTMC with a one-center model potential and IPM; other calculations: broken lines.[36,38,39] Experimental results for SI[19,20,25,27,31]: (these data correspond to SI+TI for $15 < E < 100$ keV and to SI for $E > 500$ keV).

is expected to be very small at low E so the EP cross section must be practically identical to the sum of SI and TI ones. Our improved CTMC results are in very good agreement with all measurements. They also nicely agree with the BGM results of Ref. 39, even though the maximum of the CTMC is located at slightly higher E than its BGM counterpart.

Our orientation-averaged SC cross section, σ^{SC}, is displayed in Fig. 6(a) as a function of the impact energy E. The figure also includes the experimental data of Refs. 17,27,31. As for the SI calculations shown in Fig. 5, the improvement inherent in the use of the three-center V_{mod} and the IEV clearly shows up as one compares one- and three-center CTMC results,

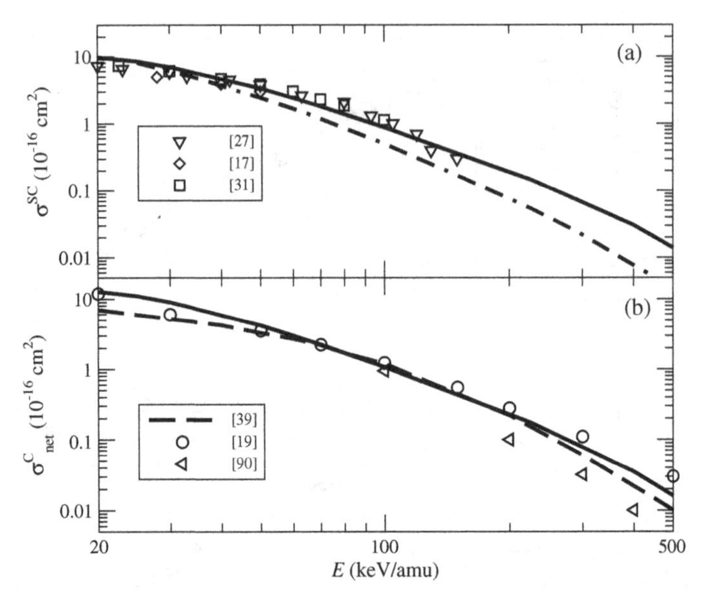

Fig. 6. Total cross sections for (a) single electron capture, σ^{SC}, and (b) net electron capture, σ^{C}_{net}, in proton collisions with water as functions of collision energy. Present results: solid lines, CTMC calculations using the three-center model potential and the IEV; dotted-dashed line, CTMC calculation using the one-center model potential and the IPM. Dashed line, calculation of Ref. 39. Experimental data for SC: Refs. 17,27,31 Experimental data for net electron capture: Refs. 19,90.

taking into account that the latter ones nicely agree with experiments over the whole impact energy range. Our computed *net* capture cross section, displayed in Fig. 6(b), is also in agreement with both the measurements[19,90] and BGM calculations.[39] The small deviations which persist between both calculations are attributed to the liabilities of both the IPM and IEV to provide very accurate cross sections for two electron processes; here the TI contribution to *net* capture has been found to be one order of magnitude smaller than the SC for $E > 50\,\text{keV/amu}$ but it can be slightly overestimated at lower E.

The dependence of the cross sections on the trajectory orientation has been discussed in detail in Ref. 45 and it will not be repeated here. A conclusion of this study is that the cross section evaluated for the t_4 projectile-target orientation is such that $\sigma_4^{\mathrm{SI}} \approx \sigma^{\mathrm{SI}}$ for the energy range of Fig. 5 and $\sigma_4^{\mathrm{SC}} \approx \sigma^{\mathrm{SC}}$ at $E < 100\,\text{keV/amu}$. Accordingly, we present in Fig. 7 the weighted monoelectronic ionization (bp_i^{ion}) and capture (bp_i^{cap}) probabilities, as functions of b, for each initial valence MO ϕ_i and for

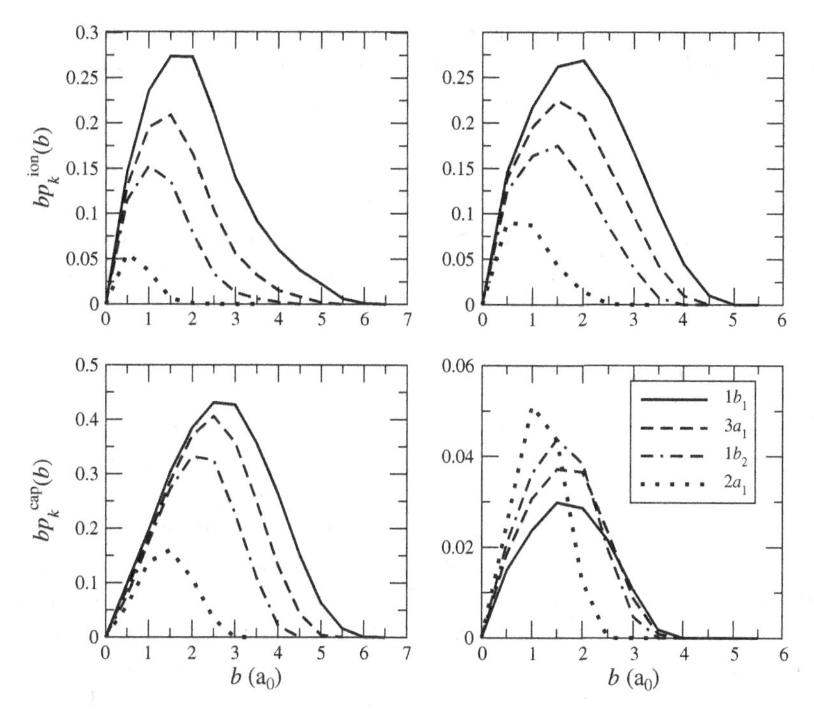

Fig. 7. One-electron transition probabilities for ionization p_k^{ion} (top panels) and capture p_k^{cap} (bottom panels), multiplied by the impact parameter, b, as functions of b for H^++H_2O collisions at $v = 1$ a.u. (left panels) and $v = 2$ a.u. (right panels), and for the trajectory orientation described in the text.

this representative trajectory orientation. We plot these probabilities for $v = 1$ a.u. ($E \approx 25$ keV/amu) and $v = 2$ a.u. ($E \approx 100$ keV/amu); the former impact velocity roughly corresponds to the maximum of the averaged SI cross section, where ionization and capture strongly compete with each other; $v = 2$ a.u. is representative of the high velocity regime, and we have explicitly checked that the trends found in Fig. 7 at $v = 2$ a.u. hold for higher v. We have also verified that t_m trajectories with $m \neq 4$ lead to similar features to those displayed in Fig. 7.

Concerning ionization, we observe that the more weakly bound the MO is, the larger the corresponding p_i^{ion}; this conforms to our intuition as well as to our experience in ion–atom collisions where it is well-known that it is generally easier to pull out an electron from a highly excited orbital. Furthermore, the weighted probabilities bp_i^{ion} peak at larger b when the initial MO is more diffuse; once again, such a behavior is intuitive and can be traced back to the spatial extension of the MO. The capture probabilities $bp_i^{\rm cap}$ behave as their ionization counterparts at small v; in this velocity regime, the first steps of ionization and capture mechanisms are the same[91–93] so that it is reasonable to obtain similar trends for capture and ionization as a function of b. As v increases, we find that the contribution to the electron capture of the inner MO, $2a_1$, increases, becoming the main contribution for $v > 2.0$ a.u, ($E > 100$ keV/amu). In fact, electron capture occurs at small impact parameters, where inner shell processes are also known to be important in ion-atom systems.

4.2. *Semiclassical Calculations*

The total electron capture cross sections have been computed using methods MI and MII, and the results are compared in Fig. 8 with the CTMC results and with the experimental data. It can be noted the reasonable agreement between the results of both semiclassical calculations. In general, the MI method leads to a capture cross section in better agreement with the experiments than MII, because of the larger basis set employed in the former calculation. At $E > 20$ keV/amu, the ionization cross section is of the same order of magnitude as the capture one, and the separation of capture and ionization in the close-coupling calculation is difficult. In addition, the translation factor is not appropriate to describe ionizing

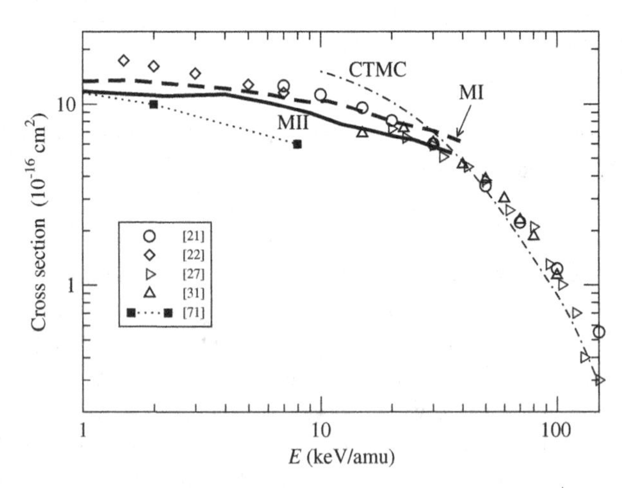

Fig. 8. SC total cross section as a function of the impact energy for $H^+ + H_2O$ collisions. Dashed line, present results using the semiclassical method MI; full line, present results using the semiclassical method MII; dashed-dotted line, present CTMC results; other theoretical data: Ref. 71. Experimental data: Refs. 21, 22, 27, 31.

transitions (see, e.g. Ref. 94). The comparison of semiclassical and CTMC results shows a behavior similar to that found in ion-atom collisions: both methods yield similar cross sections at $E \approx 25\,\text{keV/amu}$, but a smooth joining of the cross sections curves is not easy because of the above-mentioned difficulty of the semiclassical calculation to disentangle capture and ionization fluxes. In addition, the classical method starts to be inappropriate at $E < 25\,\text{keV/amu}$, where, in general, it overestimates the capture cross section.

5. He^{2+}+H_2O Collisions

We display in Fig. 9 the cross section for electron production as a function of the impact energy E. We compare our results, obtained by means of the CTMC method with the IEV and the three-center model potential, with the experimental data of Refs. 21,95,96. Good agreement is found for $E \geq 50\,\text{keV/amu}$, but discrepancies appear at lower E, where the underestimation of the experimental EP cross section can be attributed to the inaccurate computation, by means of the IEV, of the many-electron probabilities that contribute to EP. On the other hand, the SI cross section, included in Fig. 9, largely falls down as E decreases from 50 to 10 keV/amu. Besides the IEV deficiency, the SI contribution to EP can be underestimated at low E, due to the use of microcanonical phase-space initial conditions

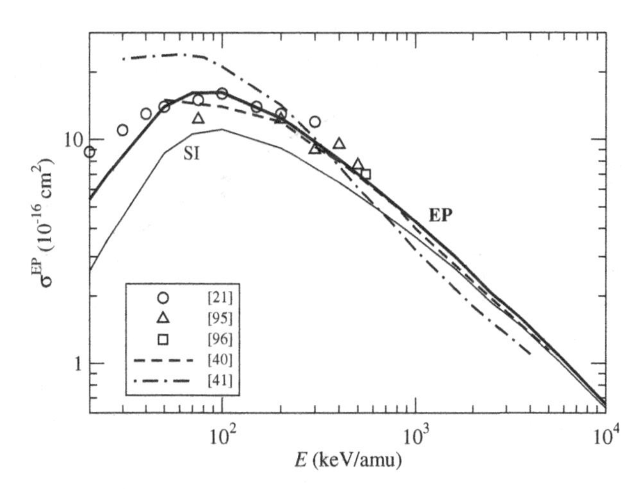

Fig. 9. Electron production cross sections as a function of the impact energy for He^{2+}+H_2O collisions. Present calculations curves labeled EP and SI. Other theoretical results: Refs. 40 and 41. Experimental results for EP: Refs. 21, 95, 96.

in the CTMC calculations; such a problem is widely documented for low-E multicharged ion-atom collisions[78,79]: in these (simple) collisional systems, it has been soon realized[78] that accurate calculations of inelastic cross sections at low and intermediate E require the use of improved initial conditions, beyond the microcanonical framework. On the other hand, the over-barrier based model of Ref. 41, which does not account for target anisotropy, works worse than the present CTMC approach at any E. The FB calculations[40] behave better at $E \geq 200$ keV/amu and coalesce, in this energy range, with the present CTMC EP cross section.

Since SC total cross sections for He^{2+} impact have not been measured at $E > 25$ keV/amu, we plot in Fig. 10 the sum of SC and TI cross sections as a function of E, and compare with the measurements of Rudd et al.[21] and the over-barrier calculations of Abbas et al.[41] The shape of the experimental cross section is well reproduced by our calculations; nevertheless the measured and computed cross sections differ in magnitude (up to a factor of ~2 at low E). The TI contribution to the SC+TI cross section is found to be small over the whole impact energy range, since $\sigma_{SC} \approx \sigma_{SC} + \sigma_{TI}$ in Fig. 10. The over-barrier model,[41] which yields a badly shaped cross section, does not allow us to ascertain the experimental or the present CTMC-IEV values of the SC+TI cross section. New experimental investigations of both SC and TI cross sections, together

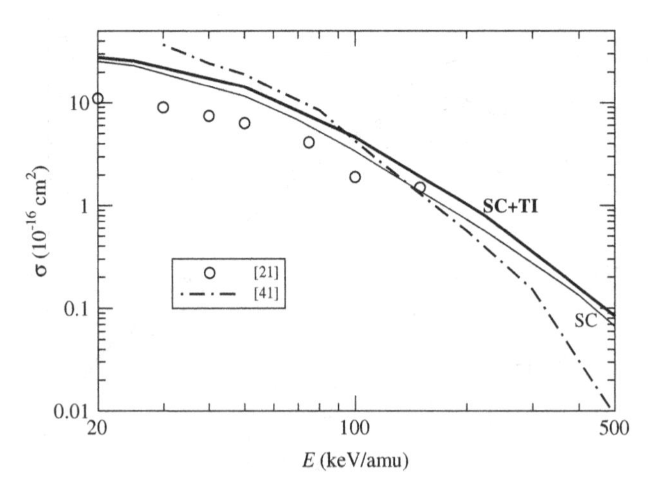

Fig. 10. Single capture plus transfer ionization (SC+TI) cross sections for $He^{2+} + H_2O$ collisions as a function of the impact energy: solid lines, present results; dotted-dashed line, classical calculations from Ref. 41. Experimental data: Ref. 21.

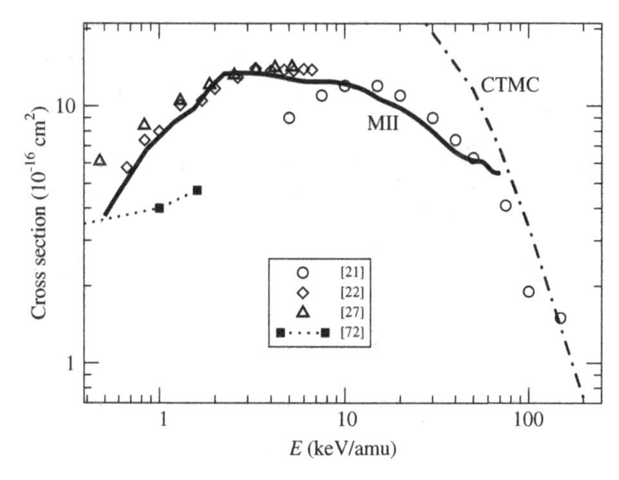

Fig. 11. SC total cross section as a function of the impact energy for $He^{2+} + H_2O$ collisions. Full line, present results using the semiclassical method MII; dashed-dotted line, present CTMC results; other theoretical data: Ref. 72. Experimental data: Ref. 21,22,27.

with the implementation of improved calculations, would allow eliciting the validity and limitations of the IEVM at low E.

We have also evaluated the capture cross section by applying the semiclassical method MII (Fig. 11). In this result we do not find contamination by ionization at relatively high energies ($E > 10\,\text{keV/amu}$), where our cross section agrees with the experimental values of Rudd et al.,[21] with the exception of the oscillation at $E \approx 50\,\text{keV/amu}$ that indicates the high-energy limit of the method. At $E < 10\,\text{keV/amu}$ our cross section agrees with the experimental data of Refs. 22, 27, but they do not agree with the calculation of Ref. 72. The semiclassical calculations, together with the experimental results, suggest that the CTMC method overestimates the capture cross section at $E < 60\,\text{keV/amu}$.

6. $C^{6+} + H_2O$ collisions

Drawing from the satisfactory implementations of the three-center CTMC model for H^+ and He^{2+}+H_2O collisions, we have also considered C^{6+} ion impact because of its paramount importance in ion beam cancer therapy. In Fig. 12, we report our computed EP and SI cross sections as functions of E, lying in the wide range $10\,\text{keV/amu} \leq E \leq 10\,\text{MeV/amu}$. Perturbative calculations by Bernal and Liendo[42] and Dal Cappello et al.[43] are also

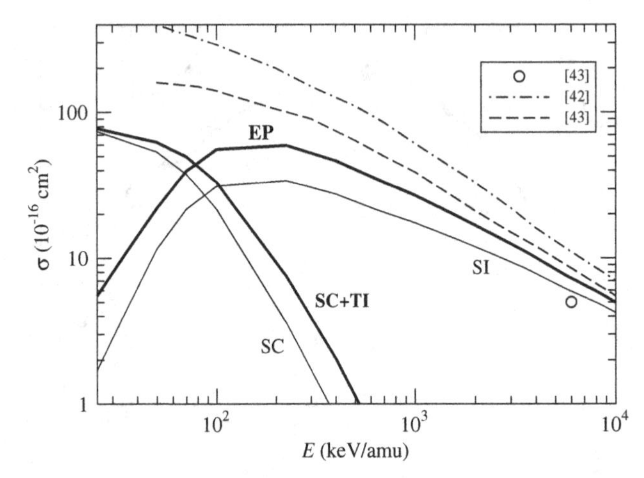

Fig. 12. Cross section as a function of the impact energy for C^{6+} + H_2O collisions. Present calculations: curves labaled EP, SI, SC and SC + TI; electron production from other calculations Ref. 42,43. Experimental point at $E = 6$ MeV/amu.[43]

included in the figure. It is clear that the present CTMC and perturbative cross sections shall coalesce for $E > 10\,\mathrm{MeV/amu}$. In this respect, it has to be noted that the lower bound of validity of perturbative calculations increases with the projectile charge,[83] so that particular care has to be taken before including perturbative cross sections in the Monte Carlo track structure codes that aim at describing the dynamics induced by charged particles passing through biological environments.[15,16] We safely venture that our non-perturbative CTMC results are more accurate than those of Refs. 42, 43; the former lie closer to the unique experimental point at $E = 6$ MeV/amu than the latter ones. We also plot in Fig. 12 our cross sections for SC and SC+TI. As for He^{2+} collisions, we expect that the main limitation of our method is the description of two- and three-electron processes, although the estimate of σ^{TI} from Fig. 12 indicates that this process is probably not very relevant at the lowest energies of our calculations.

7. Scaling Laws and Fragmentation

As stated in the Introduction, the collisional database required to model radiation damage and interaction of charged particles with biological environment is huge. In this respect, our CTMC calculations of cross sections for three ion-H_2O systems can be used to extract some scaling laws, with respect to projectile charge and impact energies, that would avoid the need for further ion-H_2O calculations. We additionally consider in this section the fragmentation processes subsequent to the primary ion-H_2O collisions. Since we have found that in general the three-center model potential approach is more accurate than the one-center model potential approximation, we limit the presentation of this section to the results of the former calculation. The results of the CTMC one-center calculation can be found in Ref. 44.

7.1. *Scaling Laws for Ion-H_2O Collisions*

Most of the scaling laws that have been derived in atomic collisions stem from first-order perturbative calculations of inelastic cross sections. For instance, FB theory indicates that ionization cross sections behave, in the high velocity regime, as $Z_{\rm P}^2 \ln(E)/E$, whereas for capture, cross sections scale as $Z_{\rm P}^5/v^{12}$.[76] These scaling relations are valid, and useful, provided the requirements for first-order perturbative conditions are fulfilled. In this respect, the perturbation strength $Z_{\rm P}/v$ is the important parameter and FB (and related laws) apply provided $Z_{\rm P}/v \ll 1$.

At the energy range considered in this work, and in usual radiation damage applications, the FB validity criterion is not fulfilled for high projectile charges. As an example, the previous scaling laws are expected to fail for C^{6+} impact if $E < 1\,{\rm MeV/amu}$, where term orders higher than one are necessary, within perturbative expansions, to account for strong interactions between target and projectile. Introducing high-order terms in Born-type expansions is generally not a practical solution to derive simple scaling laws (non-unitarity problems can further show up), and non-perturbative calculations hide simple relations between cross sections, $Z_{\rm P}$ and v. One is thus led to derive semi-empirical relations *a posteriori* from calculations involving various Z_P and v for a given target. Such explicit calculations in ion–atom collisions[97,98] have shown that ionization cross sections fall in an universal curve, in both the $Z_{\rm P}/v < 1$ and $Z_{\rm P}/v > 1$ regimes, if the cross section and the impact energy are linearly scaled with Z_P. In this work, we show in Fig. 13 that the same applies for EP cross sections for $E/Z_{\rm P} \geq 100\,{\rm keV/amu}$, where $\sigma^{EP} \approx \sigma^{SI}$. The relation between $\sigma^{EP}/Z_{\rm P}$ and $E/Z_{\rm P}$ must not be confused with the usual Born scaling; it further allows reproducing the strong interaction effects on the cross section when $v < Z_P$. In order to apply the scaling laws to obtain fragmentation cross sections it is useful to check that they also hold for SI cross sections

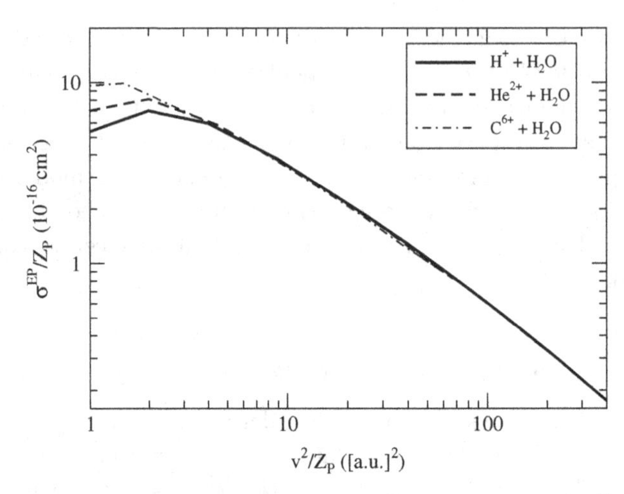

Fig. 13. Electron production cross sections over $Z_{\rm P}$ as a function of $v^2/Z_{\rm P}$ for H^+, He^{2+} and C^{6+} + H_2O collisions.

from individual MOs, $\sigma_k^{\rm SI}$; they are obtained by using equations (21)–(22) with the ionization probabilities from individual MOs: $P_k^{\rm SI} = 2p_k^{\rm ion}p_k^{\rm el}$ [see equation (16)]. This point is illustrated in Fig. 14, where we have plotted the orientation-averaged total cross sections for SI from the valence-shell MOs ($k = 2, 3, 4, 5$).

Concerning electron capture, we show in Fig. 15 that, for E ranging from $\sim 100\,{\rm keV/amu}$ to $\sim 1\,{\rm MeV/amu}$, SC cross sections for multicharged ion impact can be simply, and quite accurately, scaled using H^+ reference data through $\sigma^{\rm SC}(Z_{\rm P}, v) = Z_{\rm P}^2\sigma^{\rm SC}(1, v)$. The latter relation is in sharp disagreement with the FB capture scaling that would apply for $E \gg 1\,{\rm MeV/amu}$. For energies lower than $100\,{\rm keV/amu}$, it seems that a linear scaling with respect to $Z_{\rm P}$ is better than the proposed $Z_{\rm P}^2$ one; this agrees with the low-E over-barrier prediction of Knudsen et al.,[99] which leads to constant $\sigma^{\rm SC}(Z_{\rm P}, v)/Z_{\rm P}$ values for any values of $Z_{\rm P}$ and v. In other words, our proposed SC scaling fills the gap between low and high velocity regimes.

7.2. *Fragmentation Cross Sections*

Fragmentation reactions are of great importance in the radiation damage of biological systems by ion impact (see Ref. 100) because the secondary ions formed in the fragmentation can interact with the DNA. In this work, we focus on the fragmentation processes associated with the most significant SC and SI (single) electron removals. Indeed, SC and SI can

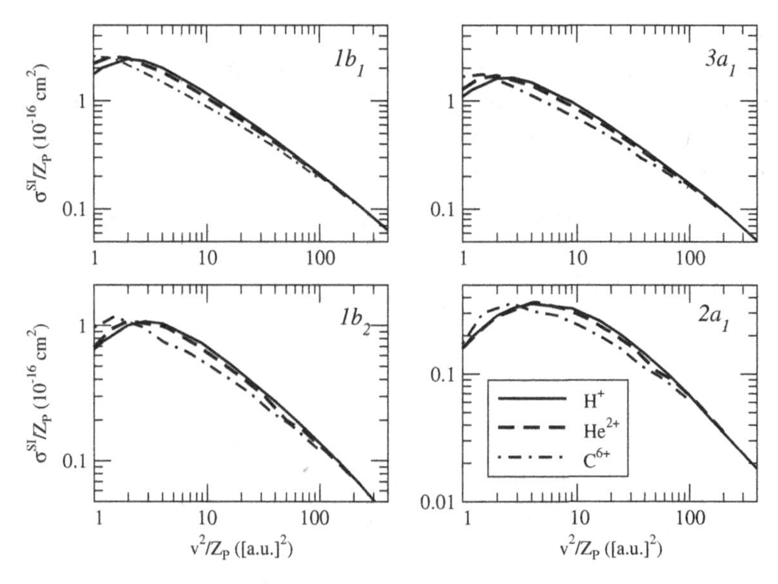

Fig. 14. Contributions to the SI cross section of the different orbitals of the valence shell of H_2O, for H^+, He^{2+} and $C^{6+}+H_2O$ collisions. As in Fig. 13, the cross sections are divided by Z_P and plotted as functions of v^2/Z_P to check the scaling law.

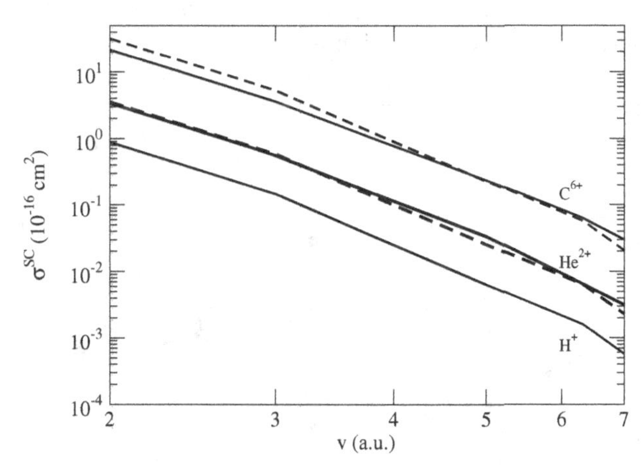

Fig. 15. Single capture cross sections as a function of the collision velocity for H^+, He^{2+} and $C^{6+}+H_2O$ collisions. Dashed lines: proton data multiplied by Z_P^2.

occur throughout a non-dissociative process, leading to a stable H_2O^+ fragment, and also through dissociative pathways. Under the assumption that dissociation takes place after the electron transitions,[29,31,36] the dissociation channels consist of (i) evaporation $H_2O^+ \rightarrow OH^+ + H$;

(ii) fission $H_2O^+ \to OH + H^+$; and (iii) break-up $H_2O^+ \to O^+ + H + H$. The cross sections associated to all these channels are computed by multiplying the branching ratios of Tan et al.,[101] for fragmentation subsequent to electron removal from identified H_2O MOs, by our MO-resolved SC and SI cross sections, previously computed:

$$\begin{aligned}
\sigma_{\mathrm{H_2O^+}}^{\mathrm{SC,SI}} &= 1.00\,\sigma^{\mathrm{SC,SI}}(1b_1) + 1.00\,\sigma^{\mathrm{SC,SI}}(3a_1) + 0.08\,\sigma^{\mathrm{SC,SI}}(1b_2),\\
\sigma_{\mathrm{OH^+}}^{\mathrm{SC,SI}} &= 0.70\,\sigma^{\mathrm{SC,SI}}(1b_2),\\
\sigma_{\mathrm{H^+}}^{\mathrm{SC,SI}} &= 0.22\,\sigma^{\mathrm{SC,SI}}(1b_2) + 0.74\,\sigma^{\mathrm{SC,SI}}(2a_1),\\
\sigma_{\mathrm{O^+}}^{\mathrm{SC,SI}} &= 0.26\,\sigma^{\mathrm{SC,SI}}(2a_1).
\end{aligned} \tag{45}$$

In practice, the fragmentation cross sections have only been computed for H^+ impact; for other projectile charges, the fragmentation cross sections can easily be derived using (45) and the scaling relations previously discussed.

In Fig. 16 we report the fragmentation cross sections after SI in the $150 \leq E \leq 3000\,\mathrm{keV/amu}$ impact energy range, and compare them to the experimental results of Luna et al.[31] and Werner et al.[25] It is worth noting that the later measurements correspond to fragmentation after either SI or TI; nonetheless, the σ^{TI} cross section is largely smaller than the σ^{SI} one in

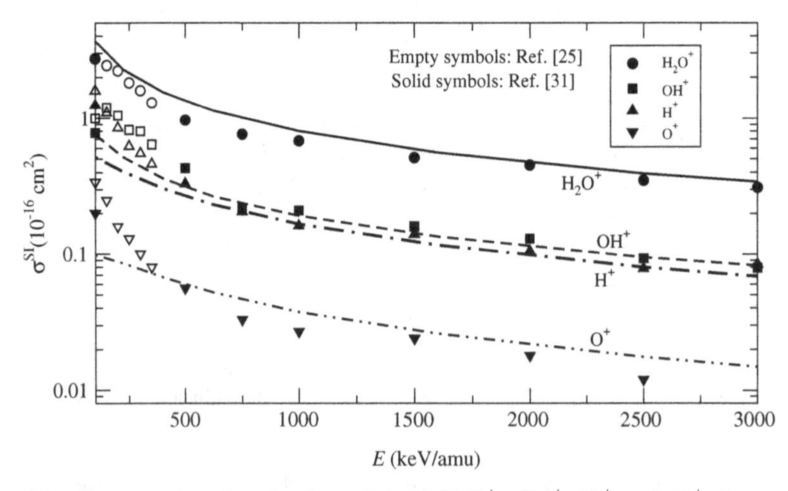

Fig. 16. Cross sections for the formation of H_2O^+, OH^+, H^+ and O^+ fragments (indicated in the figure) after single ionization in $H^+ + H_2O$ collisions. Lines: our calculations; empty symbols: experimental data of Werner et al. ;[25] solid symbols: experimental data of Luna et al..[31]

the energy range considered. It can be seen in Fig. 16 that our fragmentation cross sections are in close agreement with experiments for $E \geq 400\,\text{keV}$, but significant discrepancies appear at lower E. At the relatively high collision energies of Fig. 16, the assumption of postcollisional fragmentation applies, and these discrepancies are probably a consequence of the underestimation of the ionization probability from the orbitals $1b_2$ and $2a_1$, due to the limitations of the microcanonical initial distribution. In this respect, it has to be noted that the fragmentation cross section provides a more stringent check of accuracy of the MO-resolved SI cross sections than the total σ^{SI} to which $\sigma^{\text{SI}}(2a_1)$ little contributes (see Fig. 7).

In the energy range of Fig. 16, SI is mostly non-dissociative because it is mainly tailored by electron removal from the outer $1b_2$, $3a_1$ and $1b_1$ MOs (see Fig. 7) that favor the production of stable H_2O^+ according to (45). Furthermore, all the product cross sections present parallel shapes as functions of E for $E > 100\,\text{keV}$. As explained by Montenegro et al.,[102] this pattern shows up in the high impact velocity regime where all the MO-resolved SI cross sections exhibit the Bethe-Born $(1/I_P)\ln(E)/E$ behavior as a function of E, with I_P the ionization potential of the MO, so that according to (45), the ratios $\sigma^{\text{SI}}_{\text{H}_2\text{O}^+} : \sigma^{\text{SI}}_{\text{OH}^+} : \sigma^{\text{SI}}_{\text{H}^+} : \sigma^{\text{SI}}_{\text{O}^+}$ reduce to the constant values $[1/I_P(3a_1) + 1/I_P(1b_1) + 0.08/I_P(1b_2)] : [0.70/I_P(1b_2)] : [0.74/I_P(2a_1) + 0.22/I_P(1b_2)] : [0.26/I_P(2a_1)] = 0.1423 : 0.0368 : 0.0316 : 0.0070$.

The fragmentation cross sections (45) after single electron capture are plotted in Fig. 17 for E lying in the intermediate range 20–300 keV. Our calculations are compared to the experimental data of Luna et al.[31] and Gobet et al.[26]. As for SI, it seems that our $\sigma^{\text{SC}}(2a_1)$ cross section is underestimated for $E \leq 100\,\text{keV}$ so that the fission ($\sigma^{\text{SC}}_{\text{H}^+}$) and break-up ($\sigma^{\text{SC}}_{\text{O}^+}$) cross sections lie below the measurements, while $\sigma^{\text{SC}}_{\text{H}_2\text{O}^+}$ and $\sigma^{\text{SC}}_{\text{OH}^+}$ are in satisfactory agreement with the data of Luna et al.[31] The most conspicuous difference between fragmentation cross sections after SC from those after SI (Figs. 17 and 16, respectively) is the fact that in the former case the lines are not parallel. However, one can note in this figure that the lines for non-dissociative and evaporation reactions, and also those for fission and break up processes, are parallel. To explain this fact, one has to take into account that cross sections for fission and break up reactions are dominated by the contribution of the $2a_1$ MO, and, as explained in section 3, the SC probability from this MO increases with E, becoming clearly dominant at high E, while a similar effect is not observed in SI. Hence, the energy dependence of $\sigma^{\text{SC}}(2a_1)$ changes the slope of fission and break

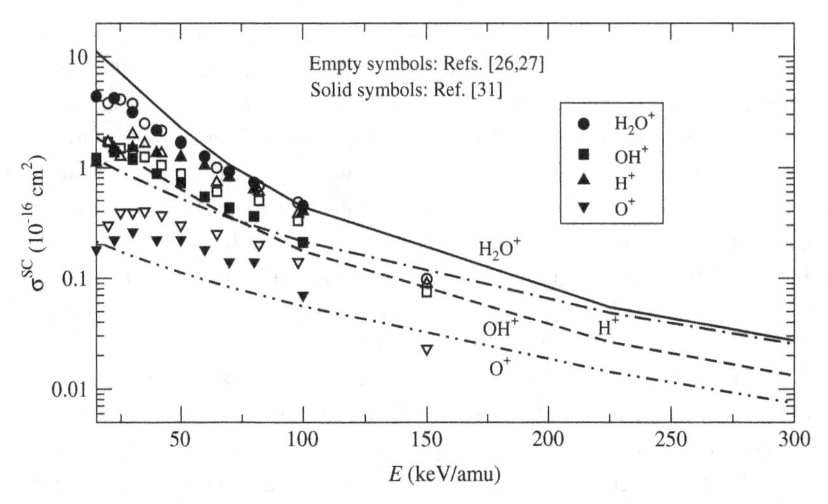

Fig. 17. Cross section for the formation of H_2O^+, OH^+, H^+ and O^+ fragments (indicated in the figure) after single electron capture in H^++H_2O collisions. Lines: our calculations; solid symbols: experimental data of Luna et al.;[31] empty symbols: experimental data of Gobet et al.[26,27]

up lines and has no affect on the other two lines of Fig. 17. This effect also explains why $\sigma^{SC}_{H^+}$ attains similar values to that of $\sigma^{SC}_{H_2O^+}$ at high energies.

8. Conclusions

We have employed an improved impact parameter classical trajectory Monte Carlo model to calculate cross sections for single ionization, single capture, and two-electron processes (transfer ionization, double capture and double ionization) in H^+, He^{2+}, and C^{6+}+H_2O collisions in an impact energy range $20 \leq E \leq 10000$ keV/amu that largely encompasses the intermediate and high energy regimes of interest for ion-based cancer therapy applications. In the framework of the independent electron approximation, our improved model employs a three-center model potential to describe the interaction of the active electron with the H_2O^+ core, beyond the usual one-center $Z_{\rm eff}$ description. This has allowed us to explicitly consider anisotropy effects related to the multi-center nature of the target. The evaluation of inelastic probabilities for the physical many-electron system has been performed by means of the independent event model which has been found to be more accurate than the usual independent particle model. Moreover, the new results with a three-center model potential lie close to the basis generator method results of Lüdde et al. [see Figs. 5(b) and 6(b)]. In order to extend the calculation to lower energies, we have carried out semiclassical calculations of total cross sections for single capture employing molecular bases and using simplified methods based on the IPM and IEV approximations. The calculated cross sections show reasonable agreement with the experiments, and the overlap between classical and semiclassical results is similar to that found for ion-atom collisions.

For He^{2+}+H_2O, the three-center CTMC description provided cross sections in better agreement with the experimental data than the over-barrier model of Abbas et al.[41] (see Figs. 9 and 10). C^{6+} impact has also been considered because of its great importance in ion beam cancer therapy, in a wide impact energy range beyond the scope of perturbative approaches (see Fig. 12).

The explicit calculations of cross sections for the three above-mentioned systems have allowed us to extract some scaling laws, with respect to projectile charge Z_P and impact energies E, for electron production and single capture cross sections (see Figs. 13 and 15). Such scaling laws are useful to fill in the gaps (in Z_P and E) which exist in the collisional database. In addition, the scaling relations that apply to bare projectile impact can be safely used for dressed projectiles at intermediate E when inner electronic structure is not important.[47,76] Finally, as a stringent test of the calculations, we have computed the fragmentation cross sections associated with the single ionization and single capture processes (see Figs. 16 and 17) using the three-center CTMC method. The non-dissociative and evaporation ($H_2O^+ \rightarrow OH^+$+H) cross sections present satisfactory agreement with measurements from low to high impact energies, whereas the fission ($H_2O^+ \rightarrow$ OH+H^+) and break-up ($H_2O^+ \rightarrow O^+$+H+H) ones are underestimated at low E because of a primary underestimation of the cross sections for electron removal from the innermost $2a_1$ MO, which is practically unnoticeable in the total cross sections.

In this work we have shown that the use of effective potentials allows us to extend the application of methods, widely employed in ion-atom collisions, to the treatment of ion collisions with complex molecules. Future work along this line will involve a detailed study of the independent electron approximation in both CTMC and semiclassical treatments, including numerical calculations of ion collisions with other molecular systems. The main goal of future work is to implement a methodology for evaluating cross sections for other ion-biomolecule collisions, relevant in radiation damage of biological systems.

Acknowledgments

This work has been partially supported by CCG08-UAM/ESP3990 project and the DGICYT project ENE2007-62934. The use of the computational facilities at the CCC of the Universidad Autónoma de Madrid is gratefully acknowledged.

References

1. U. Amaldi and G. Kraft, *Rep. Prog. Phys.* **68**, 1861–1882 (2005).
2. D. Schardt, T. Elsässer, and D. Schulz-Ertner, *Rev. Mod. Phys.* **82**, 383–425 (2010).
3. J. R. Castro, D. E. Linstadt, J.-P. Bahary, P. L. Petti, I. Daftari, J. Collier, P. H. Gutin, G. Gauger, and T. L. Phillips, *Int. J. Radiat. Oncol. Biolog. Phys.* **29**, 647–655 (1994).
4. H. Tsujii, S. Morita, T. Miyamoto, J. Mizoe, H. Katou, H. Tsuji, S. Yamada, N. Yamamoto, and H. Murata, *Progress in radio-oncology VII, Monduzzi Editore, Bologna, Italy.* p. 393 (2002).
5. D. Schulz-Ertner, A. Nikoghosyan, C. Thilmann, T. Haberer, O. Jkel, C. Karger, G. Kraft, M. Wannenmacher, and J. Debus, *Int. J. Radiat. Oncol. Biol. Phys.* **58**, 631–640 (2004).
6. M. Krämer, O. Jäkel, T. Haberer, G. Kraft, D. Schardt, and U. Weber, *Physics in Medicine and Biology.* **45**, 3299–3317 (2000).
7. M. Krämer and M. Scholz, *Phys. Med. Biol.* **45**, 3319–3330 (2000).
8. O. Jäkel, G. Hartmann, C. Karger, P. Heeg, and J. Rassow, *Med. Phys.* **27**, 1588–1600 (2000).
9. M. Scholz, A. Kellerer, W. Kraft-Weyrather, and G. Kraft, *Radiat. Environ Biophys.* **36**, 59–66 (1997).
10. F. Ballarini, D. Alloni, A. Facoetti, and A. Ottolenghi, *New Journal of Physics.* **10**, 075008-1–075008-17 (2008).
11. J. de Vries, R. Hoekstra, R. Morgenstern, and T. Schlathölter, *Phys. Rev. Lett.* **91**, 053401-1–053401-4 (2003).
12. M. C. Bacchus-Montabonel, M. Łabuda, Y. S. Tergiman, and J. E. Sienkiewicz, *Phys. Rev. A.* **72**, 052706-1–052706-9 (2005).
13. M. C. Bacchus-Montabonel, Y. S. Tergiman, and D. Talbi, *Phys. Rev. A.* **79**, 012710-1–012710-7 (2009).
14. B. Boudaïffa, P. Cloutier, D. Hunting, M. A. Huels, and L. Sanche, *Science.* **287**, 1658–1660 (2000).
15. S. Uehara, L. Toburen, and H. Nikjoo, *Int. J. Radiat. Biol.* **77**, 139–154 (2001).
16. S. Uehara and H. Nikjoo, *J. Phys. Chem. B.* **106**, 11051–11063 (2002).
17. R. Dagnac, D. Blanc, and D. Molina, *J. Phys. B: At. Mol. Opt. Phys.* **3** (9), 1239–1251 (1970).
18. M. E. Rudd, R. D. Dubois, L. H. Toburen, C. A. Ratcliffe, and T. V. Goffe, *Phys. Rev. A.* **28**, 3244–3257 (1983).

19. M. E. Rudd, T. V. Goffe, R. D. DuBois, and L. H. Toburen, *Phys. Rev. A.* **31**, 492–494 (1985).
20. M. A. Bolorizadeh and M. E. Rudd, *Phys. Rev. A.* **33**, 888–892 (1986).
21. M. E. Rudd, T. V. Goffe, and A. Itoh, *Phys. Rev. A.* **32**, 2128–2133 (1985).
22. J. B. Greenwood, A. Chutjian, and S. J. Smith, *Astrophys. J.* **529**, 605–609 (2000).
23. R. J. Mawhorter, A. Chutjian, T. E. Cravens, N. Djurić, S. Hossain, C. M. Lisse, J. A. MacAskill, S. J. Smith, J. Simcic, and I. D. Williams, *Phys. Rev. A.* **75**, 032704-1–032704-7 (2007).
24. B. Seredyuk, R. W. McCullough, H. Tawara, H. B. Gilbody, D. Bodewits, R. Hoekstra, A. G. G. M. Tielens, P. Sobocinski, D. PešićPesic, R. Hellhammer, B. Sulik, N. Stolterfoht, O. Abu-Haija, and E. Y. Kamber, *Phys. Rev. A.* **71**, 022705-1–022705-7 (2005).
25. U. Werner, K. Beckord, J. Becker, and H. O. Lutz, *Phys. Rev. Lett.* **74**, 1962–1965 (1995).
26. F. Gobet, B. Farizon, M. Farizon, M. J. Gaillard, M. Carré, M. Lezius, P. Scheier, and T. D. Märk, *Phys. Rev. Lett.* **86**, 3751–3754 (2001).
27. F. Gobet, S. Eden, B. Coupier, J. Tabet, B. Farizon, M. Farizon, M. J. Gaillard, M. Carré, S. Ouaskit, T. D. Märk, and P. Scheier, *Phys. Rev. A.* **70**, 062716-1–062716-8 (2004).
28. F. Alvarado, R. Hoekstra, and T. Schlathölter, *J. Phys. B: At. Mol. Opt. Phys.* **38**, 4085–4094 (2005).
29. H. Luna and E. C. Montenegro, *Phys. Rev. Lett.* **94**, 043201-1–043201-4 (2005).
30. P. Sobocinski, Z. D. Pešić, R. Hellhammer, N. Stolterfoht, B. Sulik, S. Legendre, and J.-Y. Chesnel, *J. Phys. B: At. Mol. Opt. Phys.* **38**, 2495–2504 (2005).
31. H. Luna, A. L. F. de Barros, J. A. Wyer, S. W. J. Scully, J. Lecointre, P. M. Y. Garcia, G. M. Sigaud, A. C. F. Santos, V. Senthil, M. B. Shah, C. J. Latimer, and E. C. Montenegro, *Phys. Rev. A.* **75**, 042711-1–042711-11 (2007).
32. M. N. Monce, S. Pan, N. L. Radeva, and J. L. Pepper, *Phys. Rev. A.* **79**, 012704-1–012704-4 (2009).
33. B. D. Goldman, S. A. Timpone, M. N. Monce, L. Mitchell, and B. Griffin, *Phys. Rev. A.* **83** (4), 042701-1–042701-7 (2011).
34. Dž. Belkić, *J. Math. Chem.* **47**, 1366–1419 (2010).
35. P. D. Fainstein, G. H. Olivera, and R. D. Rivarola, *Nucl. Instr. and Meth. B.* **107**, 19–26 (1996).
36. G. H. Olivera, C. Caraby, P. Jardin, A. Cassimi, L. Adoui, and B. Gervais, *Physics in Medicine and Biology.* **43**, 2347–2360 (1998).
37. B. Gervais, M. Beuve, G. H. Olivera, and M. E. Galassi, *Radiat. Phys. Chem.* **75**, 493–513 (2006).
38. O. Boudrioua, C. Champion, C. D. Cappello, and Y. V. Popov, *Phys. Rev. A.* **75**, 022720-1–022720-9 (2007).
39. H. J. Lüdde, T. Spranger, M. Horbatsch, and T. Kirchner, *Phys. Rev. A.* **80**, 060702-1–060702-4 (2009).

40. C. Champion, O. Boudrioua, C. Dal Cappello, Y. Sato, and D. Ohsawa, *Phys. Rev. A.* **75**, 032724-1–032724-9 (2007).
41. I. Abbas, C. Champion, B. Zarour, B. Lasri, and J. Hassen, *Physics in Medicine and Biology.* **53**, N41–N51 (2008).
42. M. A. Bernal and J. A. Liendo, *Nucl. Instr. and Meth. B.* **262**, 1–6 (2007).
43. C. D. Cappello, C. Champion, O. Boudriuoa, H. Lekadir, Y. Sato, and D. Ohsawa, *Nucl. Instr. and Meth. B.* **267**, 781–790 (2009).
44. L. F. Errea, C. Illescas, L. Méndez, B. Pons, I. Rabadán, and A. Riera, *Phys. Rev. A.* **76**, 040701-1–040701-4 (2007).
45. C. Illescas, L. F. Errea, L. Méndez, B. Pons, I. Rabadán, and A. Riera, *Phys. Rev. A.* **83**, 052704-1–052704-12 (2011).
46. R. Abrines and I. C. Percival, *Proc. Phys. Soc.* **88**, 861–872 (1966).
47. C. Illescas and A. Riera, *Phys. Rev. A.* **60**, 4546–4560 (1999).
48. L. F. Errea, C. Illescas, L. Méndez, B. Pons, A. Riera, and J. Suárez, *J. Phys. B: At. Mol. Opt. Phys.* **37**, 4323–4338 (2004).
49. L. F. Errea, C. Illescas, L. Méndez, B. Pons, A. Riera, and J. Suárez, *J. Phys. B: At. Mol. Opt. Phys.* **39**, L91–L97 (2006).
50. L. F. Errea, F. Guzmán, C. Illescas, L. Méndez, B. Pons, A. Riera, and J. Suárez, *Plasma Physics and Controlled Fusion.* **48**, 1585–1604 (2006).
51. F. Guzmán, L. F. Errea, C. Illescas, L. Méndez, and B. Pons, *J. Phys. B: At. Mol. Opt. Phys.* **43**, 144007–1–144007–8 (2010).
52. C. L. Kirschbaum and L. Wilets, *Phys. Rev. A.* **21**, 834–841 (1980).
53. J. S. Cohen, *Phys. Rev. A.* **54**, 573–586 (1996).
54. F. Guzmán, L. F. Errea, and B. Pons, *Phys. Rev. A.* **80**, 042708-1–042708-9 (2009).
55. J. H. McGuire and L. Weaver, *Phys. Rev. A.* **16**, 41–47 (1977).
56. V. A. Sidorovich, *J. Phys. B: At. Mol. Phys.* **14**, 4805–4823 (1981).
57. H. J. Lüdde and R. M. Dreizler, *J. Phys. B: At. Mol. Opt. Phys.* **18**, 107–112 (1985).
58. C. O. Reinhold and C. A. Falcón, *J. Phys. B: At. Mol. Opt. Phys.* **21**, 1829–1843 (1988).
59. L. Meng, C. O. Reinhold, and R. E. Olson, *Phys. Rev. A.* **40**, 3637–3645 (1989).
60. L. Meng, C. O. Reinhold, and R. E. Olson, *Phys. Rev. A.* **42**, 5286–5291 (1990).
61. C. Illescas and A. Riera, *J. Phys. B: At. Mol. Opt. Phys.* **31**, 2777–2793 (1998).
62. S. Otranto, R. E. Olson, and P. Beiersdorfer, *J. Phys. B: At. Mol. Opt. Phys.* **40**, 1755–1766 (2007).
63. S. Otranto and R. E. Olson, *Phys. Rev. A.* **77**, 022709-1–022709-9 (2008).
64. C. Champion and R. D. Rivarola, *Phys. Rev. A.* **82**, 042704-1–042704-8 (2010).
65. D. S. F. Crothers and R. McCarroll, *J. Phys. B: At. Mol. Phys.* **20**, 2835–2842 (1987).

66. R. K. Janev, E. A. Solov'ev, and D. Jakimovski, *J. Phys. B: At. Mol. Opt. Phys.* **28**, L615–L620 (1995).
67. L. A. Wehrman, A. L. Ford, and J. F. Reading, *J. Phys. B: At. Mol. Opt. Phys.* **29**, 5831–5842 (1996).
68. J. Bradley, R. J. S. Lee, M. McCartney, and D. S. F. Crothers, *J. Phys. B: At. Mol. Opt. Phys.* **37**, 3723–3734 (2004).
69. L. Méndez, L. F. Errea, C. Illescas, I. Rabadán, B. Pons, and A. Riera, *AIP Conf. Proc.* **1080**, 51–58 (2008).
70. H. Getahun, L. F. Errea, C. Illescas, L. Méndez, and I. Rabadán, *Eur. Phys. J. D.* **60**, 45–49 (2010).
71. S. Mada, K. N. Hida, M. Kimura, L. Pichl, H.-P. Liebermann, Y. Li, and R. J. Buenker, *Phys. Rev. A.* **75**, 022706-1–022706-7 (2007).
72. R. Cabrera-Trujillo, E. Deumens, Y. Ohrn, O. Quinet, J. R. Sabin, and N. Stolterfoht, *Phys. Rev. A.* **75**, 052702-1–052702-13 (2007).
73. O. L. Polyansky, A. G. Császár, S. V. Shirin, N. F. Zobov, P. Barletta, J. Tennyson, D. W. Schwenke, and P. J. Knowles, *Science.* **299**, 539–542 (2003).
74. T. Rajamäki, A. Miani, and L. Halonen, *J. Chem. Phys.* **118**, 10929–10938 (2003).
75. L. F. Errea, A. Macías, L. Méndez, I. Rabadán, A. Riera, A. Rojas, and P. Sanz, *Phys. Rev. A.* **63**, 062713-1–062713-9 (2001).
76. B. H. Bransden and M. H. C. McDowell, *Charge Exchange and the Theory of Ion-Atom Collisions.* Oxford, Clarendon (1992).
77. I. Rabadán, L. F. Errea, P. Martínez, and L. Méndez, *AIP Conf. Proc.* **1080**, 98–103 (2008).
78. D. J. W. Hardie and R. E. Olson, *J. Phys. B: At. Mol. Phys.* **16**, 1983–1996 (1983).
79. G. Peach, S. L. Willis, and M. R. C. McDowell, *J. Phys. B: At. Mol. Phys.* **18**, 3921–3937 (1985).
80. J. S. Cohen, *J. Phys. B: At. Mol. Opt. Phys.* **18**, 1759–1769 (1985).
81. M. J. Raković, D. R. Schultz, P. C. Stancil, and R. K. Janev, *J. Phys. A: Math. Gen.* **34**, 4753–4770 (2001).
82. L. F. Errea, C. Illescas, L. Méndez, B. Pons, A. Riera, and J. Suárez, *Phys. Rev. A.* **70**, 52713-1–52713-12 (2004).
83. L. F. Errea, L. Méndez, B. Pons, A. Riera, I. Sevila, and J. Suárez, *Phys. Rev. A.* **74**, 012722-1–012722-17 (2006).
84. C. Illescas, I. Rabadán, and A. Riera, *J. Phys. B: At. Mol. Opt. Phys.* **30**, 1765–1784 (1997).
85. D. Elizaga, L. F. Errea, J. D. Gorfinkiel, C. Illescas, L. Méndez, A. Macías, A. Riera, A. Rojas, O. J. Kroneisen, T. Kirchner, H. J. Lüdde, A. Henne, and R. M. Dreizler, *J. Phys. B: At. Mol. Opt. Phys.* **32**, 857–875 (1999).
86. A. L. Ford, L. Wehrman, K. A. Hall, and J. F. Reading, *J. Phys. B: At. Mol. Opt. Phys.* **30**, 2889–2897 (1997).
87. L. F. Errea, J. D. Gorfinkiel, A. Macías, L. Méndez, and A. Riera, *J. Phys. B: At. Mol. Opt. Phys.* **30**, 3855–3872 (1997).

88. L. F. Errea, C. Harel, H. Jouin, L. Méndez, B. Pons, and A. Riera, *J. Phys. B: At. Mol. Opt. Phys.* **27**, 3603–3634 (1994).
89. L. Errea, J. D. Gorfinkiel, C. Harel, H. Jouin, A. Macías, L. Méndez, B. Pons, and A. Riera, *J. Phys. B: At. Mol. Opt. Phys.* **33**, 3107–3122 (2000).
90. L. H. Toburen, M. Y. Nakai, and R. A. Langley, *Phys. Rev.* **171**, 114–122 (1968).
91. C. Illescas, I. Rabadán, and A. Riera, *Phys. Rev. A.* **57**, 1809–1820 (1998).
92. L. F. Errea, C. Harel, C. Illescas, H. Jouin, L. Méndez, B. Pons, and A. Riera, *J. Phys. B: At. Mol. Opt. Phys.* **31**, 3199–3214 (1998).
93. E. A. Solov'ev, *J. Phys. B: At. Mol. Opt. Phys.* **38**, R153–R194 (2005).
94. L. F. Errea, C. Harel, H. Jouin, L. Méndez, B. Pons, A. Riera, and I. Sevila, *Phys. Rev. A.* **65**, 022711-1–022711-9 (2002).
95. L. H. Toburen, W. E. Wilson, and R. J. Popowich, *Radiation Research.* **82**, 27–44 (1980).
96. P. S. Rudolph and C. E. Melton, *J. Chem. Phys.* **45**, 2227–2235 (1966).
97. A. S. Schlachter, K. H. Berkner, W. G. Graham, R. V. Pyle, P. J. Schneider, K. R. Stalder, J. W. Stearns, J. A. Tanis, and R. E. Olson, *Phys. Rev. A.* **23**, 2331–2339 (1981).
98. H. Knudsen, L. H. Andersen, P. Hvelplund, G. Astner, H. Cederquist, H. Danared, L. Liljeby, and K. G. Rensfelt, *J. Phys. B: At. Mol. Opt. Phys.* **17**, 3545–3564 (1984).
99. H. Knudsen, H. K. Haugen, and P. Hvelplund, *Phys. Rev. A.* **23**, 597–610 (1981).
100. N. J. Mason, *AIP Conf. Proc.* **1080**, 3–20 (2008).
101. K. H. Tan, C. E. Brion, P. E. Van der Leeuw, and M. J. Van der Wiel, *Chem. Phys.* **29**, 299–309 (1978).
102. E. C. Montenegro, S. W. J. Scully, J. A. Wyer, V. Senthil, and M. B. Shah, *J. Electron. Spectros. Relat. Phenom.* **155**, 81–85 (2007).

Chapter 8

Proton Beam Irradiation of Liquid Water: A Combined Molecular Dynamics and Monte Carlo Simulation Study of the Bragg Peak Profile

Rafael Garcia-Molina*
Departamento de Física — CIOyN, Universidad de Murcia,
E-30100 Murcia, Spain
rgm@um.es

Isabel Abril and Pablo de Vera
Departament de Física Aplicada, Universitat d'Alacant,
E-03080 Alacant, Spain

Ioanna Kyriakou and Dimitris Emfietzoglou
Medical Physics Laboratory, University of Ioannina Medical School,
GR-45110 Ioannina, Greece

The simulation code SEICS (simulation of energetic ions and clusters through solids) has been used to calculate the depth-dose profile and the spatial distribution of proton beams incident in liquid water. The SEICS combines molecular dynamics and Monte Carlo techniques, and considers the main interaction phenomena between a swift charged particle and the target constituents: (i) the electronic stopping power due to energy loss to target electronic excitations, including stochastic fluctuations due to the energy-loss straggling; (ii) the elastic scattering with the target nuclei, with their corresponding energy loss; and (iii) the dynamical changes in projectile charge-state due to electronic capture and loss processes. An important advantage of the SEICS is that beam-induced electronic excitations in liquid water are described self-consistently using the momentum dependent energy-loss-function of the target.

*The author to whom corrspondence should be addressed.

In order to study the effect of the various energy-loss mechanisms in the energy dissipation profile of the projectiles as a function of the depth in liquid water, we have simulated the motion through the stopping medium of proton beams having incident energies of the order of several MeV. Moreover, the beam lateral-broadening has been evaluated as a function of the depth.

From our simulations we conclude the following: (i) A proper description of the excitation spectrum of liquid water, taking into account its condensed phase state, is essential for obtaining the stopping power of liquid water in a broad projectile-energy range. (ii) The position of the Bragg peak is mainly determined by the stopping power, whereas its width can be attributed to the energy-loss straggling. (iii) Multiple elastic scattering processes contribute slightly only at the distal part of the Bragg peak. (iv) The charge-state of the projectiles is practically constant along most of its path, but decreases when approaching the end of their trajectories (i.e., near the Bragg peak), where a considerable proportion of neutral hydrogen ions play a significant role in delivering energy to the target.

1. Introduction

The study of the interaction of energetic protons (and heavier charged particles) with biological materials has been given extra impetus nowadays due to its relevance to ion-beam radiotherapy[1–9] and radiation protection in manned space missions.[10–12]

In order to predict radiation damage to living tissue by energetic ions it is necessary to understand the "physical track structure", i.e., all the projectile-target interaction processes at the atomic/molecular level. The use of energetic protons (and heavier ions, like carbon) in radiation oncology is known to have important advantages against conventional photon or electron beam radiotherapy.[13–16] This is because photon or electron beams, at clinically relevant energies, deposit a substantial fraction of their energy near the surface of the irradiated tissue, thus causing undesirable damage when treating deep-seated tumours. On the other hand, high energy protons have a well-defined penetration range, do not suffer significant angular scattering, and lose most of their energy in a well-localized region near the end of their trajectories. This pattern of energy deposition (known as the Bragg curve) results in a more favourable depth-dose profile for radiotherapy purposes, allowing substantially higher doses to deep seated tumors and a sparing effect to shallow normal tissue compared to the conventional electron or photon beams.[1,17]

Several processes take part in the effective damage of tumour cells by ion beams: from the energy delivered by the incident projectile, followed by the generation and transport of secondary electrons, as well as radicals, until the damage mechanism affecting DNA molecules. All the above-mentioned processes take part at different scales in energy, time, or space, therefore a multiscale approach should be used to deal with the problem of ion-beam cancer therapy.[18] However, in this work we will only pay attention to the initial step in this process (i.e., the propagation of the incident ions through the target), because an accurate knowledge of the energy delivered as a function of the depth (the so-called Bragg curve or depth-dose profile), as well as the spatial evolution of the beam, is extremely important for optimizing treatment planning, as a prescribed dose must be successfully delivered at a given volume with minimum dose to healthy tissue surrounding the tumour.

Several general-purpose Monte Carlo simulation codes have been developed over the last years, like GEANT4,[19,20] PENELOPE,[21] PARTRAC,[22] SHIELD-HIT[23] or FLUKA,[24] which provide essential data for radiotherapy applications; see Nikjoo *et al.*[25] for a review. Along these lines, efforts have been made to develop Monte Carlo codes for the full-slowing-down simulation of proton and secondary electron tracks in water medium.[26–30]

In this work we do not attempt a detailed description of the electron exchange processes taking place in the Bragg peak region, whose comprehensive study requires more sophisticated methodologies, like the continuum distorted wave approximation.[31–33]

The purpose of the present chapter is to study by means of simulation the main physical phenomena that influence the energy deposition by proton beams as a function of depth in soft tissue. The latter is approximated here by liquid water at a density of $1\,\text{g/cm}^3$. More specifically, we focus on the proton-water interactions around the Bragg peak, knowledge of which is essential for radiotherapy since this is the region where the dose reaches its maximum and the particles exhibit their highest relative biological effectiveness (RBE).[34] For this reason, we deal with relatively low projectile energies (around a few MeV) instead of the higher incident beam energies commonly used in proton radiotherapy (~70–250 MeV). Liquid water is chosen as the target medium because human cells are mainly composed of water (50–80%) and, therefore, proton beams must travel through this stopping medium, although the energy they deliver to subcellular structures (most notably to the DNA) is also relevant from a biophysical perspective.[18,35–39] It is important to note that an advantage of the present methodology is that it can be effectively applied to other condensed-phase targets of biological interest.[40,41]

The main tool we use for this study is the simulation code SEICS, which is based on a combination of molecular dynamics and Monte Carlo techniques to follow the projectile motion through the target.[42,43] This simulation code includes in a detailed manner both the interaction of the projectile with the target electrons (i.e., the stopping force, which is mainly responsible for the energy lost by the projectile) and the interaction with the target nuclei (i.e., the elastic scattering, which is mainly responsible for the beam angular spread). Given that the stopping force depends on the charge-state of the projectile, the electron-capture and -loss processes by the projectile when it moves through the target have been also implemented in the simulation. Taking into account all these processes, the SEICS code dynamically follows the motion of each projectile, providing its position, velocity and charge-state at any instant. From these data very useful magnitudes for radiotherapy can be evaluated, namely, the spatial profile of energy deposition, the energy distribution and the angular spread of the beam at a given depth, the penetration range, the average charge of the projectile as a function of the depth, etc.

In Section 2 we present the procedure to calculate the main inelastic magnitudes (i.e., electronic stopping power and energy-loss straggling), whose values essentially determine the slowing down pattern of protons in liquid water. We use the dielectric formalism together with a suitable method to characterize the electronic excitation spectrum of liquid water. In Section 3 we describe in detail the simulation code SEICS which is used to evaluate the proton depth-dose profile, as well as its energy distribution and charge state in liquid water. The contribution of each process considered in the simulation will be discussed in detail in Section 4. Finally, our conclusions are presented in Section 5.

2. Inelastic Energy-Loss Magnitudes of Swift Projectiles

When a swift charged particle moves through a medium (liquid water, in our case), it loses energy mainly by producing excitations and ionizations of the target electrons (i.e., inelastic collisions). Moreover, the charge state of the projectile can change due to electronic capture and loss processes with the target electrons until reaching a dynamical equilibrium.

The most important magnitudes that describe the electronic slowing down of the projectile are the stopping power (or stopping force) and the energy-loss straggling, both depending on the velocity v and charge-state q of the projectile, as well as on the target characteristics.[44,45]

2.1. *Dielectric Formalism*

To calculate the electronic stopping magnitudes, we apply the dielectric formalism, which is based on the plane-wave (first) Born approximation and consistently accounts for the condensed-phase properties of liquid water.[46,47] In this framework, a projectile with charge-state q, atomic number z, and mass m moving with a velocity v through a medium characterized by its dielectric function, $\epsilon(k,\omega)$, experiences a stochastic slowing down. The average energy loss per unit path length is the so-called stopping power S_q, which is given by

$$S_q(v) = \frac{2e^2}{\pi v^2} \int_0^\infty \frac{\mathrm{d}k}{k} \rho_q^2(k) \int_0^{kv} \mathrm{d}\omega\, \omega\, \mathrm{Im}\left[\frac{-1}{\epsilon(k,\omega)}\right]. \tag{1}$$

The energy-loss straggling Ω_q^2, which is related to the width of the energy loss distribution, can be obtained as

$$\Omega_q^2(v) = \frac{2\hbar e^2}{\pi v^2} \int_0^\infty \frac{\mathrm{d}k}{k} \rho_q^2(k) \int_0^{kv} \mathrm{d}\omega\, \omega^2\, \mathrm{Im}\left[\frac{-1}{\epsilon(k,\omega)}\right]. \tag{2}$$

Here e is the elemental charge, and $\rho_q(k)$ is the Fourier transform of the projectile charge density for the charge-state q, which is calculated according to the modified Brandt-Kitagawa model.[48,49] The only magnitude in the above expressions dependent on the target is the energy-loss function (ELF), $\mathrm{Im}[-1/\epsilon(k,\omega)]$, which accounts for the probability that the projectile loses energy producing an electronic excitation with momentum transfer $\hbar k$ and energy transfer $\hbar\omega$ to the target.

After the projectile charge-state reaches dynamic equilibrium, the stopping power, S, and the energy-loss straggling, Ω^2, can be expressed as a weighted sum over these magnitudes for the different charge-states q of the projectile, namely,

$$S(v) = \sum_{q=0}^{z} \phi_q(v) S_q(v), \tag{3}$$

$$\Omega^2(v) = \sum_{q=0}^{z} \phi_q(v) \Omega_q^2(v), \tag{4}$$

with $\phi_q(v)$ being the probability of finding the projectile with a charge-state q when having a velocity v, which varies with the energy and nature of the projectile and with the target. For a proton beam incident on water, $\phi_q(v)$ is determined from a parameterization to experimental data.[50]

It follows from the above discussion that the key magnitude for computing the electronic energy loss of a swift projectile is the ELF, which formally depends on the initial and final state many-electron wavefunctions of the target. Thus, apart from the homogeneous free-electron gas and the atomic hydrogen, where closed analytic forms exist, the calculation of the ELF from first-principles requires a formidable computational effort. In the present work, we adopt an alternative, more practical, approach whereby the dependence of the ELF upon energy transfer is determined from experimental data at $k = 0$ (or optical limit) and extended to $k \neq 0$ by suitable extension (or dispersion) algorithms. Such a hybrid approach, based on so-called extended-optical-data models, is expected to provide a computationally simple, yet sufficiently accurate, representation of the ELF over the whole $k - \omega$ plane (i.e., the Bethe surface). Moreover, by taking advantage of available optical data (that can be checked for internal consistency through various sum rules) it overcomes the necessity of making theoretical predictions of the strength and energy distribution of the various electron excitations of the system.

In what follows, we discuss the procedure used to obtain the ELF of liquid water based on extended-optical-data models; further details can be found in recent work by our group.[51,52]

2.2. *Energy Loss Function from Extended Optical-Data Models*

The target electronic-excitation spectrum over the whole $k - \omega$ plane is contained in its ELF, $\mathrm{Im}[-1/\epsilon(k,\omega)]$, which can be obtained by different methods. In this work we distinguish between the contributions to the ELF from outer-electron excitations and inner-shell electrons, respectively,

$$\mathrm{Im}\left[\frac{-1}{\epsilon(k,\omega)}\right] = \mathrm{Im}\left[\frac{-1}{\epsilon(k,\omega)}\right]_{\text{outer}} + \mathrm{Im}\left[\frac{-1}{\epsilon(k,\omega)}\right]_{\text{inner}}. \tag{5}$$

Due to the large difference in binding energies, the description of the inner-shell contribution to the ELF can be safely assumed to be independent of that of the outer-shells. For the latter, condensed phase effects are expected to play an important role and the dielectric theory to be most justified.

2.2.1. *Inner-shell electrons contribution*

As the inner-shell electrons preserve their atomic character, they can be properly described by their generalized oscillator strengths (GOS) in the hydrogenic approach[53–55]

$$\mathrm{Im}\left[\frac{-1}{\epsilon(k,\omega)}\right]_{\mathrm{inner}} = \frac{2\pi^2\mathcal{N}}{\omega}\sum_j \alpha_j \sum_{n\ell} \frac{\mathrm{d}f^j_{n\ell}(k,\omega)}{\mathrm{d}\omega}\Theta(\omega-\omega^j_{\mathrm{ionz},n\ell}), \qquad (6)$$

where $\mathcal{N}$ is the molecular density of the target, $\mathrm{d}f^j_{n\ell}(k,\omega)/\mathrm{d}\omega$ is the hydrogenic GOS corresponding to the (n,ℓ)-subshell of the jth-element, $\hbar\omega^j_{\mathrm{ioniz},n\ell}$ is the ionization energy of the (n,ℓ)-subshell, and α_j indicates the stoichiometry of the jth-element in the compound target. For a liquid water target j refers to oxygen, therefore $\alpha_{\mathrm{O}}=1/3$, $(n=1,\ell=0)$ and $\mathcal{N}=3.34\times10^{22}$ molecules/cm^3 (mass-density 1 g/cm^3). The K-shell electrons of oxygen in the liquid water molecule are treated as inner electrons with an ionization energy $\hbar\omega_{\mathrm{ioniz},n=1,\ell=0}=540\,\mathrm{eV}$.

2.2.2. *Outer-shell electrons contribution*

The first step in any extended optical data model is the description of the target ELF at the optical limit ($k=0$). The most recent set of experimental data for the ELF of liquid water at $k\approx0$ was provided by the Sendai group,[56] by using inelastic X-ray scattering spectroscopy (IXSS) to measure the generalized oscillator strength (GOS) of liquid water at nearly vanishing momentum transfer. The IXSS data extend from 6 to 160 eV excitation energy, providing a near complete knowledge of the dielectric response properties of the valence-shells of liquid water, whose ELF is then obtained from

$$\mathrm{Im}\left[\frac{-1}{\epsilon(k\approx0,\omega)}\right]_{\mathrm{experimental}} = \frac{\pi\omega_{\mathrm{pl}}^2}{2\omega}\left.\frac{\mathrm{d}f(k\approx0,\omega)}{\mathrm{d}\omega}\right|_{\mathrm{experimental}}, \qquad (7)$$

with $\mathrm{d}f(k\approx0,\omega)/\mathrm{d}\omega\,|_{\mathrm{experimental}}$ being the experimental GOS at $k\approx0$. The nominal plasmon energy of the material, $\hbar\omega_{\mathrm{pl}}$, is determined from the relationship $\hbar\omega_{\mathrm{pl}}=4\sqrt{n_{\mathrm{e}}\pi a_0^3}\mathrm{Ry}$, where n_{e} is the target electronic density, $a_0=0.529$ Å is the Bohr radius and $\mathrm{Ry}=13.606\,\mathrm{eV}$ is the Rydberg energy. As the number of electrons in the water molecule is $Z=10$, then we have $n_{\mathrm{e}}=\mathcal{N}Z=3.34\times10^{23}\,\mathrm{cm}^{-3}$ and $\hbar\omega_{\mathrm{pl}}=21.4\,\mathrm{eV}$.

All the extended optical data models discussed below account for the outer-shell electrons contribution to the ELF (at $k = 0$) by fitting the experimental ELF using a sum of Drude-type ELF as follows:

$$\mathrm{Im}\left[\frac{-1}{\epsilon(k=0,\omega)}\right]_{\text{experimental}} = \sum_i \frac{A_i}{W_i^2}\mathrm{Im}\left[\frac{-1}{\epsilon_{\mathrm{D}}(k=0,\omega;W_i,\gamma_i)}\right], \quad (8)$$

with each Drude-type ELF

$$\mathrm{Im}\left[\frac{-1}{\epsilon_{\mathrm{D}}(k=0,\omega;W_i,\gamma_i)}\right] = \frac{W_i^2\omega\gamma_i}{(W_i^2-\omega^2)^2+(\omega\gamma_i)^2} \quad (9)$$

being characterized by the position W_i and width γ_i of its peak,[57] whose intensity is quantified by the constant A_i. These parameters are obtained through a fitting of equations (8) and (9) to the available optical ELF data (i.e., at $k = 0$).

Besides a satisfactory agreement with the available optical data, the consistency of the fitting procedure must be checked by the fulfilment of physical constraints, such as several sum rules.[58,59]

The f-sum rule gives the effective number of electrons per molecule that can be excited, Z, and ensures a good behaviour of the ELF at high energy transfers:

$$\frac{m_{\mathrm{e}}}{2\pi^2 e^2 \mathcal{N}}\int_0^\infty \mathrm{d}\omega\,\omega\left\{\mathrm{Im}\left[\frac{-1}{\epsilon(k=0,\omega)}\right]_{\text{outer}} + \mathrm{Im}[\epsilon(k=0,\omega)]_{\text{K-shell}}\right\} = Z; \quad (10)$$

in this equation we have used the relation $\mathrm{Im}[-1/\epsilon(k,\omega)] = \mathrm{Im}[\epsilon(k,\omega)]$ for inner shells.

The Kramers-Kronig, or perfect-screening, sum rule provides an important test for the accuracy of the ELF at low energy transfer:

$$\frac{2}{\pi}\int_0^\infty \frac{1}{\omega}\mathrm{Im}\left[\frac{-1}{\epsilon(k=0,\omega)}\right]\mathrm{d}\omega + n^{-2}(\omega=0) = 1, \quad (11)$$

where $n(\omega = 0)$ represents the refractive index at the static limit.

Proceeding according to equations (5)–(9), our fitting to the optical ELF of liquid water satisfies better than 99% both sum rules, equations (10) and (11).

In Fig. 1 we show the fitting curve (solid line) resulting from applying equation (8) to the experimental optical ELF derived from the IXSS data,[56] for the outer-shell electron excitations of liquid water. The right panel of Fig. 1 corresponds to higher energy transfer, where the contribution of the

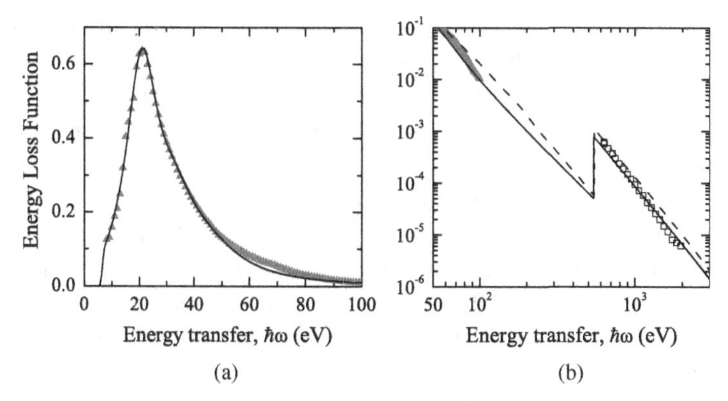

Fig. 1. Energy loss function of liquid water in the optical limit ($k = 0$). (a) Triangles are experimental IXSS data from Hayashi *et al.*[56] whereas the solid line represents the corresponding extended Drude model fitting, as described in the text. (b) Optical ELF at high energy transfers; the solid line represents the extended Drude model fitting to the IXSS data, to which we have added the contribution from the oxygen K-shell electrons through their GOS after the K-shell binding energy ($E_K = 540\,\text{eV}$). We also show the results provided by the FFAST database of NIST for the water molecule (squares)[60] and from the application of the Bragg rule to the X-ray scattering factors of H and O (dashed line)[61].

inner-shell electrons to the optical ELF of liquid water, equation (6), has been added to the valence contribution, equation (8), for energy transfers greater than the K-shell binding energy of oxygen ($E_K = 540\,\text{eV}$). For comparison purposes, at high energy transfers we have included the optical ELF values from the FFAST database of NIST for the water molecule,[60] and also the optical ELF obtained by applying the Bragg rule to the experimental X-rays scattering factors of H and O.[61]

The fitted ELF curve in Fig. 1 satisfactorily reproduces the main trends of the experimental optical ELF of liquid water. This is one of the main advantages of the approach based on using optical data specific to the material under consideration, which automatically accounts for electronic-structure effects in a realistic manner not always possible within electron gas models.

2.2.3. *Extension algorithms at $k \neq 0$*

In order to use equations (1)–(4) for calculating the stopping power or the energy-loss straggling, the ELF must be known for arbitrary momentum-transfer (i.e., $k \neq 0$). However, experimental data for the ELF of liquid water at $k \neq 0$ is only available from the Sendai group[62–64] in the range $0.19 \leq k \leq 3.59$ a.u.

A very convenient method that overcomes the theoretically formidable problem of describing the ELF over the complete Bethe surface is to use a suitable extension algorithm for extrapolating the optical data at any momentum transfer. In the simplest case, an extension algorithm is equivalent to a dispersion relation, i.e., an analytic expression of energy-transfer as a single-valued function of momentum-transfer. In what follows we summarise three of the most used extension algorithms for calculating the ELF at finite k. All of them provide the outer-shell electron contribution to the ELF at $k=0$ by fitting the experimental optical ELF from 6 to 160 eV through a sum of Drude-type ELF, but use different approaches to extend the optical ELF at $k \neq 0$. Furthermore, they guarantee the fulfilment of the sum rules for any momentum transfer, provided it is satisfied at $k = 0$.

Extended Drude (ED) model. Ritchie and Howie[65] proposed a very simple scheme for extending the ELF to the whole $k - \omega$ plane by incorporating non-zero momentum-transfers to the ELF by means of k-dependent Drude parameters $W_i(k)$ and $\gamma_i(k)$:

$$\text{Im}\left[\frac{-1}{\epsilon(k,\omega)}\right]_{\text{ED}} = \sum_i \frac{A_i}{W_i^2(k)}\text{Im}\left[\frac{-1}{\epsilon_{\text{D}}\left(k=0,\omega;W_i(k),\gamma_i(k)\right)}\right]. \tag{12}$$

The dependence of the parameters $W_i(k)$ and $\gamma_i(k)$ on momentum transfer is obtained from physically motivated dispersion relations, such as the observation that at $k=0$ (and $W_i \approx \omega_{\text{pl}}$) the Drude-type ELF of equation (12) with $\gamma=0$ coincides with the ELF of the homogeneous free-electron gas in the random phase approximation (RPA). Therefore, according to the RPA, the functional form of $W_i(k)$ follows the quadratic relation:

$$W_i(k) = W_{i,0} + \alpha_{\text{RPA}}\frac{k^2}{2m_{\text{e}}} \simeq W_{i,0} + \frac{k^2}{2m_{\text{e}}}, \tag{13}$$

where m_e is the electron mass, and $\alpha_{\text{RPA}} = 6\omega_{\text{F}}/(5\omega_{\text{pl}}) = 0.981 \simeq 1$ for liquid water, whose free-electron Fermi and plasmon energies are $\hbar\omega_{\text{F}} = 17.5\,\text{eV}$ and $\hbar\omega_{\text{pl}} = 21.4\,\text{eV}$, respectively. In this scheme, no dispersion is assumed for the damping coefficient, i.e., $\gamma_i(k) = \gamma_{i,0}$.

Despite its simplicity, the quadratic dispersion relation equation (13) is adequate for sufficiently fast projectiles by virtue of its correct limiting form at $k \to 0$ and $k \to \infty$.[66,67] The latter ensures that, at high k, single-particle effects are accounted for in an approximate manner by the term $k^2/2m_{\text{e}}$, which represents a free-electron like response and reproduces the characteristic Bethe ridge.[68] However, for not-too-fast projectiles (e.g. with energies around the Bragg peak region) improvements upon the quadratic RPA dispersion must be considered[69–72] since, in this case, the details of the Bethe ridge (position, width) will determine the fraction of the ELF that falls within the integration limits of equations (1) and (2).

Improved extended Drude (IED) model. In an effort to resolve the discrepancy between the IXSS data at $k \neq 0$ and the extended-Drude model with the RPA dispersion, Emfietzoglou and co-workers[73,74] proposed the following extension algorithms:

$$W_i(k) = W_{i,0} + [1 - \exp(-c\, k^d)] \frac{k^2}{2m_e}, \tag{14}$$

and

$$\gamma_i(k) = \gamma_{i,0} + ak + bk^2, \tag{15}$$

with a, b, c, and d being empirical parameters whose values are given by Emfietzoglou *et al.*[73,74] These dispersion relations account in an empirical way for the shifting and broadening of the Bethe ridge as observed experimentally for liquid water[62,63] and predicted by the many-body local-field factor of the electron-gas theory. Note that the term $[1 - \exp(-ck^d)]$ provides a reduction of $W_i(k)$ at not-too-large k, which results in shifting the ELF to lower excitation energies, while the k-dependence of the damping constant leads to the broadening of the ELF.

Mermin-Energy Loss Function (MELF) model. The Lindhard theory[46,57,75] provides an analytic expression for the dielectric response function of the homogeneous free-electron gas, where plasmons are undamped electronic excitations (i.e., having infinite lifetime or zero linewidth) up to a critical momentum transfer when they decay to single-electron excitations. However, a large body of experimental evidence indicates a strong damping mechanism at all k for most materials.[76]

Mermin[77] solved the problem of an undamped plasmon in the RPA by providing a phenomenological modification to the Lindhard dielectric function that includes plasmon damping through phonon-assisted electronic transitions. By a consistent account of the finite plasmon lifetime and the associated non-zero width of the plasmon peak in the ELF, the Mermin dielectric function provides a more realistic extension to finite momentum transfers than the Lindhard function.

Due to the equivalence between the Mermin-type ELF and the Drude-type ELF at the optical limit ($k = 0$), the complex structure of the ELF for real materials can be described in a similar manner to equation (8). Therefore, a linear combination of the MELF, $\mathrm{Im}[-1/\epsilon_{\mathrm{M}}(k,\omega)]$, is fitted to the experimental optical data of the target[55,78]

$$\mathrm{Im}\left[\frac{-1}{\epsilon(k,\omega)}\right]_{\mathrm{MELF}} = \sum_i \frac{A_i}{W_i^2}\mathrm{Im}\left[\frac{-1}{\epsilon_{\mathrm{M}}(k,\omega;W_i,\gamma_i)}\right]\Theta(\omega-\omega_{\mathrm{th},i}). \qquad (16)$$

As in equation (8), the fitting parameters A_i, W_i and γ_i are chosen in such a way that the MELF reproduces the main trends of the experimental ELF and satisfies the sum rules at $k = 0$, equations (10) and (11); $\omega_{\mathrm{th},i}$ is a threshold energy, which is 7 eV for liquid water.

An important advantage of the MELF model is that, unlike the Drude models, a prescription for the extension of the optical data to $k \neq 0$ is built into the model through the analytic properties of the Mermin dielectric function. The MELF model has been successfully applied to both elemental and compounds targets,[78–81] as well as biomaterials like liquid water[82,83] and dry DNA.[40,74]

In Fig. 2 we compare the results of the three extension algorithms (ED, IED and MELF) with the experimental ELF for two different values of the momentum transfer. The first feature to be noted is the experimental

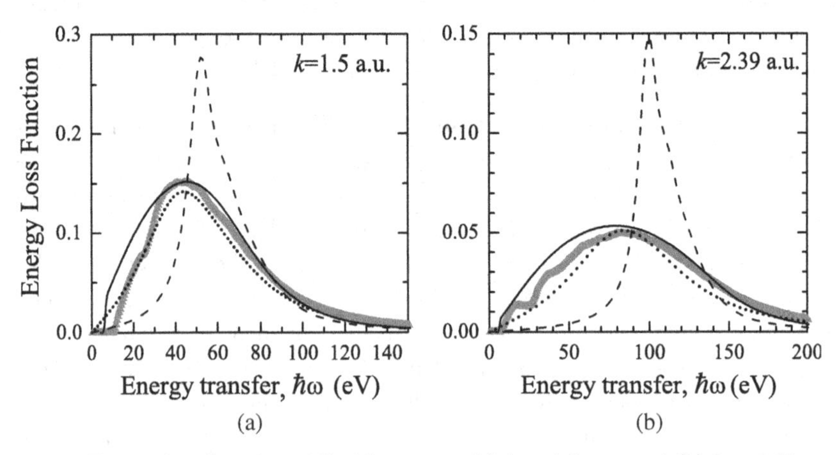

Fig. 2. Energy loss function of liquid water at (a) $k = 1.5$ a.u. and (b) $k = 2.39$ a.u. Thick grey lines correspond to experimental data.[62,64] Lines represent the results of different ELF models: solid line comes from the MELF model,[55,78] dashed line corresponds to the extended Drude model,[65] and dotted line stems from the improved extended-Drude model.[73]

broadening of the ELF as the momentum transfer increases,[64] which agrees with the theoretical expectation that single-particle excitations should gradually prevail over collective excitations for large momentum transfers. The second feature is the shift of the ELF to higher energy transfers.

Although the results of the three extension algorithms when $k \neq 0$ predict the shift to high energy transfers, only the improved extended Drude model[73,82] and the MELF model[40,55,78] reproduce the correct broadening and reduction in intensity as k increases.

2.3. *Stopping Power and Energy Loss Straggling*

The stopping power represents the most fundamental magnitude in the clinical dosimetry protocols for proton beam radiotherapy. Thus, any uncertainty in the stopping power directly translates to uncertainties in radiotherapy dosimetry.[84]

By using the previous optical data models for the outer-shell electron excitations and the hydrogenic GOS approximation for the oxygen K-shell contribution to the ELF, we can evaluate analytically the stopping power and the energy-loss straggling of liquid water for protons.

In order to illustrate the charge-state dependence of the stopping power, we have depicted in Fig. 3 the stopping power of liquid water for protons and neutral hydrogen projectiles as a function of their incident energy, as obtained from the MELF-GOS model.

The inset in Fig. 3 represents the energy dependence of the equilibrium charge-state fractions ϕ_q of H^+ and H^0 in water, as obtained by applying the Bragg rule to a parameterization of experimental data.[50]

Notice that in order to compare with experimental data, the total stopping power corresponding to each energy must be obtained by properly averaging each charge fraction with its corresponding stopping power, according to equations (1) and (3). In Fig. 4 we show the calculated stopping power of liquid water together with available experimental data. Black lines represent the results obtained from the different extension algorithms for the ELF, whereas symbols are experimental data for liquid water[85–87] and ice.[88–90] For comparison purposes, we also depict as a grey dashed line the semiempirical results provided by the SRIM code[91] and as a grey solid line the stopping power compiled in the ICRU Report 49.[92] The predictions at high energies of all three models agree (among themselves and) with the newest experimental data in liquid water.[87] However, at lower energies the predictions of the three models clearly disagree among

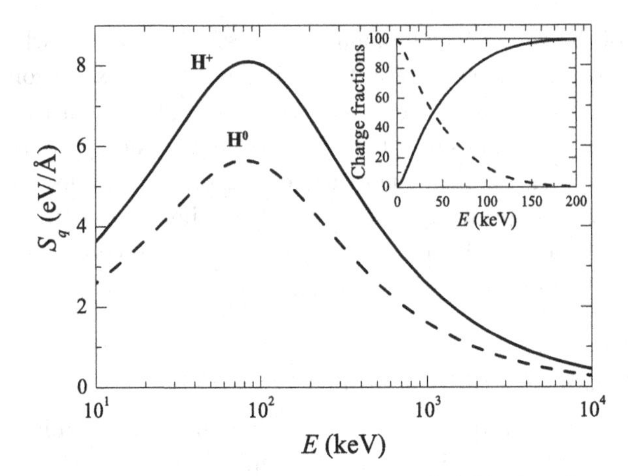

Fig. 3. Stopping power of liquid water for protons, S_+ (solid line), and neutral hydrogen projectiles, S_0 (dashed line), as a function of the incident energy. The results were obtained from the MELF-GOS model. The inset shows the charge fractions of protons, ϕ_+ (solid line), and neutral hydrogen projectiles, ϕ_0 (dashed line), in liquid water as derived from a semiempirical fitting procedure.[50]

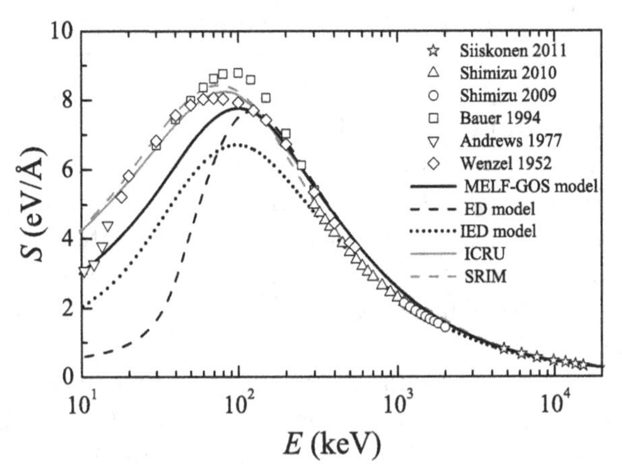

Fig. 4. Stopping power of liquid water for an incident proton beam as a function of its energy. Symbols are experimental data for liquid water[85–87] and ice.[88–90] Solid line corresponds to the MELF-GOS model, the dashed line results from the extended Drude model,[65] the dotted line is from the improved extended-Drude model.[73] Results from semi-empirical models, such as SRIM[91] (grey dashed line) and ICRU[92] (grey solid line), are also depicted.

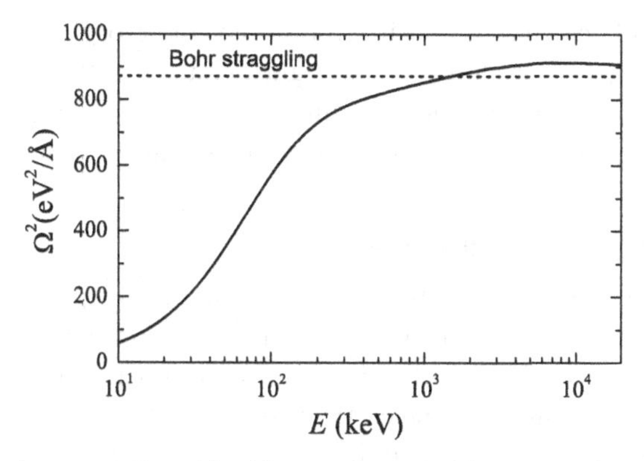

Fig. 5. Energy-loss straggling of liquid water for an incident proton beam, as a function of its energy, calculated from the MELF model. The horizontal dashed line represents the Bohr energy-loss straggling $\Omega^2_{\rm Bohr} = 4\pi z^2 Z e^4 \mathcal{N}$.

themselves and depart from the available experimental data, although these are not for liquid water but for ice. Both SRIM and ICRU curves show a good agreement with the experimental data for ice, because they use a parameterization to these available experimental data.

As the MELF-GOS model provides the stopping power that is closest to the experimental data in a broader region of incident energies, in our simulations (to be described in what follows) we will use the stopping magnitudes derived from this model.

We depict in Fig. 5 the energy-loss straggling for a proton beam incident on liquid water, calculated by the MELF-GOS model for a wide range of the projectile incident energy. Unfortunately, there is no experimental data against which to compare our calculated values. We also depict by a horizontal dashed line the Bohr energy-loss straggling, valid at high energies, which is given by $\Omega^2_{\rm Bohr} = 4\pi z^2 Z e^4 \mathcal{N}$, where $z = 1$ and $Z = 10$ in the case of liquid water.

3. Simulation of Swift Particles Moving Through Condensed Matter

To calculate the spatial and the energy distributions of a proton beam incident in liquid water as a function of depth we apply the SEICS code.[42,43]

In what follows we describe in detail this code, which is based on molecular dynamics and Monte Carlo techniques to follow in detail the motion of a charged projectile through a medium.[42,93] As a result, we obtain the position and velocity at any time along the trajectory of each projectile through the target. Therefore, it is straightforward to get (as a function of the depth) the beam energy and spreading as well as the spatial distribution of deposited energy, or the average charge-state, among other magnitudes.

3.1. *Electronic Stopping Force*

First of all we discuss the force that acts on the projectile due to the inelastic collisions with the target electrons. This produces the so-called electronic stopping force, which depends on the projectile charge-state q and velocity v. We use the dielectric framework,[46,47] together with the MELF-GOS model,[55,78] to calculate this electronic stopping force.

The position $\vec{r}$ and velocity $\vec{v}$ of the projectile, with mass m, are obtained by numerically solving its equation of motion at discrete time intervals Δt. For this purpose we use the velocity variant of Verlet's algorithm,[94] with an *ad hoc* correction for high (relativistic) velocities,

$$\vec{r}(t+\Delta t) = \vec{r}(t) + \vec{v}(t)\Delta t + \frac{\vec{F}(t)}{2m}(\Delta t)^2[1-(v(t)/c)^2]^{3/2}, \tag{17}$$

$$\vec{v}(t+\Delta t) = \vec{v}(t) + \frac{\vec{F}(t)+\vec{F}(t+\Delta t)}{2m}\Delta t[1-(v(t)/c)^2]^{3/2}, \tag{18}$$

where c is the velocity of light.

There are fluctuations in the force felt by the projectile due to the stochastic nature of the interaction with the target electrons. Then in the simulation we take the modulus of the electronic stopping force felt by the projectile (with charge state q) from a Gaussian distribution with mean value S_q, and standard deviation

$$\sigma = \sqrt{\Omega_q^2/\Delta s}, \tag{19}$$

where S_q and Ω_q^2 are given respectively by equations (1) and (2); $\Delta s = v\Delta t$ is the distance travelled by the projectile (with velocity v) in a time step Δt. Therefore, according to the Box-Müller procedure to generate a Gaussian distribution,[95] the electronic stopping force acting on the projectile is written as

$$\vec{F} = -[S_q + (\Omega_q/\sqrt{\Delta s})\sqrt{-2\ln\xi_1}\cos(2\pi\xi_2)]\hat{v}, \tag{20}$$

where $\hat{v}$ is the unit vector of the instantaneous projectile velocity $\vec{v}$. The symbols ξ_i (with $i=1,2,\ldots$) refer to random numbers uniformly distributed between 0 and 1,[96] with the value of the subscript i denoting each time a random number ξ_i is used in the simulation.

To speed up the simulation, at the higher projectile velocities ($v \gtrsim 20$ a.u.) we use the Bohr energy-loss straggling and the Bethe-Bloch stopping power formula,[97] with a mean excitation energy $I = 79.4\,\text{eV}$ for liquid water.[83]

3.2. *Elastic Scattering*

Elastic scattering due to collisions between the projectile and the target atomic cores[98,99] are simulated in such a manner that it provides the projectile scattering angle and the corresponding elastic energy loss at each collision. The path length L of the projectile between two successive elastic collisions with the target atoms is given by

$$L = -\frac{\ln \xi_3}{\sum_i \mu_i}, \tag{21}$$

where μ_i is the projectile inverse mean free path for having an elastic interaction with the i-atom of the target compound. Therefore $\sum_i \mu_i$ is the total macroscopic cross section for having an elastic collision with a target atom.

If we assume that each target atom is an effective scattering centre having a spherical volume with radius $r_0 = 1/(2\mathcal{N}^{1/3})$, where $\mathcal{N}$ is the target molecular density, then we can write

$$\mu_i = \mathcal{N}_i \pi r_0^2 = \frac{\pi}{4\mathcal{N}^{2/3}} \mathcal{N}_i, \tag{22}$$

with $\mathcal{N}_i$ being the atomic density of the i-atom type in the target compound. For a liquid water target we have $\mathcal{N}_{H_2O} = \mathcal{N}_O = \mathcal{N}_H/2$.

To determine the type of the target atom that undergoes the collision with the projectile, we suppose that the collision probability P_i with the i-atom type is proportional to the fractional contribution made by each atom to the total cross section[100]

$$P_i = \frac{\mu_i}{\sum_j \mu_j}. \tag{23}$$

Therefore a collision of the projectile with the i-atom type of the target takes place when the following condition is fulfilled

$$\sum_{j=1}^{i-1} P_j \leq \xi_4 < \sum_{j=1}^{i} P_j. \tag{24}$$

The polar scattering angle ϑ relative to the projectile direction of motion is given as a function of the parameter η (calculated using equations (28) and (29)) through

$$\cos\vartheta = \left(1 - \frac{2M_i\eta^2}{(m+M_i)\varepsilon_i^2}\right)\left[1 - \frac{4mM_i\eta^2}{(m+M_i)^2\varepsilon_i^2}\right]^{-1/2}, \quad (25)$$

where m is the projectile atomic mass and M_i is the atomic mass of the i-atom type of the target compound; ε_i is the corresponding reduced energy defined by

$$\varepsilon_i = \frac{a_{\mathrm{U},i}M_i}{zZ_i(m+M_i)}E, \quad (26)$$

with E being the projectile instantaneous kinetic energy, z and Z_i are the projectile and i-atom type target atomic numbers. The scattering between the projectile and the target atom is described by a screened Coulomb potential; here we use the universal interatomic potential[91] with the universal screening length

$$a_{\mathrm{U},i} = \frac{0.8853}{z^{0.23} + Z_i^{0.23}}a_0, \quad (27)$$

where a_0 is the Bohr radius. The value of the parameter η in equation (25) is calculated using the scattering cross section in reduced units,

$$\mathcal{J}(\eta) = \mathcal{J}(\varepsilon_i) + \frac{1-\xi_5}{4\mathcal{N}^{2/3}a_{\mathrm{U},i}^2}. \quad (28)$$

$\mathcal{J}(\eta)$ can be evaluated using

$$\mathcal{J}(\eta) = \mathcal{J}(\eta_0) + \int_{\eta_0}^{\eta} \mathrm{d}\eta' \frac{f(\eta')}{\eta'^2}, \quad (29)$$

with $f(\eta)$ being a function given by Meyer[101] and assuming $\eta_0 = 10^{-4}$ as a fixed lower integration limit. Notice that equation (28) is independent of η_0, because both $\mathcal{J}(\eta)$ and $\mathcal{J}(\varepsilon_i)$ depend on η_0 through equation (29). We have evaluated the function $\mathcal{J}(\eta)$ according to equation (29) and tabulated its values in order to obtain the value $\mathcal{J}(\varepsilon_i)$ required in equation (28) to calculate $\mathcal{J}(\eta)$; then, the value of η, used in equation (25) to determine the scattering angle ϑ, is obtained by interpolation of the tabulated function $\mathcal{J}(\eta)$.

The azimuthal scattering angle φ relative to the projectile direction of motion is simply

$$\varphi = 2\pi\xi_6. \tag{30}$$

The energy loss in the scattering process is

$$T = \frac{4mzZ_i}{a_{U,i}\varepsilon_i(m + M_i)}\eta^2, \tag{31}$$

then the modulus of the projectile velocity after the elastic collision takes place will be

$$v' = \sqrt{v^2 - \frac{2T}{m}}. \tag{32}$$

We represent schematically in Fig. 6 the geometry of the scattering process. The projectile direction of motion after the $(n-1)$-collision is defined by the polar angle Θ_{n-1} and the azimuthal angle Ψ_{n-1} in the laboratory frame of reference. The path length L until the next elastic collision is determined by using equation (21); then the n-collision takes place and we determine the i-type of target atom that is involved in the collision according to equation (24). The scattering angles ϑ and φ, with respect to the direction of motion, are obtained using equations (25) and

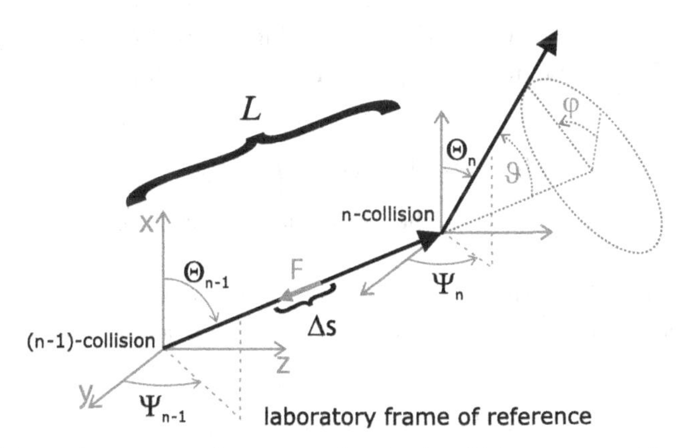

Fig. 6. Schematic geometry of two successive elastic collisions, separated by a distance L, given by equation (21). The angles in the laboratory frame of reference determining the projectile direction of motion before $(\Theta_{n-1}, \Psi_{n-1})$ and after (Θ_n, Ψ_n) the n-collision are related to the angles (ϑ, φ) with respect to the instantaneous velocity through equations (33) and (34). During the path L between two successive elastic collisions the projectile experiences a retarding force F that depends on its instantaneous velocity along each infinitesimal segment Δs; see text for more details.

(30), respectively, and the new direction of the projectile after the n-collision is obtained through[99]

$$\cos\Theta_n = -\sin\vartheta\cos\varphi\sin\Theta_{n-1} + \cos\vartheta\cos\Theta_{n-1}, \tag{33}$$

$$\cos\Psi_n = \frac{1}{\sin\Theta_n}(\sin\vartheta\cos\varphi\cos\Theta_{n-1}\cos\Psi_{n-1} - \sin\vartheta\sin\varphi\sin\Psi_{n-1} + \cos\vartheta\sin\Theta_{n-1}\cos\Psi_{n-1}). \tag{34}$$

The modulus of the projectile velocity after the n-collision is obtained using equation (32). Once the velocity (modulus and direction) is obtained after the n-collision, the next collision takes place following the same steps that have been explained in the preceding paragraphs.

3.3. *Electron Capture and Loss*

We have also included in our simulation code the capture and loss of electrons by the projectile.[42] In the model we are using, the path length of the projectile between two successive electronic capture or loss events is given by

$$L_{\mathrm{C\&L}} = -\frac{\ln\xi_7}{\mu_{\mathrm{C}} + \mu_{\mathrm{L}}}, \tag{35}$$

where μ_{C} and μ_{L} are the inverse mean free paths for electron capture and electron loss, respectively, both depending on the charge-state q of the projectile. The inverse mean free path for electron loss μ_{L} can be evaluated through

$$\mu_{\mathrm{L}}(q \to q+1) = \mathcal{N}\sigma_{\mathrm{L}}(q \to q+1), \tag{36}$$

where $q \to q+1$ denotes that the projectile changes its charge-state from q to $q+1$.

We assume that the electron-loss cross section σ_{L} is proportional to both the geometrical cross section σ of the projectile and its bound electrons,

$$\sigma_{\mathrm{L}}(q \to q+1) = (z-q)\sigma, \tag{37}$$

where we estimate the geometrical cross section of the projectile using

$$\sigma = \pi\langle r\rangle^2 = 4\pi\Lambda^2, \tag{38}$$

with $\langle r\rangle = 2\Lambda$ being the average distance between the bound electrons and the atomic nucleus in the modified Brandt-Kitagawa model.[48,49]

If multiple-electron processes are neglected, the inverse mean free path for electron capture can be obtained from the equilibrium relation

$$\mu_{\mathrm{C}}(q+1 \to q) = \frac{\phi_q}{\phi_{q+1}} \mu_{\mathrm{L}}(q \to q+1), \tag{39}$$

where ϕ_q and ϕ_{q+1} are the equilibrium fractions of the q and $q+1$ charge-states, respectively. These charge-state fractions ϕ_q should be evaluated depending on the projectile atomic number. For hydrogen projectiles,

$$\phi_0 + \phi_{+1} = 1, \tag{40}$$

$$\phi_{+1} = \langle q \rangle, \tag{41}$$

where $\langle q \rangle$ is the average charge-state. For helium projectiles, the conditions are

$$\phi_0 + \phi_{+1} + \phi_{+2} = 1, \tag{42}$$

$$\phi_{+1} + 2\phi_{+2} = \langle q \rangle, \tag{43}$$

$$\langle q \rangle^2 \phi_0 + (1 - \langle q \rangle)^2 \phi_{+1} + (2 - \langle q \rangle)^2 \phi_{+2} = \sigma_q^2, \tag{44}$$

with σ_q being the standard deviation of the charge-states distribution. The charge-state fractions ϕ_q for heavier projectiles can be calculated assuming a Gaussian distribution,

$$f(q) = \frac{1}{\sqrt{2\pi\sigma_q^2}} \exp\left[-\frac{(q - \langle q \rangle)^2}{2\sigma_q^2}\right], \tag{45}$$

where ϕ_q are then evaluated using

$$\phi_q = \frac{f(q)}{\sum_{q=0}^{z} f(q)}, \tag{46}$$

in order to ensure that the charge fractions satisfy the normalization condition

$$\sum_{q=0}^{z} \phi_q = 1. \tag{47}$$

In all these cases, both the average charge-state $\langle q \rangle$ and the standard deviation σ_q are obtained through a fit to experimental data.[50]

The probabilities of electron loss P_{L} or electron capture P_{C} by a projectile with charge-state q are proportional to the corresponding inverse mean free paths,

$$P_{\mathrm{L}}(q \to q+1) = \frac{\mu_{\mathrm{L}}(q \to q+1)}{\mu_{\mathrm{C}}(q \to q-1) + \mu_{\mathrm{L}}(q \to q+1)}, \tag{48}$$

$$P_{\mathrm{C}}(q \to q-1) = 1 - P_{\mathrm{L}}(q \to q+1). \tag{49}$$

Therefore, in order to determine the new projectile charge-state q' we use the following condition

$$q' = \begin{cases} q+1 & \text{for } \xi_8 \leq P_{\mathrm{L}}(q \to q+1) \\ q-1 & \text{for } \xi_8 > P_{\mathrm{L}}(q \to q+1). \end{cases} \tag{50}$$

In summary, according to this model we obtain the path length $L_{\mathrm{C\&L}}$ using equation (35); after an elapsed time $L_{\mathrm{C\&L}}/v$, either an electronic capture or loss event takes place determined according to equation (50).

The classical trajectory of the projectile is followed until it reaches a threshold energy E_{th}; we use $E_{\mathrm{th}} \sim 250\,\mathrm{eV}$, although reducing this value has no consequences in the final depth-dose distributions, to be discussed in the next section.

Some of the results provided by the SEICS code are illustrated in Fig. 7, which depicts several typical trajectories corresponding to a 2 MeV proton beam incident in liquid water, whose electronic excitation spectrum in the whole $k-\omega$ plane was obtained with the MELF-GOS method.[55,78] It can clearly be seen the dose delivered by the projectile at each position along its path and how the lateral broadening of the beam increases as a function of the depth. The maximum of the dose delivered by the projectile concentrates in a small (darker in the figure) region, which corresponds to the Bragg peak, after which it decreases sharply.

Although Fig. 7 offers qualitative information, the SEICS code evaluates several useful magnitudes in detail, such as the depth-dose curve and the root mean square radius, both of which will be discussed in detail further on.

4. Results and Discussion

The simulations presented in this work have been done using as input quantities the stopping power, S_q, and the energy-loss straggling, Ω_q^2, for each charge state q, calculated by equations (1) and (2) with the MELF-GOS method, to describe realistically the energy-loss function of liquid water.

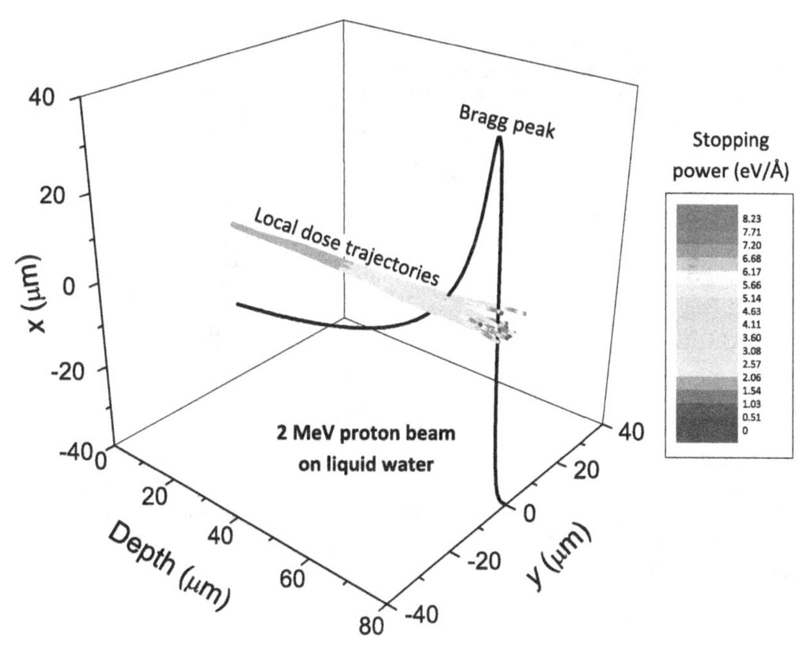

Fig. 7. Simulated trajectories (and the corresponding local dose) corresponding to a proton beam incident in liquid water with 2 MeV; only a few trajectories are displayed for clarity. The depth-dose curve, shown here qualitatively by a thin solid line, will be discussed in detail in the section 4.

The simulated depth-dose distribution of a 2 MeV proton-beam in liquid water is depicted in Fig. 8. To visualise how the elastic scattering processes and the energy-loss straggling affect the spatial distribution of the energy deposition, the solid line shows the depth-dose distribution when all the interactions are included in the simulation, whereas the dashed line and the grey dotted line are the results obtained when removing the energy-loss straggling or the elastic scattering processes, respectively. We find that the plateau of the Bragg curve is mainly determined by the electronic stopping force, with the elastic scattering and the energy-loss straggling having negligible effects before the projectile reaches the region around the Bragg peak. Removing the elastic scattering in the simulation shifts the Bragg peak distal-edge a bit deeper (around 1%), because it is the elastic scattering which is mainly responsible for the projectile slowing down at low energies, i.e., at the end of its trajectory. The removal of the fluctuations of the projectile energy loss along its full path (accounted for by the energy-loss straggling) produces a deeper (around 1–2%) and

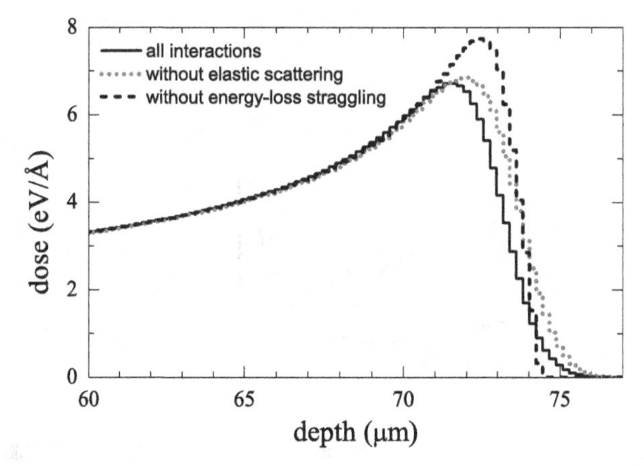

Fig. 8. Simulated depth-dose distribution obtained with the SEICS code for a 2 MeV proton beam in liquid water. Each curve represents the result obtained when removing different interactions in the simulation code SEICS: all the interactions are included (solid line), without the elastic scattering (grey dotted line) and without the energy-loss straggling (dashed line).

sharper peak (around 15%) with a steep fall-off, as can be observed in Fig. 8. Therefore, the results depicted in Fig. 8 clearly indicate that energy-loss straggling is crucial to determining the position and shape of the Bragg peak, although their contribution to the plateau of the Bragg curve is not critical.

The average charge-state for 1 MeV and 2 MeV proton-beams is plotted in Fig. 9 as a function of the depth in liquid water; for comparison purposes we also depict the corresponding depth-dose distributions as grey curves. Although electron capture and loss processes between the projectile and the target are taking place during all the projectile path, a dynamic equilibrium is quickly reached, and the projectile moves as a bare proton during most of its trajectory (for energies larger than 220 keV). The average charge of the projectile only decreases near the Bragg peak and, in particular, at the end of its trajectory. This explains the similar shape obtained for the average charge-state of 1 MeV and 2 MeV proton beams. It is worth mentioning that this behaviour can also be observed for higher energy protons, which travel as protons until the end of their trajectories, when the fraction of neutral hydrogen atoms becomes significant and, therefore, the projectile average charge decreases. Besides, neither multiple scattering nor energy-loss straggling influence the results of the simulations for the average charge-state.

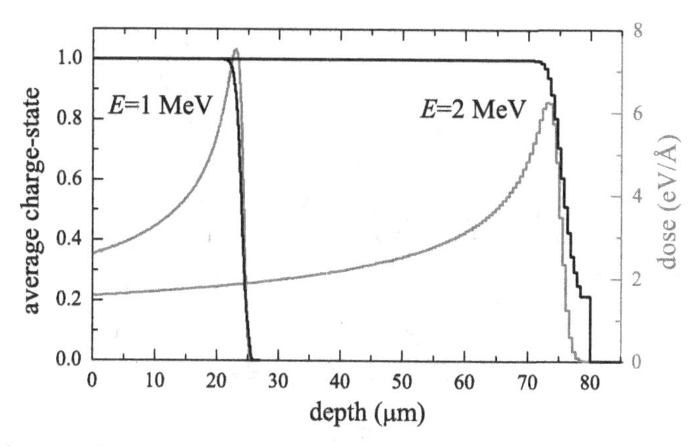

Fig. 9. (Left axis) Average charge-state of 1 MeV and 2 MeV proton beams incident in liquid water as a function of the depth, obtained by the SEICS code. (Right axis) For comparison we also represent by grey lines the corresponding depth-dose profiles.

As the energy loss magnitudes depend on the projectile charge-state, a detailed evaluation of the energy delivered by the projectile around the Bragg peak at a nano and/or micrometer scale must take into account both the electron charge-exchange processes and the fluctuations in the projectile energy-loss at each inelastic collision.

The depth-dose profile was depicted in Fig. 9 without distinguishing contributions from H^+ or H^0, but as the projectile energy decreases the proportion of neutral hydrogen overcomes the number of protons. In Fig. 10 we show separately the energy distributions for protons and neutral hydrogen projectiles around the Bragg peak, corresponding to a 1 MeV proton beam incident on liquid water. We conclude that around this important region, the number of neutral hydrogen projectiles is larger than the number of protons and with an energy distribution shifted to lower energies than protons.

We represent in Fig. 11 the average energy (dashed line) and the energy distribution (black lines) corresponding to a 10 MeV proton beam as a function of the depth in liquid water. For comparison purposes, the depth-dose distribution is represented by a grey line. As expected, the average energy of the projectiles decreases with depth, mainly due to inelastic collisions with target electrons (i.e., electronic excitations and ionizations). This figure also shows that the Bragg peak takes place when the average projectile energy is about 1.2 MeV.

The energy distribution of an initially 10 MeV monoenergetic proton beam at several depths in liquid water becomes broader as the depth

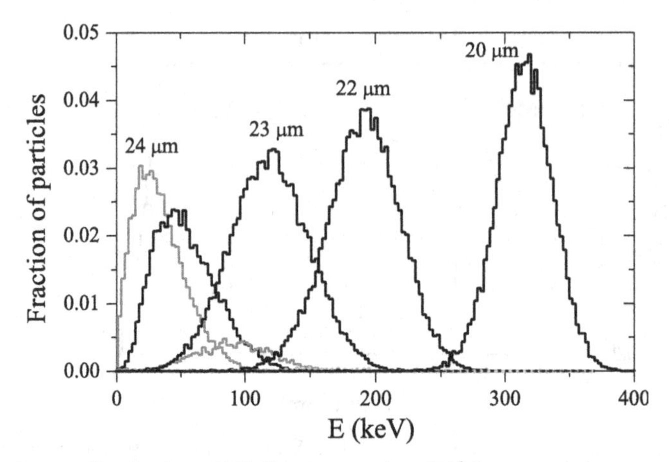

Fig. 10. Energy distribution of H^+ (black curves) and H^0 (grey curves) at several depths (indicated in the figure) around the Bragg peak for a 1 MeV proton beam in liquid water, obtained by the SEICS code.

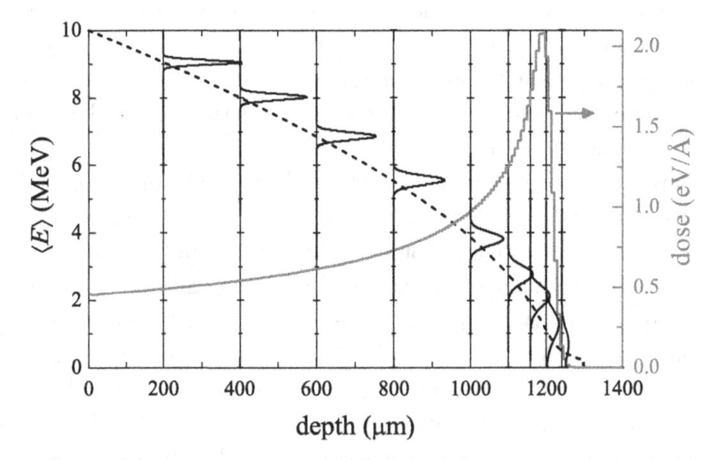

Fig. 11. (Left axis) Average energy (dashed line) for a proton-beam incident with 10 MeV in liquid water, as a function of its depth. The projectile energy-distributions are depicted by solid lines at several depths. For comparison the depth-dose distribution (right axis) is shown by a grey line.

inside the target increases, as can be seen in Fig. 11. Since the projectile energy distribution determines the generation of secondary electrons in the target, which are the main instigators of the DNA damage,[102–106] it is worth considering how the different interactions in our simulation affect the proton energy-distribution. Note that at the Bragg peak, the distribution

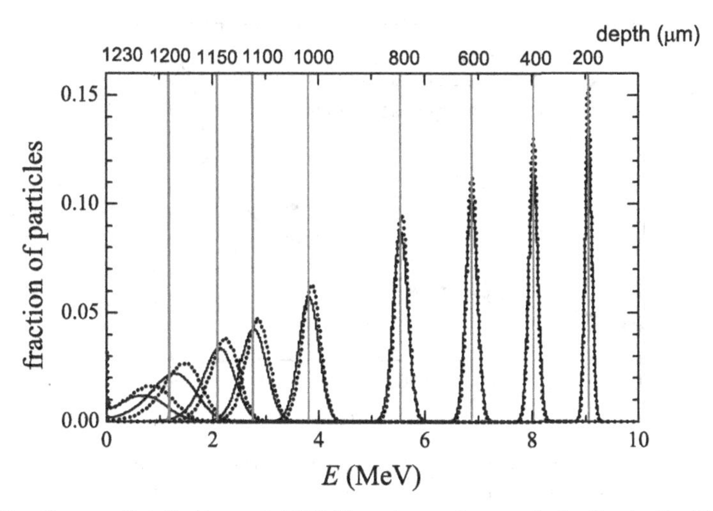

Fig. 12. Energy distributions of 10 MeV protons at several depths in liquid water. Different lines represent different interactions accounted for with SEICS: all the interactions are taken into account (black solid lines), only elastic scattering is removed (black dotted lines) and only energy-loss straggling is removed (grey lines).

of projectile energy is no longer symmetric because these are the depths where projectiles practically slow down, resulting in a larger contribution to histogram bins corresponding to lower energies.

The simulated energy-distribution for a 10 MeV proton-beam at different depths inside the liquid water target is represented in Fig. 12, with emphasis around the Bragg peak. Black solid lines represent the results obtained with SEICS when all interactions are included in the simulation, whereas the grey lines are the results obtained removing energy-loss straggling from the simulation code; the results when multiple scattering is not taken into account are plotted as black dotted lines and are almost indistinguishable from the full simulation curves (black solid lines), except at the distal part of the Bragg peak. The results show that elastic scattering processes have no sizeable influence in the projectile energy distribution, whereas the energy-loss straggling (accounting for the fluctuations in the projectile energy-loss) is mainly responsible for the broadening of the proton energy-distribution. Note that the increase in the fraction of histories observed at the low-energy side of the proton energy-distribution at 1230 μm (corresponding to the distal part of the Bragg peak) can be attributed to the significant fraction of projectiles that have almost slowed down at such depth.

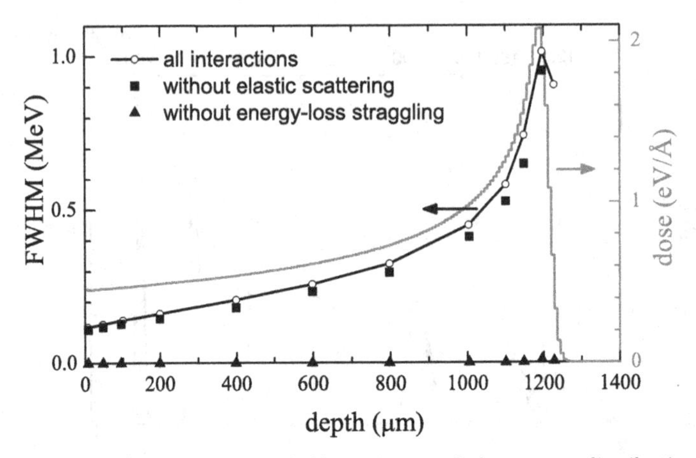

Fig. 13. (Left axis) Full width at half maximum of the energy distributions corresponding to a 10 MeV proton beam as a function of the depth in liquid water. The empty circles connected by a solid curve represent the results obtained when considering all interactions in the simulations, whereas full squares and triangles are the results obtained when elastic scattering and energy-loss straggling are removed, respectively, from the SEICS code; notice that the latter are almost over the abscissa. (Right axis) For comparison, the depth-dose distribution is shown by a grey line.

The full width at half maximum (FWHM) of the projectiles energy-distribution as a function of the depth in liquid water is plotted in Fig. 13, as well as the depth-dose distribution (grey line) for the same system as in Fig. 12. The FWHM when all the interactions are considered in the simulation is depicted by empty circles connected by a solid line, whereas full squares and full triangles are the results obtained when the simulation does not include multiple scattering and energy-loss straggling, respectively. No significant differences in the FWHM are observed regardless of whether elastic scattering is included or not in the simulations; however, a sizeable reduction of the FWHM appear when energy-loss straggling is not considered.

In order to analyse the evolution of the radial (i.e., lateral) broadening inside the target of an initial pencil beam, we have depicted in Fig. 14 the root mean square radius of a 5 MeV proton beam as a function of the depth in liquid water; the insets illustrate pictorically the broadening of the beam as it evolves inside the target. Due to multiple elastic collisions the mean radius of the beam follows an increasing trend as a function of depth travelled in the medium, which is nearly a parabolic function. The root mean square radius of the projectile beam becomes approximately constant just after the Bragg peak, and sharply falls off at the distal region;

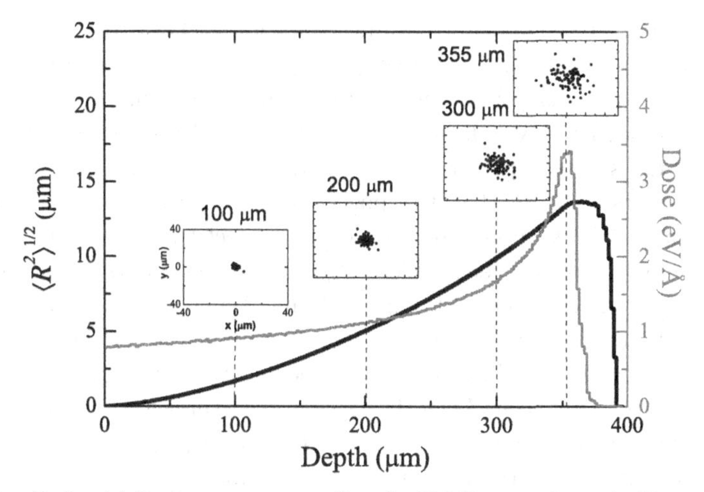

Fig. 14. (Left axis) Root mean square radius of a 5 MeV proton beam incident on liquid water, as a function of the depth. (Right axis) For comparison purposes, the depth-dose distribution is shown by a grey line. The evolution of the beam broadening is illustrated graphically in the insets, where the cross section of a beam is shown at different depths (for clarity reasons, only 100 projectile trajectories are depicted).

this fact can be explained because only very few projectiles travel in almost a straight path to reach these deep regions. As most of the projectiles deviate from their initial direction, they are stopped at lower depths.

5. Conclusions

We have presented the main ingredients of the SEICS code, which combines molecular dynamics and Monte Carlo methods to simulate the slowing down of swift projectiles in liquid water taking into account in a detailed and self-consistent manner the various energy loss and scattering processes. The classical trajectory of each projectile is calculated numerically by solving its equation of motion. The coordinates and the velocity of the projectile at each point are then determined and the deposited energy in the irradiated target is calculated, as a function of the depth.

The code includes the self-retarding stopping force (experienced by the projectile due to the target electronic interactions), with statistical fluctuations around the mean energy-loss value provided by the energy-loss straggling. The multiple elastic scattering of the projectile with the target nuclei is also accounted for through a Monte Carlo algorithm. This latter process is mainly responsible for the angular deflection of the projectile and also contributes significantly to its energy loss at low projectile energy, i.e., around the distal part of the Bragg peak. Electronic capture and loss processes are also considered along the projectile path, dynamically varying its charge-state as it moves through the target, which leads to corresponding changes of the stopping power and the energy-loss straggling.

The results indicate that the position of the Bragg peak is mainly determined by the stopping power of the projectile, whereas its shape strongly depends on the energy-loss straggling. The multiple scattering processes only modify the distal part of the Bragg peak, at the end of the projectile trajectories. Also, the average charge-state of the projectile has been obtained, revealing that the projectile travels most of the time as a bare charge, but around the Bragg peak the average charge decreases, due to a significant presence of neutral hydrogen projectiles.

We conclude that for a realistic evaluation of the Bragg peak at the microdosimetric level it is necessary to consider the condensed-phase electronic excitation properties of liquid water that essentially determine the magnitude of the main energy-loss processes.

Acknowledgements

RGM, IA and PdV thank the Spanish Ministerio de Ciencia e Innovación (Project FIS2010-17225) for financial support. PdV thanks the Conselleria d'Educació, Formació i Ocupació de la Generalitat Valenciana for its support under the VALi+d program. IK and DE acknowledge financial support from the European Union FP7 ANTICARB (HEALTH-F2-2008-201587). We acknowledge support provided by the COST Action MP 1002, Nanoscale Insights into Ion Beam Cancer Therapy.

References

1. G. Kraft, *Prog. Part. Nucl. Phys.* **45**, S473–544 (2000).
2. M. Goitein, A. J. Lomax, and E. Pedroni, *Phys. Today.* **55**, 45–51 (2002).
3. A. Brahme, *Int. J. Radiat. Oncol. Biol. Phys.* **58**, 603–616 (2004).
4. E. B. Podgoršak, ed., *Radiation oncology physics: A handbook for teachers and students*, International Atomic Energy Agency, Vienna (2005).
5. H. Nikjoo, S. Uehara, D. Emfietzoglou, and A. Brahme, *New J. Phys.* **10**, 075006 (2008).
6. P. Andreo, *Phys. Med. Biol.* **54**, N205–N215 (2009).
7. D. Schardt, T. Elsässer, and D. Schulz-Ertner, *Revs. Mod. Phys.* **82**, 383–425 (2010).
8. H. Paganetti and H. Kooy, *Expert Rev. Med. Devices.* **7**, 275–285 (2010).

9. Dž. Belkić, *J. Math. Chem.* **47**, 1366–1419 (2010).
10. F. A. Cucinotta, J. W. Wilson, R. Katz, W. Atwell, G. D. Badhwart, and M. R. Shave, *Adv. Space Res.* **18**, 183–194 (1996).
11. L. Sihver, *Z. Med. Phys.* **18**, 253–264 (2008).
12. Y. Zhang, J. Q. Clement, D. S. Gridley, L. H. Rodhe, and H. Wu, *Adv. Space Res.* **44**, 1450–1456 (2009).
13. W. P. Levin, H. Kooy, J. S. Loeffler, and T. F. DeLaney, *Brit. J. Cancer* **93**, 849–854 (2005).
14. J. Kiefer, *New J. Phys.* **10**, 075004 (2008).
15. E. A. Blakely and P. Y. Chang, *Cancer J.* **15**, 271–284 (2009).
16. M. W. McDonald and M. M. Fitzek, *Current Problems in Cancer.* **34**, 257–296 (2010).
17. A. R. Smith, *Phys. Med. Biol.* **51**, R491–R504 (2006).
18. A. V. Solov'yov, E. Surdutovich, E. Scifoni, I. Mishustin, and W. Greiner, *Phys. Rev. E.* **79**, 011909 (2009).
19. S. Agostinelli *et al.*, *Nucl. Instrum. Meth. A.* **506**, 250–303 (2003).
20. J. Allison, *IEEE Trans. Nucl. Sci.* **53**, 270–278 (2006).
21. F. Salvat, J. M. Fernández-Varea, and J. Sempau, PENELOPE — A Code System for Monte Carlo Simulation of Electron and Photon Transport. In *Workshop Proceedings Issy-les-Moulineaux*, AEN-NEA, France (Nov. 2003).
22. W. Friedland, P. Jacob, P. Bernhardt, H. G. Paretzke, and M. Dingfelder, *Radiat. Res.* **159**, 401–410 (2003).
23. I. Gudowska, N. Sobolevsky, P. Andreo, Dž. Belkić, and A. Brahme, *Phys. Med. Biol.* **49**, 1933–1958 (2004).
24. F. Sommerer, K. Parodi, A. Ferrari, K. Poljanc, W. Enghardt, and H. Aiginger, *Phys. Med. Biol.* **51**, 4385–4398 (2006).
25. H. Nikjoo, S. Uehara, D. Emfietzoglou, and F. A. Cucinotta, *Radiat. Meas.* **41**, 1052–1074 (2006).
26. C. Champion, A. L'Hoir, M. F. Politis, P. D. Fainstein, R. D. Rivarola, and A. Chetioui, *Radiat. Res.* **163**, 222–231 (2005).
27. I. Plante and F. A. Cucinotta, *New J. Phys.* **10**, 125020 (2008).
28. G. González-Muñoz, N. Tilly, J. M. Fernández-Varea, and A. Ahnesjö, *Phys. Med. Biol.* **53**, 2857–2875 (2008).
29. W. Friedland, M. Dingfelder, P. Kundrat, and P. Jacob, *Mutation Res.* **711**, 28–40 (2011).
30. T. Liamsuwan, S. Uehara, D. Emfietzoglou, and H. Nikjoo, *Int. J. Radiat. Biol.* **87**, 141–160 (2011).
31. G. H. Olivera, P. D. Fainstein, and R. D. Rivarola, *Phys. Med. Biol.* **41**, 1633–1647 (1996).
32. O. Boudrioua, C. Champion, C. Dal Cappello, and Y. V. Popov, *Phys. Rev. A.* **75**, 022720 — 1-9 (2007).
33. Dž. Belkić, I. Mančev, and J. Hanssen, *Revs. Mod. Phys.* **80**, 249–314 (2008).
34. H. Paganetti, A. Niemierko, M. Ancukiewicz, L. E. Gerweck, M. Goitein, J. S. Loeffler, and H. D. Suit, *Int. J. Radiat. Oncol. Biol. Phys.* **53**, 407–421 (2002).

35. H. Nikjoo, P. O. O'Neill, W. E. Wilson, and D. T. Goodhead, *Radiat. Res.* **156**, 577–583 (2001).
36. B. Stenerlöw, E. Höglund, and J. Carlsson, *Adv. Space Res.* **30**, 859–863 (2002).
37. H. Paganetti, *Med. Phys.* **32**, 2548–2556 (2005).
38. O. I. Obolensky, E. Surdutovich, I. Pshenichnov, I. Mishustin, A. V. Solov'yov, and W. Greiner, *Nucl. Instr. Meth. B.* **266**, 1623–1628 (2008).
39. L. Lindborg and H. Nikjoo, *Radiat. Prot. Dosim.* **143**, 402–408 (2011).
40. I. Abril, R. Garcia-Molina, C. D. Denton, I. Kyriakou, and D. Emfietzoglou, *Radiat. Res.* **175**, 247–255 (2011).
41. P. de Vera, I. Abril, and R. Garcia-Molina, *J. Appl. Phys.* **109**, 094901 — 1–8 (2011).
42. S. Heredia-Avalos, R. Garcia-Molina, and I. Abril, *Phys. Rev. A.* **76**, 012901 — 1–12 (2007).
43. R. Garcia-Molina, I. Abril, S. Heredia-Avalos, I. Kyriakou, and D. Emfietzoglou, *Phys. Med. Biol.* **56**, 6457–6493 (2011).
44. M. Nastasi, J. W. Mayer, and J. K. Hirvonen, *Ion-Solid Interactions: Fundamentals and Applications*, Ch. 5. Cambridge University Press, Cambridge (1996).
45. P. Sigmund, *Particle Penetration and Radiation Effects. General Aspects and Stopping of Swift Point Charges*. Springer Series in Solid-State Sciences Vol. 151, Berlin (2006).
46. J. Lindhard, *Dan. Vid. Selsk. Mat. Fys. Medd.* **28**, 1 (1954).
47. R. H. Ritchie, *Phys. Rev.* **106**, 874–881 (1957).
48. W. Brandt and M. Kitagawa, *Phys. Rev. B.* **25**, 5631–5637 (1982).
49. W. Brandt, *Nucl. Instr. Meth.* **194**, 13–19 (1982).
50. G. Schiwietz, and P. L. Grande, *Nucl. Instr. Methods B.* **175–177**, 125–131 (2001).
51. R. Garcia-Molina, I. Abril, I. Kyriakou, and D. Emfietzoglou, Energy loss of swift protons in liquid water: Role of optical data input and extension algorithms, Ch. 15 in *Radiation Damage in Biomolecular Systems*. G. G. Gómez-Tejedor a and M. C. Fuss, eds. Springer, Dordrecht (2012).
52. D. Emfietzoglou, R. Garcia-Molina, I. Kyriakou, I. Abril, and H. Nikjoo, *Phys. Med. Biol.* **54**, 3451–3472 (2009).
53. U. Fano, *Ann. Rev. Nucl. Sci.* **13**, 1–66 (1963).
54. R. F. Egerton, *Electron Energy-Loss Spectroscopy in the Electron Microscope*. Plenum Press, New York (1989).
55. S. Heredia-Avalos, R. Garcia-Molina, J. M. Fernández-Varea, and I. Abril, *Phys. Rev. A.* **72**, 052902 (2005).
56. H. Hayashi, N. Watanabe, Y. Udagawa, and C. C. Kao, *Proc. Nat. Acad. Sci. USA.* **97**, 6264–6266 (2000).
57. M. Dressel and G. Grüner, *Electrodynamics of Solids. Optical Properties of Electrons in Matter*. Cambridge University Press, Cambridge (2003).
58. S. Tanuma, C. J. Powell, D. R. Penn, *J. Electr. Spectrosc. Relat. Phenom.* **62**, 95–109 (1993).

59. D. Y. Smith, M. Inokuti, W. Karstens, and E. Shiles, *Nucl. Instr. Meth. B.* **259**, 1–5 (2006).
60. C. T. Chantler, K. Olsen, R. A. Dragoset, A. R. Kishore, S. A. Kotochigova, and D. S. Zucker, *X-ray form factor, attenuation and scattering tables* version 2.1. National Institute of Standards and Technology, Gaithersburg, MD (2005). `http://www.nist.gov/pml/ffast/index.cfm`
61. B. L. Henke, E. M. Gullikson, and J. C. Davis, *At. Data and Nucl. Data Tables* **54**, 181–342 (1993).
62. N. Watanabe, H. Hayashi, and Y. Udagawa, *Bull. Chem. Soc. Jap.* **70**, 719–726 (1997).
63. H. Hayashi, N. Watanabe, Y. Udagawa, and C.-C. Kao, *J. Chem. Phys.* **108**, 823–825 (1998).
64. N. Watanabe, H. Hayashi, and Y. Udagawa, *J. Phys. Chem. Sol.* **61**, 407–409 (2000).
65. R. H. Ritchie, and A. Howie, *Philos. Mag.* **36**, 463–481 (1977).
66. J. M. Fernández-Varea, R. Mayol, F. Salvat, and D. Liljequist, *J. Phys.: Cond. Matt.* **4**, 2879–2890 (1992).
67. J. M. Fernández-Varea, R. Mayol, D. Liljequist, and F. Salvat, *J. Phys.: Cond. Matt.* **5**, 3593–3610 (1993).
68. M. Dingfelder and M. Inokuti, *Radiat. Environ. Biophys.* **38**, 93–96 (1999).
69. D. J. Planes, R. Garcia-Molina, I. Abril, and N. R. Arista, *J. Electr. Spectrosc. Relat. Phenom.* **82**, 23–29 (1996).
70. Z.-J. Ding and R. Shimizu, *Surf. Sci.* **222**, 313–331 (1989).
71. Z. J. Ding and R. Shimizu, *Scanning* **18**, 92–113 (1996).
72. J.-Ch. Kuhr and H.-J. Fitting, *J. Electr. Spectrosc. Relat. Phenom.* **105**, 257–273 (1999).
73. D. Emfietzoglou, F. A. Cucinotta, and H. Nikjoo, *Radiat. Res.* **164**, 202–211 (2005).
74. I. Abril, C. D. Denton, P. de Vera, I. Kyriakou, D. Emfietzoglou, and R. Garcia-Molina, *Nucl. Instrum. Meth. B.* **268**, 1763–1767 (2010).
75. D. Pines, *Elementary excitations in solids*, 2nd printing. Benjamin, New York (1964).
76. K. Sturm, *Adv. Phys.* **31**, 1–64 (1982).
77. N. D. Mermin, *Phys. Rev. B.* **1**, 2362–2363 (1970).
78. I. Abril, R. Garcia-Molina, C. D. Denton, F. J. Pérez-Pérez, and N. R. Arista, *Phys. Rev. A.* **58**, 357–366 (1998).
79. J. C. Moreno-Marín, I. Abril, and R. Garcia-Molina, *Nucl. Instrum. Meth. B.* **193**, 30–35 (2002).
80. M. Behar, R. C. Fadanelli, I. Abril, R. Garcia-Molina, C. D. Denton, L. C. C. M. Nagamine, and N. R. Arista, *Phys. Rev. A.* **80**, 062901 (2009).
81. M. Behar, C. D. Denton, R. C. Fadanelli, I. Abril, E. D. Cantero, R. Garcia-Molina, and L. C. C. M. Nagamine, *Eur. Phys. J. D.* **59**, 209–213 (2010).
82. D. Emfietzoglou, I. Abril, R. Garcia-Molina, I. D. Petsalakis, H. Nikjoo, I. Kyriakou, and A. Pathak, *Nucl. Instr. Meth. B.* **266**, 1154–1161 (2008).
83. R. Garcia-Molina, I. Abril, C. D. Denton, S. Heredia-Avalos, I. Kyriakou, and D. Emfietzoglou, *Nucl. Instr. Meth. B.* **267**, 2647–2652 (2009).

84. H. Paul, O. Geithner, and O. Jäkel, *Nucl. Instr. Meth. B.* **256**, 561–564 (2007).
85. M. Shimizu, M. Kaneda, T. Hayakawa, H. Tsuchida, and A. Itoh, *Nucl. Instr. Meth. B.* **267**, 2667–2670 (2009).
86. M. Shimizu, T. Hayakawa, M. Kaneda, H. Tsuchida, and A. Itoh, *Vacuum.* **84**, 1002–1004 (2010).
87. T. Siiskonen, H. Kettunen, K. Peräjärvi, A. Javanainen, M. Rossi, W. H. Trzaska, J. Turunen, and A. Virtanen, *Phys. Med. Biol.* **56**, 2367–2374 (2011).
88. W. A. Wenzel and W. Whaling, *Phys. Rev.* **87**, 499–503 (1952).
89. D. A. Andrews and G. Newton, *J. Phys. D: Appl. Phys.* **10**, 845–850 (1977).
90. P. Bauer, W. Kaferbock, and V. Necas, *Nucl. Instr. Meth. B.* **93**, 132–136 (1993).
91. J. F. Ziegler, J. P. Biersak, and M. D. Ziegler, *SRIM. The Stopping and Range of Ions in Matter.* SRIM Co, Chester, MD (2008). `http://www.srim.org`
92. International Commission on Radiation Units and Measurements, *Stopping Powers and Ranges for Protons and Alpha Particles*, ICRU Report 49. ICRU, Bethesda, MD (1993).
93. S. Heredia-Avalos, *Pérdida de energía y estados de carga de proyectiles atómicos y moleculares cuando atraviesan láminas delgadas.* PhD Thesis, Universidad de Murcia, Murcia, Spain (2002). `http://bohr.inf.um.es/miembros/sha/index.html`
94. M. P. Allen and D. J. Tildesley, *Computer simulation of liquids.* Oxford University Press, Oxford (1989).
95. G. E. P. Box and M. E. Müller, *Ann. Math. Statist.* **29**, 610–611 (1958).
96. W. H. Press, S. A. Teukolsky, W. T. Vetterling, and B. P. Flannery, *Numerical Recipes in Fortran 77. The Art of Scientific Computing*, 2nd ed. Cambridge University Press, Cambridge (1997).
97. M. Inokuti, *Rev. Mod. Phys.* **43**, 297–347 (1971).
98. W. Möller, G. Pospiech, and G. Schrieder, *Nucl. Instr. Meth.* **130**, 265–270 (1975).
99. D. Zajfman, G. Both, E. P. Kanter, and Z. Vager, *Phys. Rev. A.* **41**, 2482–2488 (1990).
100. J. E. Turner, H. A. Wright, and R. N. Hamm, *Health Phys.* **48**, 717–733 (1985).
101. L. Meyer, *Phys. Stat. Sol. (b).* **44**, 253–68 (1971).
102. H. Nikjoo, S. Uehara, W. E. Wilson, M. Hoshi, and D. T. Goodhead, *Int. J. Radiat. Biol.* **73**, 355–364 (1998).
103. C. Champion, A. L'Hoir, M. F. Politis, A. Chetioui, B. Fayard, and A. Touati, *Nucl. Instr. Meth. B.* **146**, 533–540 (1998).
104. S. Uehara, L. H. Toburen, and H. Nikjoo, *Int. J. Radiat. Biol.* **77**, 139–154 (2001).
105. L. Sanche, *Eur. Phys. J. D.* **35**, 367–90 (2005).
106. H. Nikjoo, D. Emfietzoglou, R. Watanabe, and S. Uehara, *Radiat. Phys. Chem.* **77**, 1270–1279 (2008).

Index